Lecture Notes
in Computational Science
and Engineering

49

Editors

Timothy J. Barth
Michael Griebel
David E. Keyes
Risto M. Nieminen
Dirk Roose
Tamar Schlick

Benedict Leimkuhler   Christophe Chipot   Ron Elber
Aatto Laaksonen   Alan Mark   Tamar Schlick
Christophe Schütte   Robert Skeel   (Eds.)

New Algorithms
for Macromolecular
Simulation

With 90 Figures and 22 Tables

 Springer

Editors

Benedict Leimkuhler
Department of Mathematics
University of Leicester
University Road
Leicester LE1 7RH, U.K.
email: b.leimkuhler@mcs.le.ac.uk

Christophe Chipot
Institut nancéien
de chimie moléculaire
Université Henri Poincaré - Nancy I
B.P. 239
54506 Vandoeuvre-lès-Nancy, France
email: christophe.chipot@edam.uhp-nancy.fr

Ron Elber
Department of Computer Science
Cornell University
4130 Upson Hall
Ithaca, NY 14853-7501, U.S.A.
email: ron@cs.cornell.edu

Aatto Laaksonen
Arrhenius Laboratory
Division of Physical Chemistry
Stockholm University
106 91 Stockholm, Sweden
email: aatto@physc.su.se

Alan Mark
Laboratory of Biophysical Chemistry
University of Groningen
Nijenborgh 4
9747 AG, Groningen, The Netherlands
email: a.e.mark@chem.rug.nl

Tamar Schlick
Department of Chemistry
Courant Institute of Mathematical Sciences
New York University
Mercer Street 251
New York, NY 10012, U.S.A.
email: schlick@nyu.edu

Christoph Schütte
FB Mathematik und Informatik
Freie Universität Berlin
Arnimallee 2-6
14195 Berlin, Germany
email: schuette@zib.de

Robert Skeel
Department of Computer Science
Purdue University
N. University Street 250
West Lafayette, IN 47907-2066, U.S.A.
email: skeel@cs.purdue.edu

Library of Congress Control Number: 2005932829

Mathematics Subject Classification: 92C40, 92C05

ISBN-10   3-540-25542-7 Springer Berlin Heidelberg New York
ISBN-13   978-3-540-25542-0 Springer Berlin Heidelberg New York

Springer is a part of Springer Science+Business Media
springer.com
© Springer-Verlag Berlin Heidelberg 2006
Printed in The Netherlands

Typesetting: by the authors and TechBooks using a Springer LaTeX macro package

Cover design: *design & production* GmbH, Heidelberg

Printed on acid-free paper   SPIN: 11360575   46/TechBooks   5 4 3 2 1 0

Preface

This volume consists of a collection of papers given at the fourth edition of the Algorithms for Macromolecular Modelling meeting held in Leicester, UK in August 2004. The purpose of the meeting series is to foster a high level discussion of mathematical formulation, algorithmic tools, and simulation methodology for improving computational study of (primarily) biological molecules.

With the advent of supercomputers, workstation clusters, and parallel computing software, molecular modelling and simulation has become the essential counterpart to experiment. Computer simulations are helping to link the static structural information on proteins and nucleic acids obtained by X-ray crystallography and nuclear magnetic resonance with the dynamic behavior in the realistic environment of the cell. Molecular dynamics (MD) and other modelling and simulation techniques can offer an effective tool for refining experimental models and for probing systematically how molecules fold and reshape to perform the basic functions of life. Some problems that are beyond experiment can also be tackled by modelling and simulation. Indeed, experimentalists are becoming increasingly interested in modelling to decipher detailed conformational and structural aspects of macromolecular function. Yet many fundamental and practical limitations face biomolecular modelers. They include the approximate nature of the governing force fields as well as simulation protocols, the limited range of configurational sampling and relatively short trajectory times, the neglect of quantum effects in classical molecular dynamics, and the enormous computational requirements (needed to simulate a solvated macro-molecular system with full details of the environment). New algorithmic approaches, hierarchical spatial representations, and improved computing platforms are thus continuously in demand to enhance the reliability of macromolecular simulations, enhance the scope of theoretical work, and address biological problems with great specificity. In light of the the highly multidisciplinary nature of macromolecular modelling, educational efforts are also crucial for training the current generation of young biomolecular modelers.

The purpose of this volume is to help shape a dialog between developers of computational tools for biomolecular simulations and those biological and chemical scientists interested in modelling applications. In keeping with the spirit of the

AM^3 meetings, the authorship is very broad, including chemists, physicists, biologists, mathematicians and computer scientists. The book is divided into to six parts based on the content of submissions. The opening section considers novel modelling paradigms for biomolecules, including articles on challenging applications such as membrane proteins (Bond et al), enzyme simulation (Ma et al), RNA modelling (Laserson, Gan and Schlick), and sequence alignment (Joachims, Galor and Elber). Part II presents the cornerstone of most molecular modelling: the classical picture of the potential energy function and the minimization problem for molecular landscapes (Wales, Car and James), followed by an up to date overview of the protein folding problem (Scheraga et al).

Computing trajectories on the classical energy landscape is the traditional goal of molecular dynamics, although the ultimate purpose of these computations is usually to construct an efficient sampling of the low energy basins and to determine the relative likelihood of transitions amongst them; mathematicians are paying increasing attention to these issues. In Part III, we present a variety of perspectives on efficient sampling based on dynamics and stochastic-dynamics, beginning with an overview (Hampton et al) of biomolecular sampling methodology. Schemes for sampling based on extended Hamiltonians are discussed in the article by Barth, Leimkuhler and Sweet. Advances in hybrid Monte Carlo methods (Akhmatskaya and Reich) and Langevin formulation (Akkermans) are then presented, followed by a study of metastability (Huisinga and Schmidt) based on eigenstructure of the transfer operator.

Increasingly, molecular simulation is used to evaluate free energies or potentials of mean force, for example to the determine protein-ligand binding affinities, an essential challenge in de novo drug design. This topic is discussed in Part IV. In the articles of Chipot and Darve, free energy calculation methods based on evolving a system from one state to another are presented and compared. An alternative approach, based on replica-exchange methods, is considered in the article of Woods, King and Essex.

The final two parts of the volume focus on modification of the foundations of molecular simulation. In Part V, articles by Baker, Bashford and Case and Sagui, Roland , Pedersen and Darden consider alternative models of solvation, in the one case using a simplified implicit solvent model to increase efficiency and facilitate treatment of more complex macromolecules, in the other case focussing on more accurate electrostatic models based on higher order multipoles than are traditionally used. The last section, including articles by Tu and Laaksonen, and Lee, Sagui and Roland focusses on incorporation of even more accurate treatment of molecular interactions by use of quantum-mechanical calculations.

Authors have been asked to explain terminology and notation carefully and to provide complete bibliographies, enhancing the usefulness of the book as an advanced textbook for graduate study and multidisciplinary research preparation.

During the AM^3 meeting, an panel discussion was held at which certain questions regarding the direction of the field were put to a small group of leading practitioners drawn from the various fields covered by the conference. A summary of this discussion is included at the end of the book. Their candid responses, along with some additional comments and questions raised by the assembled audience, are meant to be taken in the spirit of friendly and open exchange.

In 2004, the AM^3 meeting was funded by grants from the Engineering and Physical Sciences Research Council (UK), the National Science Foundation (US), the National Institutes of Health (US), the Burroughs Wellcome Foundation (US). It was an official cooperative activity of the Society for Industrial and Applied Mathematics. A website for the meeting series is maintained at www.am-3.org.

Leicester, UK   *Benedict Leimkuhler*
Nancy, France   *Christophe Chipot*
Ithaca, New York   *Ron Elber*
Stockholm, Sweden   *Aatto Laaksonen*
Groningen, The Netherlands   *Alan Mark*
New York, New York   *Tamar Schlick*
Berlin, Germany   *Christof Schütte*
West Lafayette, Indiana   *Robert Skeel*
August 2005

Contents

**Part I Macromolecular Models: From Theories
to Effective Algorithms**

Membrane Protein Simulations: Modelling a Complex Environment
*P.J. Bond, J. Cuthbertson, S.S. Deol, L.R. Forrest, J. Johnston, G. Patargias,
M.S.P. Sansom* 3
1   Introduction – Membrane Proteins and Their Importance 3
2   Membrane Protein Environments *in Vivo* and *in Vitro* 5
3   Simulation Methods for Membranes 6
4   Using Simulations to Explore Membrane Protein Systems 6
5   Complex Solvents 7
6   Detergent Micelles 10
7   Lipid Bilayers 12
8   Self-Assembly and Complex Systems 14
References 16

**Modeling and Simulation Based Approaches
for Investigating Allosteric Regulation in Enzymes**
M.Q. Ma, K. Sugino, Y. Wang, N. Gehani, A.V. Beuve 21
1   Introduction 21
2   Modeling and Simulation of sGC 25
3   Future Work 29
References 31

**Exploring the Connection Between Synthetic and Natural RNAs
in Genomes via a Novel Computational Approach**
U. Laserson, H.H. Gan, T. Schlick 35
1   Introduction: Importance of RNA Structure and Function 35
2   Exploring the Connection between Synthetic
     and Natural RNAs 36
3   Methods 37

4   Results 41
5   Conclusions and Future Directions 48
References 53

Learning to Align Sequences:
A Maximum-Margin Approach
T. Joachims, T. Galor, R. Elber 57
1   Introduction 57
2   Sequence Alignment 58
3   Inverse Sequence Alignment 59
4   A Maximum-Margin Approach to Learning
      the Cost Parameters 60
5   Training Algorithm 63
6   Experiments 66
7   Conclusions 68
References 68

Part II Minimization of Complex Molecular Landscapes

Overcoming Energetic and Time Scale Barriers Using
the Potential Energy Surface
D.J. Wales, J.M. Carr, T. James 73
1   Introduction 73
2   Discrete Path Sampling 74
3   Basin-Hopping Global Optimisation 79
References 83

The Protein Folding Problem
H.A. Scheraga, A. Liwo, S. Oldziej, C. Czaplewski, J. Pillardy, J. Lee, D.R.
Ripoll, J.A. Vila, R. Kazmierkiewicz, J.A. Saunders, Y.A. Arnautova, K.D.
Gibson, A. Jagielska, M. Khalili, M. Chinchio, M. Nanias, Y.K. Kang, H.
Schafroth, A. Ghosh, R. Elber, M. Makowski 89
1   Introduction 89
2   Early Approaches to Structure Prediction 89
3   Global Optimization of Crystal Structures 90
4   All-atom Treatment of Protein A and the Villin Headpiece 91
5   Hierarchical Approach to Predict Structures
      of Large Protein Molecules 91
6   Performance in CASP Tests 92
7   Computation of Folding Pathways 93
8   Molecular Dynamics with the UNRES Model 95
9   Conclusions 96
References 97

Part III Dynamical and Stochastic-Dynamical Foundations for Macromolecular Modelling

**Biomolecular Sampling: Algorithms,
Test Molecules, and Metrics**
S.S. Hampton, P. Brenner, A. Wenger, S. Chatterjee, J.A. Izaguirre 103
1   Introduction 103
2   Sampling Algorithms 105
3   Test Systems, Methods, and Metrics 111
4   Simulation Results 114
5   Discussion 119
References 121

**Approach to Thermal Equilibrium
in Biomolecular Simulation**
E. Barth, B. Leimkuhler, C. Sweet 125
1   Introduction 125
2   Molecular Dynamics Formulation 128
3   Thermostatting using Nosé-Hoover Chains, Nosé-Poincaré
     and RMT Methods 130
4   Conclusions 137
References 139

The Targeted Shadowing Hybrid Monte Carlo (TSHMC) Method
E. Akhmatskaya, S. Reich 141
1   Introduction 141
2   Description of the Basic HMC Method 142
3   Störmer-Verlet Time-Stepping Method
     and Modified Hamiltonian 143
4   Targeted Shadowing HMC 144
5   Computer Experiment 148
6   Conclusion 151
References 151

The Langevin Equation for Generalized Coordinates
R.L.C. Akkermans 155
1   Introduction 155
2   Generalized Coordinates 157
3   Generalized Langevin Equation 158
4   Point Transformations 160
5   Conclusions 163
References 164

Metastability and Dominant Eigenvalues of Transfer Operators
W. Huisinga, B. Schmidt 167
1   Introduction 167
2   Markovian Molecular Dynamics 169
3   Markov Chains, Transfer Operators,
     and Metastability 170
4   Upper and Lower Bounds 172
5   Illustrative Examples 176
6   Outlook 180
References 180

Part IV Computation of the Free Energy

Free Energy Calculations in Biological Systems. How Useful Are They in Practice?
C. Chipot 185
1   Introduction 185
2   Methodological Background 186
3   Free Energy Calculations and Drug Design 194
4   Free Energy Calculations and Signal Transduction 199
5   Free Energy Calculations and Peptide Folding 201
6   Free Energy Calculations and Membrane Protein Association 204
7   Conclusion 206
References 207

Numerical Methods for Calculating the Potential of Mean Force
E. Darve 213
1   Introduction 213
2   Generalized Coordinates and Lagrangian Formulation 218
3   Derivative of the Free Energy 223
4   Potential of Mean Constraint Force 226
5   Adaptive Biasing Force 233
6   Numerical Results 238
7   Conclusion 244
References 246

Replica-Exchange-Based Free-Energy Methods
C.J. Woods, M.A. King, J.W. Essex 251
1   Introduction 251
2   Free Energy Calculations 251
3   Hydration of Water and Methane 253
4   Halide Binding to a Calix[4]Pyrrole Derivative 255
5   Conclusion 257
References 257

Part V   Fast Electrostatics and Enhanced Solvation Models

Implicit Solvent Electrostatics in Biomolecular Simulation
N.A. Baker, D. Bashford, D.A. Case 263
1   Introduction 263
2   Poisson-Boltzmann Methods 266
3   Generalized Born and Related Approximations 273
4   Applications 278
5   Conclusions 285
References 286

New Distributed Multipole Metdhods for Accurate Electrostatics in Large-Scale Biomolecular Simulations
C. Sagui, C. Roland, L.G. Pedersen, T.A. Darden 297
1   Introduction 297
2   Calculations 300
3   Results and Discussion 302
4   Conclusion 309
References 310

Part VI   Quantum-Chemical Models for Macromolecular Simulation

Towards Fast and Reliable Quantum Chemical Modelling of Macromolecules
Y. Tu, A. Laaksonen 315
1   Introduction 315
2   Extended NDDO Approximation (PART I) 317
3   An Efficient *ab initio* Tight-Binding Method (PART II) 325
4   Conclusion 338
References 338

Quantum Chemistry Simulations of Glycopeptide Antibiotics
J.-G. Lee, C. Sagui, C. Roland 343
1   Introduction 343
2   Calculations 344
3   Results and Discussion 345
4   Conclusion 350
References 350

Panel Discussion 353

List of Corresponding Authors

Reinier Akkermans
Accelrys, Ltd.
334 Cambridge Science Park
Cambridge CB4 0WN
United Kingdom
ReinierA@Accelrys.com

Nathan Baker
Dept. of Biochemistry and Molecular
Biophysics
Washington University
St. Louis, MO 63110
USA
baker@biochem.wustl.edu

Eric Barth
Department of Mathematics
Kalamazoo College
Kalamazoo, Michigan 49006
USA
barth@kzoo.edu

Christophe Chipot
Equipe de dynamique des assemblages
membranaires
UMR CNRS/UHP 7565
Université Henri Poincaré
BP 239, 54506
Vandœuvre–lès–Nancy cedex
France
Christophe.Chipot@edam.uhp-nancy.fr

Eric Darve
Institute for Computational and
Mathematical Engineering
Mechanical Engineering Department
Stanford University, CA
USA
darve@stanford.edu

Ron Elber
Department of Computer Science
Cornell University
Ithaca, New York 14853
USA
ron@cs.cornell.edu

Jonathon Essex
School of Chemistry
University of Southampton
Southampton SO17 1BJ
UK
J.W.Essex@soton.ac.uk

Wilhelm Huisinga
Free University Berlin
Department of Mathematics and
Computer Science
Arnimallee 2-6
14195 Berlin
huisinga@math.fu-berlin.de

Jesús Izaguirre
Department of Computer Science and
Engineering
University of Notre Dame
Notre Dame, Indiana 46556
USA
izaguirr@cse.nd.edu

Aatto Laaksonen
Division of Physical Chemistry
Arrhenius Laboratory
Stockholm University
106 91 Stockholm
Sweden
aatto@physc.su.se

Marc Ma
Department of Computer Science
New Jersey Institute of Technology
University Heights
Newark, New Jersey 07102
USA
qma@oak.njit.edu

Sebastian Reich
Institut für Mathematik
Universität Potsdam
Postfach 60 15 53
D-14415 Potsdam
Germany
sreich@math.uni-potsdam.de

Christopher Roland
Center for High Performance
Simulations and
Department of Physics
North Carolina State University
Raleigh, NC 27695
USA
roland@c127.chips.ncsu.edu

Celeste Sagui
Center for High Performance
Simulations and
Department of Physics
North Carolina State University
Raleigh, NC 27695
USA
sagui@unity.ncsu.edu

Mark Sansom
Department of Biochemistry
University of Oxford
Oxford OX1 3QU
UK
mark@biop.ox.ac.uk

Harold Scheraga
Baker Laboratory of Chemistry and
Chemical Biology
Cornell University
Ithaca, New York 14853
USA
has5@cornell.edu

Tamar Schlick
Courant Institute of Mathematical
Sciences
New York University
251 Mercer Street
New York, New York 10012
USA
schlick@nyu.edu

David Wales
Department of Chemistry
Lensfield Road
Cambridge CB2 1EW
UK
dw34@cam.ac.uk

Part I

**Macromolecular Models: From Theories
to Effective Algorithms**

Membrane Protein Simulations: Modelling a Complex Environment

Peter J. Bond[1], Jonathan Cuthbertson[1], Sundeep S. Deol[1], Lucy R. Forrest[2], Jennifer Johnston[1], George Patargias[1] and Mark S.P. Sansom[1]

[1] Department of Biochemistry, University of Oxford OX1 3QU, UK
[2] Department of Biochemistry and Biophysics, Columbia University, New York, NY 10032, USA

Abstract Molecular dynamics simulations of membrane proteins are reviewed, with especial attention to exploration of the interactions between membrane proteins and their environment. Environments of membrane proteins that have been studied include: (i) solvent mixtures; (ii) detergent micelles; and (iii) lipid bilayers. Mixtures of water and non-aqueous solvents mimic a membrane via local clustering of solvent molecules to provide an anisotropic micro-environment for the protein. Detergent molecules exploit interaction sites on the protein surface that *in vivo* are occupied by lipid molecules. Simulations present a dynamic picture of lipid/protein interactions, and are starting to reveal specific lipid interaction sites on the surfaces of membrane proteins. More recently it has proved possible to use MD simulations to look at large scale dynamic events, such as self-assembly of protein/detergent micelles. The future will see an expansion of simulation studies to a range of large scale dynamic membrane and membrane protein phenomena.

Key words: Membrane protein; molecular dynamics; lipid bilayer; detergent micelle

1 Introduction – Membrane Proteins and Their Importance

Membrane proteins play key roles in the biology of cells. About 25% of all genes code for integral membrane proteins [1]. For example, a recent survey of predicted integral membrane proteins from 28 fully sequenced genomes (4 archaea, 9 Gram positive bacteria, 9 Gram negative bacteria, and 4 eukaryotes) yields an unweighted mean of 22%, with a range of from 18% to 31% (Cuthbertson & Sansom, ms. in preparation). These proteins are responsible for a diverse range of functions, ranging from signalling across membranes, to transport of metabolites, energy transduction, and cellular recognition. Membrane proteins are also of biomedical importance, forming a major category of drug targets [2].

For a long while we have remained relatively ignorant concerning the structures of membrane proteins, largely as a consequence of difficulties in their expression and crystallisation for X-ray diffraction studies [3]. However, improvements in methods for the structural biology of membrane proteins mean that > 80 high resolution structures are now known (see http://blanco.biomol.uci.edu/Membrane_Proteins_xtal.html). From analysis of the rate of membrane protein structure determination is seems likely that this number will continue to grow exponentially [4]. Thus, in order to understand the relationship between structure and dynamics, molecular simulation studies of membrane proteins will become of increasing importance.

There are two major architectures for membrane proteins (Fig. 1). The vast majority of membrane proteins are made up of bundles of transmembrane (TM) α-helices. These bundles can be simple (e.g. a TM helix dimer as in glycophorin A, Fig. 1A), or complex, with multiple TM helices and re-entrant loops containing helices that incompletely span the membrane, as in the members of the aquaporin family of membrane proteins (e.g. GlpF, Fig. 1B). In Gram negative bacteria, the proteins of the outer membrane adopt a very different architecture. Outer membrane proteins (OMPs) are based upon a TM anti-parallel β-barrel structure, as seen in one of the simplest OMPs, OmpA (Fig. 1C).

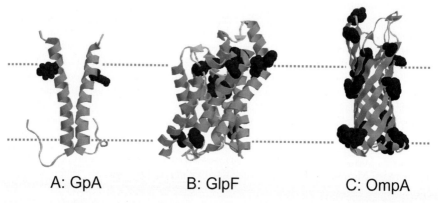

A: GpA   **B: GlpF**   **C: OmpA**

Figure 1. Structures of three membrane proteins: A GpA, a simple (two identical α-helices) α-helical membrane protein; B GlpF, a more complex (eight α-helices) α-helical membrane protein; and C OmpA, an eight-stranded β-barrel bacterial outer membrane protein. The approximate location of the lipid bilayer is indicated by the horizontal broken lines. The transmembrane secondary structure elements (α-helices or β-strands) are shown as grey ribbons. Amphipathic aromatic sidechains (Trp, Tyr) are shown in space-filling format in dark grey

Molecular simulations of membrane proteins add value to structural studies. Protein crystallography provides a static snapshot of the structure of a membrane protein, a temporal and spatial average of the structure within a crystal environment. Molecular simulations reveal the conformational dynamics of a membrane protein on a multi-nanosecond timescale [5, 6], and this timescale will be extended in the future. Simulations can also reveal details of the dynamic interactions of these proteins with their membrane environment. In this paper we consider some recent results from membrane protein simulations, with a focus on how simulations can be used to explore the interactions of these proteins with their complex environments.

2 Membrane Protein Environments *in Vivo* and *in Vitro*

The membrane environment is quite complex [7], consisting of three distinct environments to which membrane proteins are exposed: (i) an aqueous region, on either side of the membrane; (ii) a hydrophobic environment, in the centre of the lipid bilayer; and (iii) a complex interfacial environment, where the protein is exposed to a mixture of lipid headgroups and water molecules [8]. Given the spatial complexity of the membrane environment, we would expect a similar degree of complexity in the interactions of membrane proteins with this environment.

The *in vivo* membrane environment is rather complex, due to a diversity of lipid species, including e.g. cholesterol. This continues to present challenges to molecular modelling and simulation studies. As a consequence of this, the majority of membrane protein simulations correspond to the simpler environments in which membrane proteins may be studied experimentally, namely: (i) simple lipid bilayers, containing a single lipid species; (ii) detergent micelles; and (iii) solvent mixtures. Although simpler than the *in vivo* environments, these *in vitro* environments (Fig. 2) still present challenges to simulation studies.

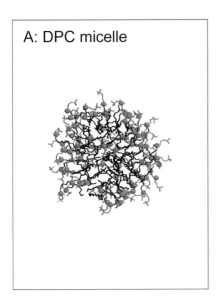
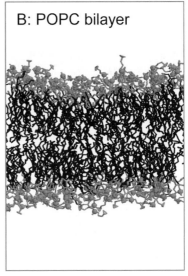

Figure 2. *In vitro* and *in vivo* environments for membrane proteins. (**A**) a detergent micelle, made up of 80 molecules of dodecylphosphocholine (DPC); and (**B**) a lipid bilayer, formed from palmitoyl oleoyl phosphatidylcholine (POPC) molecules. In both diagrams, the hydrocarbon tails are shown in *dark grey* whilst the more polar headgroups are in *pale grey*. The phosphorus atoms are shown as grey spheres

Although the native environment of a membrane protein is inserted in a lipid bilayer, in the laboratory they are frequented studied in a solubilized form, especially for NMR and related spectroscopic studies [9]. The most common way to solubilize a membrane protein is via the use of detergents [10], replacing the lipid bilayer environment by forming a mixed detergent/protein micelle. However, small membrane proteins and membrane protein fragments (e.g. single TM helices) may also be dissolved in a number of different non-aqueous solvents (e.g. TFE/water or methanol/chloroform/water) and then studied by NMR [11]. It is therefore of some interest to use simulations to characterise membrane proteins in these different *in vitro* environments, e.g. in order to estimate the extent to which the non-bilayer environments may perturb the conformational dynamics of the protein.

3 Simulation Methods for Membranes

The majority of simulations of membrane proteins use classical atomistic MD simulations, the application of which to proteins has steadily developed over the past ≈ 25 years [12]. There have been some attempts to use e.g. *ab initio* MD (and hybrid *ab initio* MD/MM) approaches for proteins, but so far these have not been widely applied to membrane proteins other than in a few special cases (e.g. the interactions of an ion channel with permeant K^+ ions [13]). In particular, the size of even a modest membrane protein/bilayer or micelle system (ca. 50,000 atoms once hydrated) currently precludes the use of such approaches.

A number of different forcefields and simulation codes have been used to simulate membrane protein systems, including CHARMM [14], GROMACS [15], GROMOS [16], and NAMD [17]. A systematic study of the dependence of membrane protein simulation behaviour on forcefield and/or simulation code would be of some interest.

One issue that arises repeatedly in simulations of membranes and membrane proteins is how to treat long-range electrostatic interactions [18]. It is generally accepted that Ewald summation based methods are preferable to simple cutoff procedures [19]. However, it must be remembered that Ewald methods are also an approximation, and periodicity-induced artefacts may occur [20, 21]. Of course, periodicity artefacts are possible for all simulations using periodic boundary conditions. There have been a number of recent studies of the effects of different treatments of electrostatic interactions on lipid bilayer simulations [22–24],the general conclusion of which is that Ewald methods give reasonable agreement with experimental data. There also has been some exploration of such effects in membrane channel simulations [25], but more systematic studies for a wider range of membrane proteins would be of interest.

4 Using Simulations to Explore Membrane Protein Systems

In this article we will focus on the use of MD simulations to explore the relationship between membrane proteins and their environment. In particular, we wish to explore how a membrane protein interacts with its lipid bilayer environment (a topic of some interest to experimentalists [26]), and how protein-lipid interactions are mimicked when the protein is relocated to a detergent micelle or complex solvent environment.

One needs to be aware of the methodological limitations of current simulation studies. A major limitation is that of scale. Typically, simulations are performed on a single membrane protein within a small patch of lipid bilayer (maybe 100 to 200 lipid molecules) or within a single detergent micelle. Simulation times generally are of the order of $\sim$10 ns, although extended simulations are becoming more common [27, 28]. From simulations of pure lipid bilayers [29] it is known that mesoscopic fluctuations may emerge when large systems are simulated for long times. Thus, there is a need for very large scale simulations of membrane/protein systems.

Our focus here will be on membrane proteins and their environment from a more general perspective. We will therefore not address in any detail the use of simulations to probe the functional properties of particular classes of membrane protein. There have been a number of reviews in this area in recent years [5, 30–32]. We will also not attempt to cover the large literatures on simulations of pure detergent micelles (see e.g. [33–36]) or pure lipid bilayers (see recent reviews, e.g. [5, 37]), but note that is by virtue of these ongoing and important studies that the membrane protein simulations described here are possible.

5 Complex Solvents

The first example is that of a (relatively) simple membrane protein studied in a complex solvent system. The F-ATP synthase (present in e.g. mitochondrial and bacterial membranes) couples H^+ translocation (down an electrochemical gradient) to synthesis of ATP, and thus is essential to cellular energy metabolism. The F-ATPase is composed of an extramembraneous F1 domain, responsible for catalysis of ATP synthesis, and a TM domain, F0, which is the H^+ translocation domain. The F0 domain is composed of a and b subunits, which interact with a ring of 9–12 c subunits. NMR structures have been determined of the monomeric c subunit dissolved in a chloroform, methanol, water (4:4:1) mixture [11, 38]. There are two TM helices per 97-residue monomer of subunit c. Comparison of the NMR structures with the 3.9 Å resolution X-ray diffraction structure of the F1-ATPase combined with a ring of 10 c subunits [39] suggests that the structure of the c subunit monomer in the solvent environment closely resembles that of the same subunit in the intact protein complex, and therefore presumably within the membrane. It is therefore of interest to use simulations to see how the complex solvent system mimics the membrane environment, and so stabilises the folded structure of the F-ATPase c subunit.

In Fig. 3 we summarise the outcome of relatively short (5 ns) MD simulations of monomer c subunit in a chloroform, methanol, water mixture (4:4:1 volume ratio). Simulations of c subunit in a lipid (POPE) bilayer (duration 10 ns) and in the three component solvents (i.e. in water, in chloroform, and in methanol alone; 3 simulations of duration 4 ns in each solvent) were also performed [40, 41].

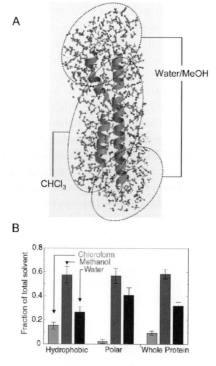

Figure 3. Simulations of the *c*-subunit from F-ATPase in a complex solvent (methanol, chloroform, water, 4:4:1 volume ratio) A Snapshot taken at the end of a 5 ns simulation, showing the clustering of the different solvent molecules around the different regions (hydrophobic helix cores, more polar helix ends) of the *c*-subunit molecule (shown in ribbons format). B Relative numbers of solvent molecules (*dark grey* = water; *mid-grey* = methanol; *pale grey* = chloroform) found within 7 Åof the *c* subunit. The bars show the mean and standard deviation (error bars) over the last 3 ns from the 5 ns simulation. Separate analyses are shown for solvents around hydrophobic, polar, and all amino acid residues of the protein

Comparison of the root mean square fluctuations about the average structure suggested that the flexibility of the protein was related to the environment as follows: POPE < solvent mixture ≈ methanol < chloroform < water. Thus it seems that the solvent mixture is able to provide a locally anisotropic environment which mimics a lipid bilayer such that the dynamic and structural properties of the protein resemble those when within a membrane. However, multiple simulations may be needed in order to more fully sample the conformational behaviour in the different environments.

One may therefore ask how the solvent mixture mimics the bilayer environment in terms of interaction with the protein. One approach to this is to examine the distribution of the different solvent molecules around the protein, looking at those solvent molecules within a certain distance of the protein (Fig. 3B). Polar amino acid residues of the protein are solvated primarily by methanol and water, with very few interactions with chloroform molecules. Hydrophobic residues are solvated primarily by methanol, with a decrease in the number of waters and an increase in the number of chloroforms forming interactions. If one examines the spatial distribution of solvent molecules around the protein it can be seen that there are clusters of water and methanol molecules around the polar helix termini and the loop between the two TM helices, with rather more chloroform molecules around the hydrophobic core of the TM helices (Fig. 3A). Thus the component molecules within the solvent mixture appear to cluster in a manner that effectively mimics the anisotropic environment provided *in vivo* by a lipid bilayer membrane. It should however be noted that these simulations are relatively short, and longer simulations are needed to explore time dependent changes in the solvation patterns. Also, careful parameterization of a range of non-aqueous solvents is required [42] in order to guarantee the accuracy of such simulations.

There have been a number of other simulations of membrane proteins and peptides in mixed solvents [43]. For example, simulations of a channel-forming α-helical peptide derived from the pore-lining M2 helix of the glycine receptor [44] in 30% TFE/water have revealed a similar pattern of clustering of the hydrophobic solvent (in this case TFE) about the hydrophobic regions of the peptide (Fig. 4). The secondary structure of the peptide in TFE/water is relatively stable, but less so than in a lipid bilayer or detergent micelle environment (Johnston and Sansom, ms. in preparation). Similar solvent clustering effects have been seen in simulations of the membrane active toxin melittin in TFE/water simulations, resulting in stabilisation of the α-helical conformation of this peptide [45].

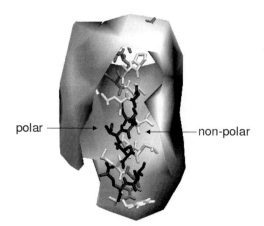

Figure 4. Simulation (20 ns) of a channel-forming peptide derived from the M2 helix of the glycine receptor [44] in 30% TFE/water (Johnston and Sansom, ms. in preparation). Polar amino acid residues are shown in *black*, non-polar residues in *white*, and charged (basic) residues in *grey*. The contour surface indicates a region of high TFE probability density around the peptide helix. The polar residues face a region of lower TFE density and the non-polar residues face the highest density of TFE

Overall, these simulation studies suggest that solvent mixtures can to some extent mimic a lipid bilayer environment by local rearrangement of the distribution of solvent molecules so as to complement the surface (hydrophobic vs. hydrophilic) presented by the membrane protein, i.e. clustering of solvent molecules provides an anisotropic micro-environment for the protein.

6 Detergent Micelles

We turn now to simulations of mixed micelles of detergents and membrane proteins. A number of such simulations have been performed for various membrane proteins: OmpA [46], a bacterial photosynthetic reaction centre [47], the ion channel KcsA (Bond and Sansom, ms. in preparation), the simple α-helix dimer glycophorin A (GpA; see below) [48], and the bacterial water and glycerol channel GlpF [49]. Here we will focus on simulations of GlpF as a representative of the major class of complex α-helical bundle membrane proteins.

Octyl glucoside (OG) is a detergent that is widely employed in experimental studies of membrane proteins, including: crystallisation for X-ray diffraction, and solubilization for NMR. MD simulations of duration 10 ns have been used to explore the conformational dynamics of GlpF, in OG micelles (Fig. 5A; [49]). Simulations of GlpF in DMPC bilayers were also performed, for comparative purposes (see below). Overall, the GlpF molecule was stable in the OG micelle: although some changes in the conformation of the loops on the surface of the protein were seen, the TM helix bundle remained largely unchanged. The mobility of the TM α-helices (assessed via analysis of residue mean square fluctuations) was $\sim$1.3 times higher in the GlpF-OG micelle than in the GlpF-DMPC bilayer simulations.

In simulations of pure OG micelles [35, 36, 49], the micelle is approximately spherical in geometry. In the mixed micelle GlpF-OG simulation, the detergent forms an irregular torus around the protein (Fig. 5A). Thus, the hydrophobic core of the protein is covered by detergent molecules whilst the polar regions of the protein are exposed to water, as they are when the protein is within a lipid bilayer. The detergent interacts tightly with the protein there is only a slightly higher penetration of water into the GlpF-OG micelle than into the pure OG micelle. Similarly, the presence of the protein resulted in only a small degree of perturbation of the dynamics of the alkyl chains in the OG micelle, namely an $\sim$15% increase in the trans-gauche($-$)-gauche($+$) transition time.

On the basis of analysis of membrane protein structures [50] and from biophysical experiments on membrane protein models it has been shown that two classes of amino acids play an important role in locking a membrane protein into a lipid bilayer [26]. Thus, amphipathic aromatic sidechains (Trp and Tyr) are thought to interact with H-bonding groups in the lipid headgroups and with water molecules in this region, whilst basic sidechains (Arg and Lys) interact with the phosphate groups of lipids. On the basis of the simulations of GlpF we may ask in detail how the protein interacts with the detergent molecules that coat its surface in the micelles.

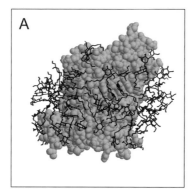

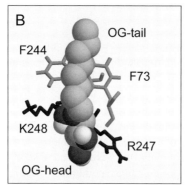

Figure 5. Simulation of a GlpF-OG micelle. A Snapshot at the end of a 10 ns simulation, showing the protein in spacefilling format (*pale grey*) and the octylglucoside (OG) molecules in bonds form (*dark grey*). B Example of the interaction between an octylglucoside molecule (in spacefill format) and selected residues (in bonds format) on the surface of GlpF. The head-group of OG interacts with the basic sidechains on R247 and K248; the alkyl tail of OG with the aromatic sidechains of F73 and F244

Examination of the interactions of aromatic sidechains with the headgroups of detergent molecules in the GlpF-OG simulation revealed two bands of such interactions, as has been observed in other simulations of membrane proteins (see below). There were also substantial interactions between basic sidechains of the protein and detergent headgroups. The detergent tails seemed to interact a little more with the hydrophobic sidechains of the protein than did the lipid tails.

To illustrate the nature of the interactions observed in these simulations, a selected interaction from the GlpF-OG simulation is shown (5B). The glucosyl headgroup forms H-bonding interactions with two basic sidechains (R247 and K248) on the protein surface at the interfacial region. The alkyl tail is sandwiched between the sidechains of two aromatic residues (F73 and F244). We note that similar, but non-identical, interactions of OG with aromatic sidechains are seen in the crystal structure of GlpF.

A difference was noticed between the GlpF-OG and GlpF-DMPC simulations in terms of the number of atomic contacts between the protein sidechains and the lipid/detergent headgroups: the OG headgroups made about four times as many contacts with the protein as did the DMPC headgroups. This difference may be explained by the fact that the octyl-glucoside molecule, containing four hydroxyl groups within the glucosyl group, has a greater potential to form hydrogen bonds with protein sidechains compared to the headgroup of DMPC.

In summary, the detergent molecules seem to exploit interaction sites on the protein surface that *in vivo* are presumably occupied by lipid molecules. Thus, the environment presented to the protein by the micelle is very similar to that in a lipid bilayer. However, there are subtle differences (see below).

7 Lipid Bilayers

There have been many simulation studies of integral membrane proteins and TM peptides embedded within lipid bilayers. This literature is reviewed by e.g. [5, 51, 52]. There have been rather fewer simulations of peripheral membrane proteins (e.g. [53, 54]). Here we will focus on simulations of a particular membrane protein, namely the bacterial outer membrane protein OmpA (see Fig. 1C).

The outer membrane of Gram-negative bacteria serves as a protective barrier against the external environment, but is rendered selectively permeable by a number of proteins including porins. Outer membrane proteins have a diversity of other functions, including active transporters, enzymes, and recognition proteins. MD simulations have been used to study a number of OMPs (reviewed in [6]) providing e.g. atomic resolution descriptions of solute permeation through porins [55]. It has also yielded insights into the dynamics of active transporters [56] and pores [57], as well as providing clues to catalytic mechanisms in outer membrane enzymes [58, 59]. Additionally, simulations are beginning to reveal common features of interactions between membrane proteins and lipids, with implications for studies of OMP folding and stability.

OmpA is composed of an N-terminal TM β-barrel domain, and a periplasmic C-terminal domain. MD simulations of the OmpA TM domain embedded within a DMPC bilayer Fig. 6A have been performed, and the results analysed in terms of possible mechanisms of pore formation by OmpA. A simulation in which putative gate region sidechains of the OmpA barrel interior were held in a non-native conformation resulted in an open pore, with a predicted conductance similar to experimental measurements [57].

OmpA is also one of just a few membrane proteins whose structure has been solved both by X-ray crystallography and by NMR. MD simulations have been used to compare the behaviour of OmpA in a detergent (DPC) micelle and in a phospholipid (DMPC) bilayer. The dynamic fluctuations of the protein structure seem to be $\sim$1.5 times greater in the micelle environment than in the lipid bilayer. This is similar to the situation seen for GlpF (see above) As a consequence of the enhanced flexibility of the OmpA protein in the micellar environment, sidechain conformational

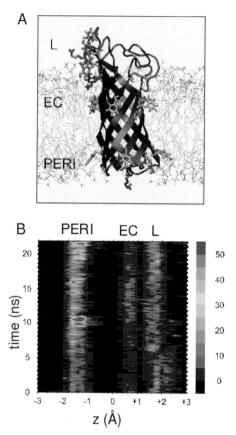

Figure 6. Lipid-protein interactions in a simulation of OmpA in a lipid bilayer. A Snapshot from a simulation of OmpA in a lipid DMPC bilayer. The aromatic residues (Trp, Tyr) in the extracellular band (EC), the extracellular loops (L), and the periplasmic band (PERI) are shown. B Number of interactions (indicated on the *greyscale*) of aromatic sidechains (Trp, Tyr) with the headgroups of lipid molecules as a function of position along the bilayer normal (z) and simulation time. The interactions corresponding to the L, EC and PERI regions are labelled

changes resulted in formation of a continuous pore through the centre of the OmpA molecule. This was used to suggest that whereas the crystal structure of OmpA corresponds to a closed conformation of the pore, the micelle may promote formation of an open conformation of the OmpA pore [46].

As mentioned above, the interactions between membrane proteins and their lipid bilayer environment play important roles in the stability and function of such proteins. MD simulations provide an opportunity to compare the interactions of different membrane proteins with lipid bilayers. For example, simulations of an α-helical protein (KcsA) and a β-barrel protein (OmpA), both in PC bilayers, have been compared [60] in terms of a number of aspects of lipid-protein interactions, including

H-bonding interactions of aromatic sidechains (Trp, Tyr) with lipid headgroups, and electrostatic interactions of basic sidechains (Lys, Arg) with lipid phosphate groups. Such interactions are seen to fluctuate on a ~1 to 5 ns timescale. There are two clear bands of interacting residues on the surface of KcsA, whereas there are three such bands on OmpA Fig. 6B. A large number of Arg/phosphate interactions are seen for KcsA; for OmpA, the number of basic/phosphate interactions is smaller and shows more marked fluctuations with respect to time. Both classes of interaction occur in clearly defined interfacial regions of width ~1 nm. Analysis of lateral diffusion of lipid molecules reveals that boundary lipid molecules diffuse at about half the rate of bulk lipid. The number of boundary lipid molecules estimated by simulation for OmpA ($\approx$14) correlates well with recent studies of the association of spin-labelled lipids with β-barrel proteins [61] which suggested there were 11 motionally restricted lipids per OmpA molecule.

Overall, these simulations present a dynamic picture of lipid/protein interactions. Simulations of related membrane proteins (e.g. OmpT; [59] and KcsA; Deol and Sansom, ms. in preparation) reveal more specific lipid interaction sites on the surfaces of certain membrane proteins. For example, in OmpT a proposed LPS binding site [62], made up of a cluster of surface-exposed lysine sidechains, forms long-lasting H-bonds with lipid (DMPC molecules). This suggests that even relatively short (10 ns) simulations may be capable of identifying specific lipid protein interactions. However, a more systematic comparison between multiple membrane protein simulations (Khalid et al., ms. in preparation) is needed to confirm this.

In attempt to further bridge the gap between our understanding of membrane proteins *in vivo* and *in vitro* we have performed an extended (50 ns) MD simulation of OmpA in a crystal unit cell (Bond & Sansom, ms. in preparation). There is good agreement between simulated and experimental (crystallographic) B-factors, providing confidence in the simulation methodology. The crystal contains detergent molecules. These form a fluid, continuous micelle-like cylinder around extended fibres of symmetry-related proteins within the crystal. The interactions of these detergents with the protein mimic those observed in both the micelle and bilayer simulations. It will be of interest to extend such studies to comparative simulations of other membrane proteins in bilayer, micelle, and crystal environments.

8 Self-Assembly and Complex Systems

The simulations that have been described above are all of systems that are presumed to be at, or close to, equilibrium. However, more recently it has proved possible to use long MD simulations to look at large scale dynamic events, such as self-assembly of detergent micelles [33] or of lipid bilayers [63]. Given the importance of dynamic events in large scale protein-mediated rearrangements of biological membranes (e.g. membrane fusion [64]) it is of interest to examine the potential for using long timescale MD simulations to explore processes of self-assembly and transition involving membrane proteins.

In order to explore the process of protein/detergent micelle self-assembly we have run simulations of the formation of a detergent micelle from a random arrangement of detergent molecules within a simulation box also containing water molecules and a membrane protein, the latter in its experimental conformation as determined by either X-ray diffraction (OmpA see below) or NMR (GpA; Fig. 7). The detergent chosen was DPC, which has been used in experimental studies of both of these proteins.

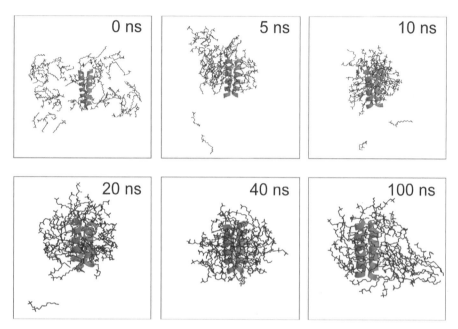

Figure 7. Self-assembly of a GpA/DPC micelle. Snapshots from a 100 ns simulation of self-assembly of a micelle from randomly positioned DPC monomers and a GpA TM helix dimer. The TM helices are shown as grey ribbons and the detergents in bonds format

Snapshots from a 100 ns self-assembly simulation of a GpA-DPC micelle are shown in Fig. 7. It can be seen that over the first few nanoseconds, individual detergent molecules rapidly fuse to form small micelle-like aggregates of ~10 DPC molecules. Thus, within ~5 ns only four or five pure detergent micelles remain. At the same time, a number of initial protein-detergent interactions are created. Subsequently, the small micelle-like aggregates fuse with one another, and with the detergent molecules already bound to the protein. After ~10 to 20 ns, the system consists of a loosely-packed protein-DPC micelle (in which detergents cover much of the protein surface) plus a weakly-interacting DPC aggregate, or globule. Micelle formation kinetics can be approximated as a single exponential process, with a time constant of ~5 ns for GpA. We note that similar events have been seen in simulations of the self-assembly of GpA-SDS micelles [48].

Micelle self-assembly has been compared for simulations of GpA-DPC and of OmpA-DPC micelles [65]. Despite differences in protein architecture (an α-helix dimer vs. a β-barrel) the simulations suggest a common mechanistic pathway for micelle formation. Both self-assembly processes exhibited exponential kinetics of detergent-protein adsorption, suggesting a simple diffusive, stochastic adsorption model for self-assembly. The simulations revealed that the end structures of the self-assembled micelles are similar to those from their pre-formed counterparts, with an approximate torus of detergent surrounding the membrane proteins (see above).

These and related simulations are of interest with respect to future directions for simulations of membranes. In particular, they demonstrate that direct simulation of larger scale dynamic reorganisation and self-assembly of membrane protein-containing structures is feasible. Recent studies have also demonstrated the value of coarse-grain approximations to lipids [66, 67] in allowing simulation of large scale dynamic events such as membrane vesicle fusion [68] or peptide-induced lipid phase transitions [69]. Thus, in the near future we are likely to see an expansion of such studies to cover a wide range of large scale dynamic membrane phenomena of biological importance.

Acknowledgements

Our thanks to all of our colleagues, especially Kaihsu Tai, for their interest in this work. Research in MSPS's laboratory is funded by the BBSRC, the EPSRC, and the Wellcome Trust.

References

[1] Wallin, E., von Heijne, G.: Genome-wide analysis of integral membrane proteins from eubacterial, archean, and eukaryotic organisms. Prot. Sci., **7**, 1029–1038 (1998)

[2] Terstappen, G.C., Reggiani, A.: *In silico* research in drug discovery. Trends Pharmacol. Sci., **22**, 23–26 (2001)

[3] Tate, C.G.: Overexpression of mammalian integral membrane proteins for structural studies. FEBS. Lett., **504**, 94–98 (2001)

[4] White, S.H.: The progress of membrane protein structure determination. Prot. Sci., **13**, 1948–1949 (2004)

[5] Ash, W.L., Zlomislic, M.R., Oloo, E.O., Tieleman, D.P.: Computer simulations of membrane proteins. Biochim. Biophys. Acta, **1666**, 158–189 (2004)

[6] Bond, P.J., Sansom, M.S.P.: The simulation approach to bacterial outer membrane proteins. Mol. Memb. Biol., **21**, 151–162 (2004)

[7] Tristram–Nagle, S., Nagle, J.F.: Lipid bilayers: Thermodynamics, structure, fluctuations, and interactions. Chem. Phys. Lipids, **127**, 3–14 (2004)

[8] Wiener, M.C., White, S.H.: Structure of a fluid dioleoylphosphatidylcholine bilayer determined by joint refinement of x-ray and neutron diffraction data. III. Complete structure. Biophys. J., **61**, 434–447. (1992)

[9] Arora, A., Tamm, L.K.: Biophysical approaches to membrane protein structure determination. Curr. Opin. Struct. Biol., **11**, 540–547 (2001)

[10] Garavito, R., Ferguson-Miller, S.: Detergents as tools in membrane biochemistry. J. Biol. Chem, **276**, 32403–32406 (2001)

[11] Girvin, M.E., Rastogi, V.K., Abildgaard, F., Markley, J.L., Fillingame, R.H.: Solution structure of the transmembrane H^+-transporting subunit c of the F1F0 ATP synthase. Biochem., **37**, 8817–8824 (1998)

[12] Karplus, M.J., McCammon, J.A.: Molecular dynamics simulations of biomolecules. Nature Struct. Biol., **9**, 646–652 (2002)

[13] Guidoni, L., Carloni, P.: Potassium permeation through the KcsA channel: A density functional study. Biochim. Biophys. Acta, **1563**, 1–6 (2002)

[14] Brooks, B.R., Bruccoleri, R.E., Olafson, B.D., States, D.J., Swaminathan, S., Karplus, M.: CHARMM: A program for macromolecular energy, minimisation, and dynamics calculations. J. Comp. Chem., **4**, 187–217 (1983)

[15] Lindahl, E., Hess, B., van der Spoel, D.: Gromacs 3.0: A package for molecular simulation and trajectory analysis. J. Molec. Model., **7**, 306–317 (2001)

[16] van Gunsteren, W.F., Kruger, P., Billeter, S.R., Mark, A.E., Eising, A.A., Scott, W.R.P., Huneberger, P.H., Tironi, I.G. 1996. Biomolecular simulation: The GROMOS96 manual and user guide. Groningen & Zurich: Biomos & Hochschulverlag AG an der ETH Zurich.

[17] Kalé, L., Skeel, R., Bhandarkar, M., Brunner, R., Gursoy, A., Krawetz, N., Phillips, J., Shinozaki, A., Varadarajan, K., Schulten, K.: NAMD2: Greater scalability for parallel molecular dynamics. J. Comp. Phys., **151**, 283–312 (1999)

[18] Sagui, C., Darden, T.A.: Molecular dynamics simulations of biomolecules: Long-range electrostatic effects. Ann. Rev. Biophys. Biomol. Struct., **28**, 155–179 (1999)

[19] Tobias, D.J.: Electrostatics calculations: Recent methodological advances and applications to membranes. Curr. Opin. Struct. Biol., **11**, 253–261 (2001)

[20] Hunenberger, P.H., J.A., M.: Effect of artificial periodicity in simulations of biomolecules under Ewald boundary conditions: A continuum electrostatics study. Biophys. Chem., **78**, 69–88 (1999)

[21] Weber, W., Hunenberger, P.H., McCammon, J.A.: Molecular dynamics simulations of a polyalanine octapeptide under Ewald boundary conditions: Influence of artificial periodicity on peptide conformation. J. Phys. Chem. B, **104**, 3668–3675 (2000)

[22] Anezo, C., de Vries, A.H., Holtje, H.D., Tieleman, D.P., Marrink, S.J.: Methodological issues in lipid bilayer simulations. J. Phys. Chem. B, **107**, 9424–9433 (2003)

[23] Patra, M., Karttunen, M., Hyvonen, M.T., Falck, E., Lindqvist, P., Vattulainen, I.: Molecular dynamics simulations of lipid bilayers: Major artifacts due to truncating electrostatic interactions. Biophys. J., **84**, 3636–3645 (2003)

[24] Patra, M., Karttunen, M., Hyvonen, M.T., Falck, E., Vattulainen, I.: Lipid bilayers driven to a wrong lane in molecular dynamics simulations by subtle changes in long–range electrostatic interactions. J. Phys. Chem. B, **108**, 4485–4494 (2004)

[25] Bostick, D.L., Berkowitz, M.L.: The implementation of slab geometry for membrane-channel molecular dynamics simulations. Biophys. J., **85**, 97–107 (2003)

[26] Killian, J.A., von Heijne, G.: How proteins adapt to a membrane-water interface. Trends Biochem. Sci., **25**, 429–434 (2000)

[27] Crozier, P.S., Stevens, M.J., Forrest, L.R., Woolf, T.B.: Molecular dynamics simulation of dark-adapted rhodopsin in an explicit membrane bilayer: Coupling between local retinal and larger scale conformational change. J. Mol. Biol., **333**, 493–514 (2003)

[28] Feller, S.E., Gawrisch, K., Woolf, T.B.: Rhodopsin exhibits a preference for solvation by polyunsaturated docosohexaenoic acid. J. Am. Chem. Soc., **125**, 4434–4435 (2003)

[29] Lindahl, E., Edholm, O.: Mesoscopic undulations and thickness fluctuations in lipid bilayers from molecular dynamics simulations. Biophys. J., **79**, 426–433 (2000)

[30] Roux, B.: Theoretical and computational models of ion channels. Curr. Opin. Struct. Biol., **12**, 182–189 (2002)

[31] Giorgetti, A., Carloni, P.: Molecular modeling of ion channels: Structural predictions. Curr. Opin. Chem. Biol., **7**, 150–156 (2003)

[32] Efremov, R.G., Nolde, D.E., Konshina, A.G., Syrtcev, N.P., Arseniev, A.S.: Peptides and proteins in membranes: What can we learn via computer simulations? Curr. Med. Chem., **11**, 2421–2442 (2004)

[33] Marrink, S.J., Tieleman, D.P., Mark, A.E.: Molecular dynamics simulation of the kinetics of spontaneous micelle formation. J. Phys. Chem. B, **104**, 12165–12173 (2000)

[34] Tieleman, D.P., van der Spoel, D., Berendsen, H.J.C.: Molecular dynamics simulations of dodecylphosphocholine micelles at three different aggregate sizes: Micellar structure and chain relaxation. J. Phys. Chem. B, **104**, 6380–6388 (2000)

[35] Bogusz, S., Venable, R.M., Pastor, R.W.: Molecular dynamics simulations of octly glucoside micelles: Structural properties. J. Phys. Chem. B., **104**, 5462–5470 (2000)

[36] Bogusz, S., Venable, R.M., Pastor, R.W.: Molecular dynamics simulations of octyl glucoside micelles: Dynamic properties. J. Phys. Chem. B., **105**, 8312–8321 (2001)

[37] Tieleman, D.P., Marrink, S.J., Berendsen, H.J.C.: A computer perspective of membranes: Molecular dynamics studies of lipid bilayer systems. Biochim. Biophys. Acta, **1331**, 235–270 (1997)

[38] Rastogi, V.K., Girvin, M.E.: Structural changes linked to proton translocation by subunit c of the atp synthase. Nature, **402**, 263–268 (1999)

[39] Stock, D., Leslie, A.G.W., Walker, J.E.: Molecular architecture of the rotary motor in atp synthase. Science, **286**, 1700–1705 (1999)

[40] Forrest, L.R. 2000. Simulation studies of proton channels and transporters [D. Phil.]: University of Oxford. 224 p.

[41] Forrest, L.R., Groth, G., Sansom, M.S.P.: Simulation studies on subunit c from f0f1-atpase in different solvents. Biophys. J., **80**, 212 (2001)

[42] Fioroni, M., Burger, K., Mark, A.E., Roccatano, D.: A new 2,2,2-trifluoroethanol model for molecular dynamics simulations. J. Phys. Chem. B, **104**, 1234712354 (2000)

[43] Kovacs, H., Mark, A.E., Johansson, J., van Gunsteren, W.F.: The effect of environment on the stability of an integral membrane helix:- molecular dynamics simulations of surfactant protein c in chloroform, methanol and water. J. Mol. Biol., **247**, 808–822 (1995)

[44] Cook, G.A., Prakash, O., Zhang, K., Shank, L.P., Takeguchi, W.A., Robbins, A., Gong, Y.X., Iwamoto, T., Schultz, B.D., Tomich, J.M.: Activity and structural comparisons of solution associating and monomeric channel-forming peptides derived from the glycine receptor M2 segment. Biophys. J., **86**, 1424–1435 (2004)

[45] Roccatano, D., Colombo, G., Fioroni, M., Mark, A.E.: Mechanism by which 2,2,2-trifluoroethanol/water mixtures stabilize secondary-structure formation in peptides: A molecular dynamics study. Proc. Nat. Acad. Sci. USA, **99**, 12179–12184 (2002)

[46] Bond, P.J., Sansom, M.S.P.: Membrane protein dynamics vs. Environment: Simulations of OmpA in a micelle and in a bilayer. J. Mol. Biol., **329**, 1035–1053 (2003)

[47] Ceccarelli, M., Marchi, M.: Simulation and modeling of the rhodobacter sphaeroides bacterial reaction center: Structure and interactions. J. Phys. Chem. B, **107**, 1423–1431 (2003)

[48] Braun, R., Engelman, D.M., Schulten, K.: Molecular dynamics simulations of micelle formation around dimeric glycophorin a transmembrane helices. Biophys. J., **87**, 754–763 (2004)

[49] Patargias, G., Bond, P.J., Deol, S.D., Sansom, M.S.P.: Molecular dynamics simulations of GlpF in a micelle vs. in a bilayer: Conformational dynamics of a membrane protein as a function of environment. J. Phys. Chem. B., **109**, 575–582 (2005)

[50] Lee, A.G.: Lipid-protein interactions in biological membranes: A structural perspective. Biochim. Biophys. Acta, **1612**, 1–40 (2003)

[51] Forrest, L.R., Sansom, M.S.P.: Membrane simulations: Bigger and better? Curr. Opin. Struct. Biol., **10**, 174–181 (2000)

[52] Domene, C., Bond, P., Sansom, M.S.P.: Membrane protein simulation: Ion channels and bacterial outer membrane proteins. Adv. Prot. Chem., **66**, 159–193 (2003)

[53] Nina, M., Bernèche, S., Roux, B.: Anchoring of a monotopic membrane protein: The binding of prostaglandin H2 synthase–1 to the surface of a phospholipid bilayer. Eur. Biophys. J., **29**, 439–454 (2000)

[54] Jensen, M.O., Mouritsen, O.G., Peters, G.H.: Simulations of a membrane-anchored peptide: Structure, dynamics, and influence on bilayer properties. Biophys. J., **86**, 3556–3575 (2004)

[55] Im, W., Roux, B.: Ions and counterions in a biological channel: A molecular dynamics simulation of ompf porin from *Escherichia coli* in an explicit membrane with 1 M KCl aqueous salt solution. J. Mol. Biol., **319**, 1177–1197 (2002)

[56] Faraldo-Gómez, J.D., Smith, G.R., Sansom, M.S.P.: Molecular dynamics simulations of the bacterial outer membrane protein FhuA: A comparative study of the ferrichrome-free and bound states. Biophys. J., **85**, 1–15 (2003)

[57] Bond, P.J., Faraldo-Gómez, J.D., Sansom, M.S.P.: OmpA – a pore or not a pore? Simulation and modelling studies. Biophys. J., **83**, 763–775 (2002)

[58] Baaden, M., Meier, C., Sansom, M.S.P.: A molecular dynamics investigation of mono- and dimeric states of the outer membrane enzyme OMPLA. J. Mol. Biol., **331**, 177–189 (2003)

[59] Baaden, M., Sansom, M.S.P.: OmpT: Molecular dynamics simulations of an outer membrane enzyme. Biophys. J., **87**, 2942–2953 (2004)

[60] Deol, S.S., Bond, P.J., Domene, C., Sansom, M.S.P.: Lipid-protein interactions of integral membrane proteins: A comparative simulation study. Biophys. J., **87**, 3737–3749 (2004)

[61] Ramakrishnan, M., Pocanschi, C.L., Kleinschmidt, J.H., Marsh, D.: Association of spin-labeled lipids with β-barrel proteins from the outer membrane of *Escherichia coli*. Biochem., **43**, 11630–11636 (2004)

[62] Vandeputte-Rutten, L., Kramer, R.A., Kroon, J., Dekker, N., Egmond, M.R., Gros, P.: Crystal structure of the outer membrane protease ompt from escherichia coli suggests a novel catalytic site. EMBO J., **20**, 5033–5039 (2001)

[63] Marrink, S.J., Lindahl, E., Edholm, O., Mark, A.E.: Simulation of the spontaneous aggregation of phospholipids into bilayers. J. Am. Chem. Soc., **123**, 8638–8639 (2001)

[64] Brunger, A.T.: Structure of proteins involved in synaptic vesicle fusion in neurons. Ann. Rev. Biophys. Biomol. Struct., **30**, 157–171 (2001)

[65] Bond, P.J., Cuthbertson, J.M., Deol, S.D., Sansom, M.S.P.: MD simulations of spontaneous membrane protein/detergent micelle formation. J. Am. Chem. Soc., **126**, 15948–15949 (2004)

[66] Shelley, J.C., Shelley, M.Y., Reeder, R.C., Bandyopadhyay, S., Klein, M.L.: A coarse grain model for phospholipid simulations. J. Phys. Chem. B, **105**, 4464–4470 (2001)

[67] Murtola, T., Falck, E., Patra, M., Karttunen, M., Vattulainen, I.: Coarse-grained model for phospholipid/cholesterol bilayer. J. Chem. Phys., **121**, 9156–9165 (2004)

[68] Marrink, S.J., Mark, A.E.: The mechanism of vesicle fusion as revealed by molecular dynamics simulations. J. Am. Chem. Soc., **125**, 11144–11145 (2003)

[69] Nielsen, S.O., Lopez, C.F., Ivanov, I., Moore, P.B., Shelley, J.C., Klein, M.L.: Transmembrane peptide-induced lipid sorting and mechanism of l-α-to-inverted phase transition using coarse-grain molecular dynamics. Biophys. J., **87**, 2107–2115 (2004)

Modeling and Simulation Based Approaches for Investigating Allosteric Regulation in Enzymes

Marc Q. Ma[1,2], Kentaro Sugino[1], Yu Wang[1], Narain Gehani[1], and Annie V. Beuve[3]

[1] Department of Computer Science, New Jersey Institute of Technology, University Heights, Newark, NJ 07102, USA
[2] Center of Applied Mathematics and Statistics, New Jersey Institute of Technology, University Heights, Newark, NJ 07102, USA
[3] Department of Pharmacology and Physiology, University of Medicine and Dentistry of New Jersey, 185 S. Orange Avenue, Newark, NJ 07103, USA

Summary. Understanding complex biological processes such as regulation of enzymes requires development and application of new mathematical and computational models, tools and/or protocols, including homology–based modeling (HM) and molecular dynamics (MD) simulations. We demonstrate how these *in silico* methods can be used to advance our understanding of complex processes using the allosteric regulation of soluble guanynyl cyclase (sGC) as an example. Future directions for the sGC research are also identified.

Key words: homology–based modeling, molecular dynamics simulations, allosteric activation/regulation, soluble guanylyl cyclase, GTP, cGMP, YC-1

1 Introduction

Understanding complex biological processes such as regulation of enzymes requires development and application of new mathematical and computational models, tools and/or protocols, including homology-based modeling (HM) and molecular dynamics (MD) simulations. We briefly introduce these *in silico* methods (this section), illustrate how they can be used to advance our understanding of complex biological processes using the allosteric regulation of soluble guanynyl cyclase (sGC) as an example (Sect. 2), and identify future directions (Sect. 3) for the sGC research.

1.1 Homology–Based Modeling

Homology-based modeling (HM) uses the sequence information and public sequence and structure databases to make predictions on the most likely native 3D structures of proteins when there exist close homologues whose 3D structures are known. Homology does not necessarily imply sequence similarity. Homology means sharing a common evolutionary origin. For a set of proteins hypothesized to be homologous, their 3D structures are conserved to a greater extent than their sequences. There are software packages and websites that provide the functionalities for protein structure prediction based on HM such as Insight II (Accelrys, San Diego, CA), 3D-JIGSAW [2, 3, 8] and MODELLER [33]. SCWL [5] can be used for rebuilding the side–chains to improve accuracy of predictions.

The basic idea of HM is to compile the structure from fragment libraries with standard geometries based on the following observations:

- The structure of a protein is uniquely determined by its amino acid sequence [11].
- During evolution, the structure is more stable and changes much slower then the associated sequence. Thus, similar sequences adopt practically identical structure, and distantly related sequences still fold into similar structure [6, 37].
- As long as the length of two sequences and the percentage of identical residues fall in the region marked as "safe", the two sequences are practically guaranteed to adopt a similar structure.

1.2 Classical Molecular Dynamics

Classical molecular dynamics (MD) is the most widely used technique in biomolecular modeling, in which starting with the atoms' coordinates, connectivity and force field parameters, we compute trajectories, viz. collections of the time evolution of the Cartesian coordinates for each atom in 3D space. To do this we solve the Newton's equations of motion, a system of 2nd order ordinary differential equations (1),

$$\mathbf{M}\frac{d^2}{dt^2}\mathbf{q}(t) = -\nabla U\left(\mathbf{q}(t)\right) , \qquad (1)$$

where $\mathbf{q(t)}$ is the position vector, $\mathbf{M}$ is the diagonal mass matrix, d is the differential operator, t is time, $U\left(\mathbf{q}(t)\right)$ is the potential energy, and $-\nabla U(\cdot)$ is the force. The energy typically consists of contributions from harmonic potentials that model the covalent bonds, dihedral and improper potentials that model the torsion by 4 atoms, Lennard–Jones (LJ) potentials that models a van der Waals attraction and hard core repulsion, and the electrostatic potentials that model the Coulomb interactions.

One can more efficiently solve (1) using the multiple time stepping (MTS) Verlet-I [14]/r-RESPA [45]/Impulse algorithm, represented by

$$\mathbf{M}\frac{d^2}{dt^2}\mathbf{q} = \mathbf{F}^{\text{fast}} - \sum_{n'=-\infty}^{\infty} \Delta t\, \delta(t - n'\Delta t)\nabla U^{\text{slow}}(\mathbf{q}) , \qquad (2)$$

which splits the potentials into fast (bond, angle, dihedral and improper, and short–range Lennard–Jones and electrostatic) and slow (long–range electrostatic) components, and evaluates the former more frequently than the latter. Here δ is the Dirac delta function, Δt is the small time step by the inner integrator, $n'\Delta t$ is the large time step for the outer integrator, and n' is an integer.

MTS integrators allow larger time steps (in outer integrators) than their single time stepping (STS) counterparts such as Verlet (or Leapfrog) and its variants, thus reducing the time for computing the long–range electrostatic forces, which are the most compute–intensive among all the forces. The Impulse algorithm has good long–time energy behavior. Due to the strong nonlinearity of non-bonded forces and extreme stiffness of (2), the time steps allowed are restricted to less than 3.3 femtoseconds (fs) (1 fs $= 10^{-15}$ second, i.e., one quadrillionth of a second or one thousandth of a nanosecond) in the Impulse integrator for most biological systems, so as to obtain stable solutions over a long period of simulated time [31]. The restriction of time steps in MTS integrators is due to 3:1 nonlinear overheating, cf. [31].

Among all MTS integrators such as Impulse, mollified Impulse (MOLLY) [19], Langevin MOLLY [17], Targeted MOLLY (TM) [30] and LN (Langevin Normal mode) [1, 38], Impulse is the most well–studied and the most suitable for long production simulations for both Newtonian dynamics and conformational sampling.

Besides using MTS integrators, modeling practitioners also use parallel computing and fast electrostatics, *e.g.*, Particle Mesh Ewald (PME) [9, 26] and Multi-Grid [40], and a well designed simulation software package such as ProtoMol [34], NAMD [25] or ORAC [36] to speed up their simulations.

1.3 Conformational Space Sampling

It may be necessary to explore the conformational space of the biological molecules of interest in order to understand better the structural basis of certain functions. Classical MD can be used to sample the conformational space by itself or in conjunction with other techniques such as Monte Carlo (MC) simulations, which leads to the Hybrid MC (HMC) method. HMC uses a short MD trajectory to generate a global MC move and then uses the Metropolis criterion to accept or reject the move.

A novel biased HMC method, termed as Shadow HMC or SHMC, has been reported with improved efficiency of sampling of the conformational space of biomolecules [18] and implemented in ProtoMol. SHMC allows larger time steps and system sizes in the molecular dynamics (MD) step and achieves an order of magnitude speedup in sampling efficiency for medium sized proteins (30) by sampling from all conformational space using high–order approximations to a shadow or modified Hamiltonian exactly integrated by a symplectic MD integrator. SHMC satisfies microscopic reversibility and is a rigorous sampling method.

Langevin dynamics can also be used for conformational space sampling. LN [1, 38], which uses a non-symplectic, stochastic and extrapolative method and basically permits time steps to be chosen on the basis of accuracy considerations only, is a valid and competitive method for solving the Langevin dynamics equations. Due to the choice of larger time steps, LN samples conformational space in extremely efficient manner. In this sense, LN is the best quasi-multiscale integrator for conformational space sampling.

1.4 Soluble Guanylyl Cyclase, an Allosterically Regulated Enzyme

The allosteric regulation [13] of catalytic reaction within soluble guanynyl cyclase (sGC) is a complex biological process and there have been rich experimental studies on sGC. The sGC contains in its catalytic core a binding site for the substrate guanosine 5'-triphosphate (GTP) and a potential regulatory site for binding of allosteric modulator [15] such as YC-1($C_{19}H_{16}N_2O_2$) [41], a benzylindazole derivative (Fig. 1).

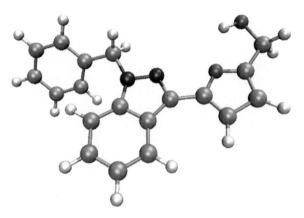

Figure 1. The structural details of YC-1 ($C_{19}H_{16}N_2O_2$)

It is generally accepted that the conformation of the active site changes when an activator binds, which results in either higher affinity for the substrate or higher velocity of the reaction or both. The mammalian sGC is a 150 kDa heterodimer, consisting of an α_1 (74-82 kDa) subunit and a β_1 (69-74 kDa) subunit [4, 32, 44]. The cyclization of GTP to guanosine 3',5'-cyclic monophosphate (cGMP) is catalyzed in the catalytic center of sGC. cGMP is a second messenger that regulates many biological processes, including vasodilation, kidney function, smooth muscle activity and light perception in the retina. The mechanisms regulating the catalytic activity of sGC remain unclear despite the extensive experimental studies [7, 27, 28, 42]. In addition, no crystal structure of sGC is currently available.

Recently, we conducted a mutational analysis of the wild–type (WT) sGC to study the impact of single–point mutation on catalytic activity and YC-1 allosteric activation of sGC [28]. Our data provided some evidence that YC-1 may adopt two orientations in its putative binding pocket. We postulated that the hydroxymethyl group of YC-1 may either face towards the inside of its binding pocket, a binding mode termed as "Normal" or the opposite direction, termed as "Flip." With different binding modes, YC-1 interacts with the protein and the substrate differently and therefore sGC's ability to catalyze the formation of cGMP should be different. The modeling and simulation work on sGC would provide additional insight on the molecular mechanisms of allosteric regulation.

2 Modeling and Simulation of sGC

2.1 Predicting the structure of sGC

We predict the structure of the catalytic center of sGC using both commercial and free software packages that implement HM methods. We used Insight II (Accelrys, San Diego, CA) and 3D-JIGSAW [2, 3, 8] for structure prediction and SCWL [5] for rebuilding the side–chains. Our predictions made using different packages are very similar and important features such as binding pockets are reproduced.

We used the following parts of the rat sGC sequences: α_1 V480-L625, β_1 V420-L485 and β_1 H492-E576, which correspond to the catalytic center of the sGC. The template molecule is adenylyl cyclase (AC, PDB entry 1AZS), which is functionally similar to sGC. The sequence identity that the three sequences of sGC share with AC is 34%, 36% and 47%, respectively. Based on previous studies, if the sequence identity is greater than 30%, no matter what kind of alignment method is used (*e.g.,* pairwise, profile or threading), correctness of the alignment result generally will be greater then 80% [23]. The center region of the predicted structure is shown in Fig. 2 We also docked GTP, YC-1 and two magnesium ions (Mg^{2+}) into the predicted structure using VMD [16]. The two magnesium ions were included for two reasons: first, the substrate for sGC is the complex [Mg^{2+}–GTP] and not GTP alone; second, the catalytic reaction, *i.e.,* formation of cGMP, is a two–ion mechanism [46].

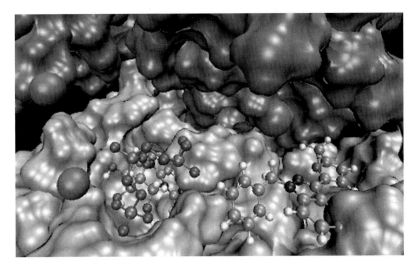

Figure 2. The center region of the predicted structure of the catalytic center of sGC in complex with GTP (in the *left* pocket) and two magnesium ions (isolated spheres) and YC-1 (in the *right* pocket). The α_1 subunit (*top*) has been lifted and rotated to reveal more structural details of the binding sites. The β_1 subunit is at the *bottom*

2.2 Constructing Model Systems for MD Simulations

Three Model Systems

We construct three model systems for MD simulations. In the first system, there is no YC-1 binding. In the second system, YC-1 initially binds with the "Normal" mode, *i.e.,* the hydroxymethyl group of YC-1 initially facing towards the inside of its binding pocket. In the third system, YC-1 binds with the "Flip" mode, *i.e.,* opposite to the orientation in the "Normal" mode.

Parameterization of GTP and YC-1

There are no standard parameters in CHARMM format directly for GTP or YC-1. Parameterization of GTP is done by replacing adenosine in ATP with guanine and using the parameters of guanine and the phosphorus groups of ATP. For novel molecules such as YC-1, one can use semi-empirical molecular orbital programs, such as MOPAC [43], to generate the parameters. MOPAC calculates optimized geometry (bond length, angle, dihedral angle), electron density, atomic charge, electro static potential (ESP) and other molecular properties. MOLDEN, a free z-matrix editor, which can convert PDB format file into z-matrix format, is used to generate the z-matrix which is used as input for MOPAC.

Adding Counter Ions and Explicit Solvation

The [sGC–GTP–YC-1–2 Mg^{2+}] system carries -1 net charge, which should be balanced by adding counteracting ions such as K^+ or Na^+. We use "autoionize" program, a plug-in for VMD, to add counter ions. One sodium (Na^+) ion was added.

To obtain the dynamics of the protein systems accurately, explicit solvation is needed to mimic the biological setting — proteins function in water environment. Implicit solvation may be used if the main purpose of the simulations is to sample the phase space efficiently. For our simulations, we use TIP3P [24] rigid water model. We used 3, 700 TIP3P water molecules as the bulk water box (65Å × 54Å × 54Å). The resulting system contains roughly 16, 000 atoms.

2.3 Three 1-Nanosecond Simulations

Simulation Protocol

The solvated sGC (before docking GTP, YC-1 and magnesium ions) undergoes a minimization, heating and equilibration process. In the minimization process (100,000 steps), the water molecules are let move while keeping protein's configuration unchanged. Then both protein and water molecules are let move for further minimization. After this process, the system is heated gradually to room temperature using the standard protocols. After the system reaches room temperature (300K), a 300ps equilibration process is performed to bring the system into an equilibrated state. At this stage, GTP, two magnesium ions and one counter ion are added to the system. Additional 6000 steps of minimization is performed. Then, we dock the YC-1 and perform additional minimization. Minimization is done using conjugate gradient method of NAMD which is used to eliminate illegal close contacts of atoms. After these steps, one 1-nanosecond MD simulation is performed on each of the three systems NAMD. For these simulations, we use CHARMM27 force field [12], which is augmented by the new parameters that we determined for YC-1 and GTP, for our model systems, Impulse [14, 45] MTS integrator with 3fs outer step size for good efficiency and stability in solving the governing equations of motion, and PME [9, 26] for fast electrostatics calculation. Extreme care is used in choosing parameters for the above protocol so that we can get the truthful dynamics and transient behaviors of GTP, YC-1 and sGC.

Dynamics of GTP

Formation of cGMP from substrate GTP proceeds through the nucleophilic attack of the α phosphorus (P_α) by the 3′ hydroxyl (oxygen O3) of the ribose ring of GTP. Shorter distance between these two atoms indicates higher probability to complete such a process.

Figure 3 shows the distance between P_α and O3 in the 1ns simulations of the three model systems. For the "Normal" binding mode of YC-1, this distance is kept low, hovering at around 3.6Å. For the "Flip" binding mode of YC-1 and for the system without YC-1 binding, this distance rises to and maintains at 4.9Å (after 200ps for YC-1 "Flip"–mode simulation and after 800ps for the no–YC-1–mode simulation). These results suggest that different initial binding mode of YC-1 makes a significant difference in terms of YC-1's effect on GTP catalysis by sGC, and initial "Normal"–mode of binding positively correlates with more chances of cyclization.

Dynamics of YC-1

YC-1 does not always stay extended in its binding pocket after initial binding. By interacting with the environment, it folds to a hairpin structure with the hairpin opening facing towards the inside of the binding pocket. Therefore, the hydroxymethyl group does not directly interact with the GTP or magnesium ions. It seems that the hydroxymethyl group maintain close interaction with the sulfur of CYS594, an amino acid located deep inside of the allosteric pocket.

The progress of hairpin formation by YC-1 is measured by the distance between the oxygen (O23) in the hydroxymethyl group and a carbon (C19) on the opposite side of YC-1. When YC-1 is fully stretched, this distance is the largest. When YC-1 bends to a hairpin structure, this distance becomes small. The measurement of this distance is shown in Fig. 4. For "Normal"-mode of binding, the YC-1 quickly forms the hairpin structure (in less than 100ps). For "Flip"–mode of binding, it takes YC-1 almost 900ps to reach stable hairpin structure.

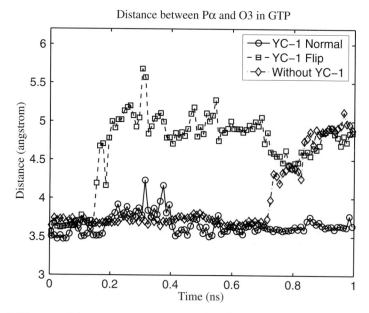

Figure 3. Distance of the α phosphorus (P_α) and the 3′ hydroxyl (oxygen O3) of the ribose ring of GTP measured from 3 1-ns simulations of sGC

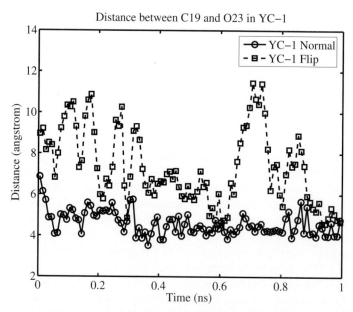

Figure 4. Distance between the oxygen (O23) in the hydroxymethyl group and a carbon (C19) on the opposite side of YC-1, indicating YC-1 forms a hairpin structure after initial binding

Summary of Observations

Key observations from these MD simulations are summarized as follows: (1) The distance between the α phosphorus (P_α) and the 3' hydroxyl (oxygen O3) of the ribose ring of GTP hovers at a low value with the "Normal" binding mode, whereas this distance keeps at a much higher value with the "Flip" binding mode or without YC-1 binding. (2) YC-1 folds to a hairpin structure after initial binding to the protein rather than staying fairly extended. And the folding times are one order of magnitude different when the initial binding modes of YC-1 are different.

These observations suggest that YC-1 has different capabilities of activating the catalysis in sGC due to different initial binding modes. The "Normal"–mode of binding allows YC-1 to increase catalysis by sGC, whereas the "Flip"–mode of binding does not. Thus, YC-1 activation of sGC is probably mainly due to the "Normal"–mode of binding. More analysis is needed to fully understand the cascading events — YC-1 folds to hairpin structure and interacts with the surrounding amino acids, which probably causes conformational changes in the catalytic core where GTP binds.

3 Future Work

It is very promising to apply a set of *in silico* methods to help unravel the molecular mechanisms of allosteric regulation of enzymes as indicated above. It is possible to gain even deeper and more thorough understanding of this and other complex biological processes, which can be achieved through better design of model systems, simulation protocols and more simulations. Some of the future work presented here is exclusively suitable for the study of sGC system, whereas the rest is suitable for many other complex processes as well.

3.1 Simulation of Mutants

Current investigation on the mechanisms of allosteric activation in sGC will be further strengthened after more simulations are performed on some experimentally well–studied mutants which will be constructed through *in silico* mutagenesis. Many more simulations, for the purpose of studying dynamics and/or sampling, on the wild–type sGC and mutants will be performed and all the trajectories will be studied from the "ensemble" point of view in addition to the detailed analysis of each individual trajectories for the transient and dynamical behaviors of the systems.

3.2 Binding Affinity Predictions

Binding affinities indicate how difficult it is to dissociate a ligand from a receptor protein [39]. The binding mode selectivity may be closely related to the difference of binding affinities of YC-1 due to different initial binding modes. We will predict the binding affinities associated with each binding mode of YC-1 in different mutant or WT sGC systems, for which we are going to use steered MD and Jarzynski's equality (SMD/JE) method [20, 21], a promising new technique for fast and accurate binding free energy prediction of ligand–receptor systems [35]. In sGC system, a natural choice of the reaction coordinate is the distance from the atom in YC-1 that is being pulled by the SMD force, *e.g.*, C19, to the bottom of the allosteric pocket.

3.3 The Feedback Loop

These simulations (dynamics and/or sampling) will be used to derive a detailed molecular model of allosteric regulation of sGC. Based on this model, predictions will be made on which sites are critical and whether the allosteric effects will get abolished or enhanced upon mutation. Wet lab mutagenesis studies will be performed to check if the predictions match with the experimental data (feedback). We may need to iterate a few times before a satisfactory model is established.

3.4 A Database for Storing and Managing all Trajectories

These MD simulations are very time–consuming: to compute 1ns trajectory of any one of our three model systems, it takes roughly 1 week of computer time using one CPU on a SunBlade 2000 workstation (1.2 GHz processor, 2GB RAM), or 24 hours using 8 CPUs on a SunFire 6900 parallel computer (24 dual–core 1.2 GHz processor, 2GB RAM for each CPU). In the current and future studies, many 100ns trajectories will be generated, which will require tremendous amount of computing time using the state of the art computers or computer clusters. These MD trajectories are extremely valuable and should be stored and managed carefully.

A relational database is needed for this purpose. This database will store all the input/output files for all MD simulations including the DCD files (a popular binary trajectory format that contains all the coordinates of the system simulated frame by frame), and other files such as the analysis scripts, raw analysis results, graphics, and movies. We are going to implement a MySQL database, a popular open-source database, and build a web interface using PHP, a widely-used general-purpose scripting language that is especially suited for Web development and can be embedded into HTML. This system will also be used to distribute the newest results.

3.5 Novel Multiscale Integrators

We will explore new ideas of developing multiscale integrators that are more efficient in time stepping than Impulse, allowing shorter turn-around time for simulations. We will use ProtoMol as the testbed of these new integrators.

It is possible to develop new Integrators similar to the mollified Impulse method (MOLLY) [19]. We can make these integrators symmetric and reversible, a property that is enough for certain applications such as hybrid Monte Carlo simulations. Computationally, these reversible but non-symplectic integrators may be more efficient than their reversible and symplectic counterparts due to the reduced overhead in computing the averaged positions and mollified slow forces.

Meanwhile, the *Projective Thermostatting Dynamics* (PTD) is a powerful new method for modeling and simulations, in which fast modes can be selectively and effectively filtered out to ensure the stability of treating slow modes with large time steps [22]. It may be possible to construct MTS integrators that allow larger time steps when we combine MOLLY [19] with PTD.

3.6 Other Ensembles and Force Fields

Results shown in this paper were obtained from simulations of microcanonical (NVE) ensemble. Isothermal-isobaric (NpT) ensemble simulations probably are more appropriate for the real biological setting in which the temperature and pressure are almost constant. NpT simulations will be performed and the results will be compared against those from the NVE simulations, which will give insights on ensemble–related numerical artifacts.

The FF03/AMBER force field [10] arguably performs better than CHARMM in balancing between the α-helix and β-strand conformations for extended-time simulations of proteins whose structure contains both α and β structural elements [29]. Choosing a more appropriate force field will undoubtedly reduce the numerical artifacts. We will need to run more simulations using the FF03/AMBER force field.

Acknowledgments

This research is funded by NSF (MRI-DMS 0420590), NIH (RO1-GM 067640-01) and NJIT. Simulations are performed on a SunFire 6900 system at UMDNJ and the Hydra Linux cluster at NJIT. We are grateful to the following people for their help in this research and manuscript preparation: Joshua Berlin, Yong Duan, Jesús Izaguirre, Ben Leimkuhler, Piero Procacci, Tamar Schlick, Martin Tenniswood, Carol Venanzi and Ruhong Zhou.

References

[1] E. Barth and T. Schlick. Overcoming stability limitations in biomolecular dynamics. I. Combining force splitting via extrapolation with Langevin dynamics in LN. *J. Chem. Phys.*, 109(5):1617–1632, August 1998.

[2] P. A. Bates, L. A. Kelley, R. M. MacCallum, and M. J. E. Sternberg. Enhancement of protein modelling by human intervention in applying the automatic programs 3D-JIGSAW and 3D-PSSM. *PROTEINS: Struc., Func., and Genetics*, Suppl 5:39–46, 2001.

[3] P. A. Bates and M. J. E. Sternberg. Model building by comparison at CASP3: Using expert knowledge and computer automation. *PROTEINS: Struc., Func., and Genetics*, Suppl 3:47–54, 1999.

[4] E. Bischoff, J.-P. Stasch, F. Mullershausen, M. Russwurm, A. Friebe, and D. Koesling. Effects of the sGC stimulator BAY 41-2272 are not mediated by phosphodiesterase 5 inhibition. *Circulation*, 110:e320–e321, 2004.

[5] A. A. Canutescu, A. A. Shelenkov, and J. R. L. Dunbrack. A graph theory algorithm for protein side-chain prediction. *Protein Science*, 12:2001–2014, 2003.

[6] C. Chothia and A. M. Lesk. The relation between the divergence of sequence and structure in proteins. *EMBO J.*, 5:823–836, 1986.

[7] P. Condorelli and S. C. George. In vivo control of soluble guanylate cyclase activation by nitric oxide: a kinetic analysis. *Biophys. J.*, 80:2110–2119, 2001.

[8] B. Contreras-Moreira and P. A. Bates. Domain Fishing: A first step in protein comparative modeling. *Bioinformatics*, 18:1141–1142, 2002.

[9] T. Darden, D. York, and L. Pedersen. Particle mesh Ewald: An N log(N) method for Ewald sums in large systems. *J. Chem. Phys.*, 98(12):10089–10092, 1993.

[10] Y. Duan, C. Wu, S. Chowdhury, M. C. Lee, G. Xiong, W. Zhang, R. Yang, P. Cieplak, R. Luo, T. Lee, J. Caldwell, J. Wang, and P. Kollman. A point-charge force field for molecular mechanics simulations of proteins based on condensed-phase quantum mechanical calculations. *J. Comp. Chem.*, 24:1999–2012, 2003.

[11] C. J. Epstein, R. F. Goldberger, and C. B. Anfinsen. The genetic control of tertiary protein structure: Studies with model systems. *Cold Springs Harb. Symp. Quant. Biol.*, 28:439, 1963.

[12] N. Foloppe and A. D. MacKerell, Jr. All-atom empirical force field for nucleic acids: I. parameter optimization based on small molecule and condensed phase macromolecular target data. *J. Comp. Chem.*, 21:86–104, 2000.

[13] A. Friebe, M. Russwurm, E. Mergia, and D. Koesling. A point-mutated guanylyl cyclase with features of the YC-1-stimulated enzyme: Implications for the YC-1 binding site? *Biochemistry*, 38:15253–15257, 1999.

[14] H. Grubmüller, H. Heller, A. Windemuth, and K. Schulten. Generalized Verlet algorithm for efficient molecular dynamics simulations with long-range interactions. *Molecular Simulation*, 6:121–142, 1991.

[15] A. J. Hobbs. Soluble guanylate cyclase: The forgotten sibling. *Trends Pharmacol. Sci.*, 18:484–491, 1997.

[16] W. F. Humphrey, A. Dalke, and K. Schulten. VMD – Visual Molecular Dynamics. *J. Mol. Graphics*, 14:33–38, 1996.

[17] J. A. Izaguirre, D. P. Catarello, J. M. Wozniak, and R. D. Skeel. Langevin stabilization of molecular dynamics. *J. Chem. Phys.*, 114(5):2090–2098, Feb. 1, 2001.

[18] J. A. Izaguirre and S. S. Hampton. Shadow Hybrid Monte Carlo: An efficient propagator in phase space of macromolecules. *J. Comp. Phys.*, 200:581–604, 2004.

[19] J. A. Izaguirre, S. Reich, and R. D. Skeel. Longer time steps for molecular dynamics. *J. Chem. Phys.*, 110(19):9853–9864, May 15, 1999.

[20] C. Jarzynski. Nonequilibrium equality for free energy differences. *Physical Review Letters*, 78:2690–2693, 1997.

[21] C. Jarzynski. Nonequilibrium work theorem for a system strongly coupled to a thermal environment. *J. Stat. Mech., Theory and Exp.*, page Online P09005, 2004.

[22] Z. Jia and B. Leimkuhler. A projective thermostatting dynamics technique. *Multiscale Model. Simul., A SIAM Interdisciplinary Journal*, 4(2):563–583, 2005.

[23] M. S. Johnson and J. P. Overington. A structural basis for sequence comparisons: An evaluation of scoring methodologies. *J. Mol. Biol.*, 233:716–738, 1993.

[24] W. L. Jorgensen, J. Chandrasekhar, J. D. Madura, R. W. Impey, and M. L. Klein. Comparison of simple potential functions for simulating liquid water. *J. Chem. Phys.*, 79:926–935, 1983.

[25] L. Kalé, R. Skeel, M. Bhandarkar, R. Brunner, A. Gursoy, N. Krawetz, J. Phillips, A. Shinozaki, K. Varadarajan, and K. Schulten. NAMD2: Greater scalability for parallel molecular dynamics. *J. Comp. Phys.*, 151:283–312, 1999.

[26] K. Kholmurodov, W. Smith, K. Yasuoka, T. Darden, and T. Ebisuzaki. A smooth-particle mesh Ewald method for DL_POLY molecular dynamics simulation package on the Fujitsu VPP700. *Journal of Computational Chemistry*, 21(13):1187–1191, 2000.

[27] M. Koglin, J. P. Stasch, and S. Behrends. BAY 41-2272 activates two isoforms of nitric oxide-sensitive guanylyl cyclase. *Biochem. Biophys. Res. Commun.*, 292:1057–1062, 2002.

[28] M. Lamothe, F.-J. Chang, R. Shirokov, and A. Beuve. Functional characterization of nitric oxide and YC-1 activation of soluble guanylyl cyclase: Structural implication for the YC-1 binding site? *Biochemistry*, 43:3039–3048, 2004.

[29] M. C. Lee, J. Deng, J. M. Briggs, and Y. Duan. Large scale conformational dynamics of the HIV-1 integrase core domain and its catalytic loop mutants. *Biophysical J. BioFAST*, page doi:10.1529/biophysj.104.058446, 2005.

[30] Q. Ma and J. A. Izaguirre. Targeted mollified Impulse – a multiscale stochastic integrator forlong molecular dynamics simulations. *Multiscale Model. Simul., A SIAM Interdisciplinary Journal*, 2(1):1–21, Nov 2003.

[31] Q. Ma, J. A. Izaguirre, and R. D. Skeel. Verlet-I/r-RESPA/Impulse is limited by nonlinear instability. *SIAM J. on Sci. Comput.*, 24(6):1951–1973, May 2003.

[32] R. Makino, H. Matsuda, E. Obayashi, Y. Shiro, T. Iizuka, and H. Hori. EPR characterization of axial bond in metal center of native and cobalt-substituted guanylate cyclase. *J. Biol. Chem.*, 274:7714–7723, 1999.

[33] M. A. Marti-Renom, A. Stuart, A. Fiser, R. Sanchez, F. Melo, and A. Sali. Comparative protein structure modeling of genes and genomes. *Annu. Rev. Biophys. biomol. Struct.*, 29:291–325, 2000.

[34] T. Matthey, T. Cickovski, S. Hampton, A. Ko, Q. Ma, T. Slabach, and J. A. Izaguirre. PROTOMOL: an object-oriented framework for prototyping novel algorithms for molecular dynamics. *ACM Trans. Math. Softw.*, 30:237–265, 2004.

[35] S. Park, F. Khalili-Araghi, E. Tajkhorshid, and K. Schulten. Free energy calculation from steered molecular dynamics simulations using Jarzynski's equality. *J. Chem. Phys.*, 119:3559–3566, 2003.

[36] P. Procacci, E. Paci, T. Darden, and M. Marchi. ORAC: A molecular dynamics program to simulate complex molecular systems with realistic electrostatic interactions. *J. Comp. Chem.*, 18:1848–1862, 1997.

[37] C. Sander and R. Schneider. Database of homology-derived protein structures and the structural meaning of sequence alignment. *PROTEINS: Struc., Func., and Genetics*, 9:56–68, 1991.

[38] A. Sandu and T. Schlick. Masking resonance artifacts in force-splitting methods for biomolecular simulations by extrapolative Langevin dynamics. *J. Comput. Phys*, 151(1):74–113, May 1, 1999.

[39] T. Simonson, G. Archontis, and M. Karplus. Free energy simulations come of age: Protein-ligand recognition. *Acc. Chem. Res.*, 35:430–437, 2002.

[40] R. D. Skeel, I. Tezcan, and D. J. Hardy. Multiple grid methods for classical molecular dynamics. *J. Comp. Chem.*, 23(6):673–684, March 2002.

[41] J. Sopkova-de Oliveira Santos, V. C. I. Bureau, and S. Rault. YC-1, an activation inductor of soluble guanylyl cyclase. *Acta Crystallographica Section C*, 56:1035–1036, 2000.

[42] J. P. Stasch, E. M. Becker, C. Alonso-Alija, H. Apeler, K. Dembowsky, A. Feurer, R. Gerzer, T. Minuth, E. Perzborn, U. Pleiss, H. Schroder, W. Schroeder, E. Stahl, W. Steinke, A. Straub, and M. Schramm. NO-independent regulatory site on soluble guanylate cyclase. *Nature*, 410:212–215, 2001.

[43] J. J. P. Stewart. MOPAC: A semiempirical molecular obital program. *J. of Computer-Aided Molecular Design*, 4:1–105, 1990.

[44] T. Tomita, T. Ogura, S. Tsuyama, Y. Imai, and T. Kitagawa. Effects of GTP on bound nitric oxide of soluble guanylate cyclase probed by resonance raman spectroscopy. *Biochemistry*, 36:10155–10160, 1997.

[45] M. Tuckerman, B. J. Berne, and G. J. Martyna. Reversible multiple time scale molecular dynamics. *J. Chem. Phys.*, 97(3):1990–2001, 1992.

[46] G. Zimmermann, D. Zhou, and R. Taussig. Mutations uncover a role for two magnesium ions in the catalytic mechanism of adenylyl cyclase. *J. Biol. Chem.*, 273:19650–19655, 1998.

Exploring the Connection Between Synthetic and Natural RNAs in Genomes: A Novel Computational Approach

Uri Laserson[1,2], Hin Hark Gan[1], and Tamar Schlick[1,2]

[1] Department of Chemistry
[2] Courant Institute of Mathematical Sciences, New York University, 251 Mercer Street, New York, New York 10012, USA

Abstract. The central dogma of biology—that DNA makes RNA makes protein—was recently expanded yet again with the discovery of RNAs that carry important regulatory functions (e.g., metabolite-binding RNAs, transcription regulation, chromosome replication). Thus, rather than only serving as mediators between the hereditary material and the cell's workhorses (proteins), RNAs have essential regulatory roles. This finding has stimulated a search for small functional RNA motifs, either embedded in mRNA molecules or as separate molecules in the cell. The existence of such simple RNA motifs in Nature suggests that the results from experimental *in vitro* selection of functional RNA molecules may shed light on the scope and functional diversity of these simple RNA structural motifs *in vivo*. Here we develop a computational method for extracting structural information from laboratory selection experiments and searching the genomes of various organisms for sequences that may fold into similar structures (if transcribed), as well as techniques for evaluating the structural stability of such potential candidate sequences. Applications of our algorithm to several aptamer motifs (that bind either antibiotics or ATP) produce a number of promising candidates in the genomes of selected bacterial and archaeal species. More generally, our approach offers a promising avenue for enhancing current knowledge of RNA's structural repertoire in the cell.

1 Introduction: Importance of RNA Structure and Function

RNA molecules play essential roles in the cellular processes of all living organisms. The wonderful capacity of RNA to form complex, stable tertiary structures has been exploited by evolution. RNA molecules are integral components of the cellular machinery for transcription regulation, chromosome replication, RNA processing and modification, and other essential biological functions [9, 41, 48]. Recent discoveries that noncoding RNAs (ncRNAs—RNA molecules that have specific functions other than directing protein synthesis through translation) make up a significant portion of the transcriptome (entire set of expressed RNA and protein transcripts of an organism) further suggest the prominent role of RNA in cellular function [16, 34]. In particular, cells employ many small ncRNAs such as microRNAs (21–23 nt) to regulate gene expression; small interfering RNAs (siRNAs), often complementary to messenger RNAs, to mediate mRNA degradation; and small nucleolar RNAs, important in post-transcriptional modification of ribosomal and other RNAs [3, 23, 24]. Thus, ncRNAs may hold the key for understanding genetic control of cell development and growth [31, 32].

With rapidly growing interest in RNA structure and function, as well as emerging technological applications of RNA to biomedicine, new scientific challenges regarding RNA are at the forefront. One important objective is to increase the functional repertoire of RNA. This can be tackled by systematically identifying and characterizing the ncRNAs in genomes, or by creating novel synthetic functional RNAs using *in vitro* selection methods. Although *in vitro* selection is a proven tool for finding novel functional RNAs, it is limited in scope to relatively small RNAs (<250 nt) [53].

Here we report development of a theoretical/experimental approach for expanding the functional repertoire of RNA using a combination of *in vitro* selection and computational methods. Alternatively, the analysis of RNA motifs as mathematical graphs, as developed recently ([14, 26, 35], and see the Appendix), may suggest candidate novel RNA topologies to direct computational/experimental searches for ncRNAs in genomes and synthetic functional RNAs in the laboratory. Before we present our approach, we describe the motivation for exploring the connection between synthetic and natural RNAs.

2 Exploring the Connection between Synthetic and Natural RNAs

In recent years, numerous target-binding nucleic acid molecules (known as *aptamers*) have been identified; the targets include organic molecules, antibiotics, peptides, proteins, and whole viruses [21, 53]. In addition, *in vitro* selection experiments have produced novel RNA enzymes (ribozymes) and led to applications in biomolecular engineering, e.g., allosteric ribozymes and biosensors [44–46].

The process of *in vitro* selection simulates evolution in the laboratory [10, 47, 51, 53]. Starting with a large pool of small, random-sequence RNA molecules, the sequences are iteratively selected for a physical or chemical property (e.g., binding affinity or catalysis) by amplifying the enriched sequence pool using the polymerase chain reaction (PCR). After ~8–15 cycles have been completed, the molecules are cloned and sequenced. The resulting artificial molecules may be target-binding RNAs (called *aptamers*) or novel catalytic RNAs. Since the process of selecting functional molecules is similar to evolution, and the process takes advantage of the PCR, common motifs among the sequences with a significant amount of sequence conservation often emerge.

The binding and catalytic properties of synthetic and natural RNAs are mediated by specific sequence and structural motifs. In 2001, Szostak's group demonstrated that the motif of the natural functional hammerhead ribozyme can be selected from random sequence pools [39], suggesting multiple origins for the ribozyme. Because a synthetic aptamer discovered by Wallace and Schroeder [52] and crystallized by the Patel group [49] binds streptomycin (which is naturally produced by bacteria) tightly and specifically, Piganeau and Schroeder reinforced the likely connection between natural and synthetic RNAs [36]. Indeed, motifs similar to aptamers may exist in natural RNAs because aptamers bind to certain targets that are prevalent in Nature and recent findings also show that metabolite-induced RNA conformational changes control gene expression in bacteria and other organisms [33, 54, 55]. As Piganeau and Schroeder conclude in their commentary on the recent article by Patel and collaborators [49] on the structure of a 40-nt aptamer binding the antibiotic streptomycin, *"We can now predict that many biosynthetic pathways will be regulated by metabolite binding 'natural aptamers,' and we might even find a structure similar to the streptomycin aptamer in a bacterium producing streptomycin"* [36].

Since *in vitro* selection is a technique that simulates an evolutionary process, it is reasonable that structures discovered through *in vitro* selection may have also evolved in the cell. The methods we develop here are meant to explore these intriguing connections between natural and synthetic RNAs in a general and systematic manner.

3 Methods

The components used in our method are standard, but the combination is novel. Namely, we combine the following techniques and tools: output from *in vitro* selection of functional molecules; RNAMotif [29], a computational tool that allows searches for RNA sequences in genomic databases that might fold into specified secondary conformations; and the Vienna RNA Package [22], which can predict the secondary structures of RNA molecules from their sequence, as well as calculate other properties, such as "sub-optimal foldings" and heat-capacity curves.

3.1 Aptamer Search Algorithm

Our aptamer search method has three major steps:

1. *Create motif descriptors* by extracting the critical structural features from the experimental aptamer molecule that confer onto the molecule its special physical or chemical property (e.g., binding affinity);
2. *Search the genomes* of selected organisms using the RNAMotif tool for the aptamer structure specified by the descriptor we have created;
3. *Assess the quality* of the candidate sequences (e.g., nature and stability of fold, energetic stability, statistical significance, etc.).

The first step involves the analysis of motif data from *in vitro* selection experiments and, if available, any structural studies of specific motifs. Important structural information includes overall qualitative structure (e.g., loops, bulges, hairpins, lengths of stems, etc.) as well as any specific sequence information that may be critical to the hydrogen-bonding scheme (Fig. 1).

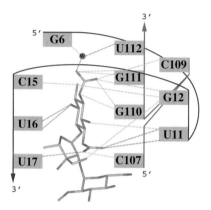

Figure 1. The hydrogen-bonding scheme of the streptomycin aptamer developed by the Schroeder group [52]. The dotted lines represent hydrogen bonds between the streptomycin molecule (center) and the nucleotides in the binding pocket. This type of information is ideal for pinpointing the specific nucleotides that account for the binding specificity. The secondary structure of the molecule is shown in Fig. 2d. (Adapted from [49])

Figure 1 illustrates this process for the streptomycin aptamer. The hydrogen-bonding scheme in the streptomycin-aptamer complex (discovered by the Schroeder group [52]) was crystallized and structurally characterized by the Patel group [49]. The corresponding secondary structure is shown in Fig. 2d. This motif has three helices (each of variable length) and two asymmetric bulges. Because the bases shown in Fig. 1 are known experimentally to be important to the binding specificity, we define the target sequence motif in the first asymmetric bulge to be GNANNUG. Likewise, we retain bases associated with other bulges since they contribute to the binding affinity.

After collecting the relevant structural information, we search the genomes of numerous organisms using RNAMotif as described below. This yields for each aptamer a pool of sequences that could potentially fold into the specified experimental aptamer structure. To filter this pool further, we "fold" each candidate sequence using the 2D prediction tool in the Vienna RNA Package. We retain the predicted structure if it is similar to the experimental aptamer structure but discard it otherwise; the similarity is determined by inspection: the sequence is retained if the general topology of the predicted structure is similar to the experimental structure. This process greatly reduces the size of the candidate sequence pool.

Finally, we subject the remaining candidates (sequences that fold as desired) to further tests that allow us to assess their potential as significant matches based on energetic and statistical measurements for significance and stability.

3.2 RNAMotif Scanning Tool: Searching for Secondary Structural Motifs

Sequence and secondary structural motifs can be searched for in genomes using the scanning tool RNAMotif [29]. The qualitative topological and secondary structural elements are specified in a "descriptor" (specifying the structural connectivity and the length of helices, loops, bulges, etc.) as well as any specific sequence information (such as a GNRA loop). The program searches for sequences in genomes that could potentially fold into the specified secondary structure based on Watson-Crick base pairing rules. Additionally, the sequences are ranked based on the energy of folding them into the specified secondary structure.

3.3 RNA Folding Algorithms: Secondary Structure Prediction

Available 2D RNA folding algorithms can predict the patterns of base pairing, the presence of base pair mismatches, and regions with unpaired bases (e.g., loops, bulges, and junctions). For RNA tree structures, the 2D folding algorithm bundled with the Vienna RNA package (P. Schuster and coworkers [22]), is widely used. (One of the limitations of the 2D prediction algorithms is that pseudoknots[3] are not accounted for since they are difficult to predict, but efforts include pseudoknot prediction [37].) This shortcoming in 2D RNA folding algorithms limits our RNA folding applications to tree structures, which nonetheless represent a large set of structures to explore.[4] The current algorithms also do not account for the effect of magnesium ions, which have been shown to be critical for RNA folding. However, the Vienna RNA package is reasonably accurate for short sequences (<100 nt), and in addition to predicting the minimum energy 2D structure, it can calculate suboptimal structures and melting curves (specific heat curves, as described next).

3.4 Melting Curve Analysis of Candidate Sequences

Biological molecules possess stable structures at physiological temperatures. The stability of RNA secondary structures with respect to temperature change can be analyzed using heat-capacity or melting curves. These curves represent the amount of energy absorbed per unit change in temperature (e.g. $\partial H/\partial T$ versus T, where H is the RNA's conformational enthalpy). For each candidate sequence, we compute the theoretical melting curve using tools available in the Vienna RNA Package. Additionally, to discriminate random from functional RNA sequences, we shuffle each candidate sequence 1,000 times and compute the corresponding melting curves. The melting temperature, T_m, is defined as the temperature of the highest peak; it may also be interpreted as the transition temperature at which the RNA molecule's secondary structure experiences significant disruption.

---

[3] A *pseudoknot* forms when consecutive single-stranded regions **a**, x, **b**, y, **c**, z, **d** (where x, y, and z are the connecting regions) fold on each other such that **a** hydrogen-bonds with **c** and **b** with **d**. A pseudoknot is technically part of a molecule's tertiary structure though it is often convenient to consider it a secondary structural element since it involves base-pairing interactions [41].

[4] Note that general 3D prediction algorithms for RNAs are not yet available and constitute a significant challenge. This gap between 2D and 3D structure prediction is best addressed at present by subjecting any predictions based on 2D folds to experimental tests.

Additionally, for each sequence (the candidate and the randomly shuffled ones), we plot the melting temperature (T_m) against the free energy (F) as a further indicator of stability. The T_m versus F plot maps the global physical characteristics of RNA secondary folds. Using principal component analysis, we compute a 90% confidence ellipse, and check to see whether the candidate sequence falls outside the ellipse (i.e., is significantly stable).

3.5 Conformational Energy Landscape Analysis

We also calculate the conformational energy landscape of the secondary RNA fold of each candidate sequence. The shape of this energy landscape offers an alternative approach for discriminating RNA-like from non-RNA-like molecules. To assess different secondary structures, we plot for each suboptimal structure (up to 10 kcal/mol above the global minimum) its energy (E) against a "distance" (D) from the minimum energy structure. We define D to be the base pair dissimilarity between the minimum energy structure and each suboptimal structure. This E versus D plot for optimal and suboptimal structures defines the conformational energy landscape of an RNA sequence.

Furthermore, we quantify the conformational energy landscape of each sequence by computing the "Valley index," V, as described in [27]. The Valley index is a Boltzmann-weighted average of the distance between all pairs of structures (minimum-energy and sub-optimal). A large Valley index implies many low energy competing structures that have very different conformations.

Similar to the melting curve analysis, each sequence is randomly shuffled 1,000 times, and the Valley index is plotted against the free energy of the sequence. A 90% confidence ellipse is computed, and the candidate sequence is tested to see whether it falls outside of the ellipse.

3.6 Statistical Analysis

Each secondary structure motif can be considered as a "word" in the 4-letter nucleotide alphabet. For example, the probability of finding any given letter in a random sequence of nucleotides is $1/4$, so the probability of finding any specific sequence of length N is $1/4^N$. Likewise, a complicated descriptor has an associated expected frequency. However, given the complexity of secondary structure descriptors, this number is difficult to compute analytically. Therefore, these frequencies are estimated using a Monte Carlo method, in which the number of matches to a given descriptor in a random 1 Mb sequence is averaged over 1 million tries. Provided the genomes of interest have a nucleotide distribution that is close to uniform,[5]

---

[5]The assumption that that a genome is uniformly distributed in the four bases is strong. Many genomes deviate from uniform distributions, and often it is more fruitful to consider di- or trinucleotide distributions. However, our uniform distribution assumption makes the analyses simple and can later be refined. We also performed tests in which we biased the base distribution to mimic the *Streptomyces avermitilis* genome (the genome we used with

our computation provides an estimate of the number of matches to the descriptor that we expect by probability alone. With this information, we can estimate whether a secondary structure motif is over or under-represented in a given genome.

3.7 Computational Performance

Creating the descriptor definition in terms of skeletal motifs requires biological intuition and experimentation. Once defined, we employ the simple "programming language" of Macke and colleagues [29] to describe an RNA secondary structure of any complexity.

The genome searches using RNAMotif are very rapid. The search involving the most complicated descriptor (streptomycin) and largest genome (*Streptomyces avermitilis*) can be completed in less than one minute. In general, the computational speed depends on genome size and complexity of the descriptor.

Subsequently, the candidate sequences are folded. We use the RNAfold program as part of the Vienna RNA package. For a sequence of less than 100 nucleotides, the predicted secondary structure is computed in a matter of seconds. Generating sub-optimal structures, however, is relatively lengthy, because sequences of similar lengths can have drastically different numbers of sub-optimal structures (e.g., ranging from ∼150 for some sequences to ∼200,000 for others). However, the generation of each suboptimal structure is extremely fast (even the case where over 200,000 structures were generated took only 2 minutes to finish).

The energetic and statistical analyses are the most computationally expensive. For the scatter plots, we compute the Valley index and melting curve for 1,000 random permutations of a given sequence. For the statistical evaluation, we search a 1 Mb sequence 1 million times. The combined calculations take several weeks, with the majority of the time in the Monte Carlo evaluation.

All computations were performed on an SGI 300 MHz MIPS R12000 IP27 processor with 4 GB of memory.

4 Results

4.1 Candidate Sequences in Bacterial and Archaeal Genomes: Initial Search Results

We began searches for three aptamers that showed binding affinity to antibiotics (chloramphenicol [6], streptomycin [49, 52], and neomycin B [25]) in bacterial genomes, as well as for an aptamer that showed affinity for ATP [40] in archaeal genomes. The descriptors are shown in Fig. 2. We use representative species for searching: since the *Streptomyces* family accounts for many of the known antibiotics (including those used here), we use all available *Streptomyces* genomes. Additionally, we choose *E. coli* genomes since they are among the best characterized

the largest deviation from uniformity) and computed the expected frequency of matches; the results were similar to the case of uniform distribution.

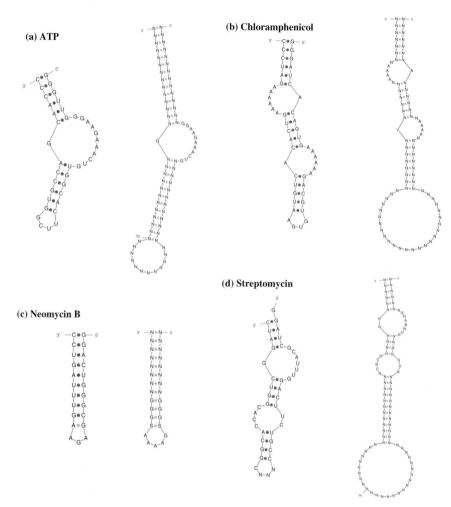

Figure 2. The four experimental aptamers (left side of each pair) used in this study along with our constructed motif descriptors (right). Regions with many 'N's are variable in length. The nucleic acid base symbols are defined as follows: N: any base; V: A, C, or G; S: C or G. We search the genome of an organism for any sequence that fits into the descriptor consensus (allowing for slight mismatches). The experimental papers associated with the aptamers are: ATP [40], Chloramphenicol [6], Neomycin B [25], and Streptomycin [49, 52]

bacterial species (and a staple of many biological studies). Other bacterial genomes were selected randomly. For the searches involving the ATP aptamer, we used all the available archaeal genomes at the time of the study from the National Center for Biotechnology Information (www.ncbi.nlm.nih.gov). The search results are displayed in Tables 1 and 2.

Table 1. Number of initial candidate sequences for the ATP-binding aptamer in selected archaeal genomes. Bold entries exhibit significant deviations from the expected number of matches

Genome	Size (Mb)	ATP Observed	Expected
Aeropyrum pernix	1.7	**6**	**1.6**
Archaeoglobus fulgidis DSM 4304	2.2	**6**	**2.1**
Halobacterium sp. NRC-1	2.6	4	2.5
Methanobacterium thermoautotrophicum str. ΔH	1.8	**6**	**1.7**
Methanococcus jannaschii	1.7	4	1.6
Methanopyrus kandleri AV19	1.7	3	1.6
Methanosarcina acetivorans C2A	5.8	**12**	**5.5**
Methanosarcina mazei Goe1	4.1	5	3.9
Pyrobaculum aerophilum	2.3	2	2.2
Pyrococcus abyssi	1.8	3	1.7
Pyrococcus furiosus DSM 3638	1.9	**8**	**1.8**
Pyrococcus horikoshii	1.8	3	1.7
Sulfolobus solfataricus	3.0	2	2.8
Sulfolobus tokodaii	2.7	4	2.5
Thermoplasma acidophilum	1.6	2	1.5
Thermoplasma volcanium	1.6	2	1.5
Total		72	

Table 2. Number of initial candidate sequences for the three antibiotic-binding aptamers in selected bacterial genomes. Bold entries exhibit significant deviations from the expected number of matches

Genome	Size (Mb)	Chloramph. Obs.	Exp.	Streptomycin Obs.	Exp.	Neomycin B Obs.	Exp.	Tot.
Streptomyces avermitilis	9.2	0	3.1	2	0.7	**6**	**19.8**	8
Streptomyces coelicolor	8.8	0	2.9	1	0.7	**1**	**18.9**	2
E. coli K12	4.7	7	**1.6**	1	0.4	11	10.1	19
E. coli O157:H7	5.6	7	**1.9**	1	0.4	**17**	**12.0**	25
E. coli O157:H7 EDL933	5.6	7	**1.9**	1	0.4	**19**	**12.0**	27
E. coli CFT073	5.3	7	**1.8**	0	0.4	**17**	**11.4**	24
Neisseria meningitidis MC58	2.3	1	0.8	0	0.2	9	4.9	10
Neisseria meningitidis Z2498	2.2	**5**	**0.7**	0	0.2	9	4.7	14
Sinorhizobium meliloti	3.7	1	1.2	1	0.3	**3**	**7.9**	5
Chlamydia trachomatis	1.1	**4**	**0.4**	0	0.1	1	2.4	5
Tot.		39		7		93		139

Tables 1 and 2 describe our results for these four aptamer targets. First, we note that the neomycin B candidate pool is much larger than both the chloramphenicol and streptomycin pools, because it is simpler (12–17 nt, with a simple loop motif; see Fig. 2c).

Second, we see that the average number of matches to neomycin B in *Strepto-myces* genomes is four, while the average number in *E. coli* genomes is four times greater. Since the average *Streptomyces* genome size is almost twice that of the average *E. coli* genome (~9 Mb versus ~5 Mb), these trends are significant because probability alone would predict the number of hits in *Streptomyces* genomes to be about *twice* as many matches as *E. coli*, not *one fourth* as we obtain. This suggests a preference for or discrimination against this aptamer, i.e., either *Streptomyces* is significantly missing this structure, or *E. coli* has a significantly large occurrence of it.

More rigorously, the Monte Carlo method described above estimates the expected frequency of each of the descriptors per 1 Mb of uniformly distributed random sequence. The Monte Carlo results are listed in Table 3 and the expected number of matches are listed in Tables 1 and 2. This information reveals several interesting trends. For streptomycin, which exhibits a number of matches similar to the computed expected number of matches, no significant findings of aptamer hits can be claimed; however, pools for neomycin B, chloramphenicol, and ATP show significant deviations from the expected behavior (e.g., much smaller for the *Strepto-myces* genomes and much larger for *E. coli* genomes for neomycin B). The archaea genomes tend to agree with the expected number of matches, except for several genomes marked in bold in Table 1.

Table 3. Computed expected frequencies of the descriptors per 1 Mb of uniformly distributed random sequence

Descriptor	Frequency ±1 Standard Error
Chloramphenicol	0.3320 ± 0.0007
Streptomycin	0.0778 ± 0.0003
Neomycin B	2.1471 ± 0.0015
ATP	0.9424 ± 0.0009

4.2 Structurally Filtered Search Results

Next, we further filter the pools of matches to eliminate candidate sequences that may not fold as intended. This step reduces our total pool of candidate sequences greatly, from 139 to 32 matches in bacterial genomes and from 72 to 5 matches in archaeal genomes. In particular, only one candidate sequence remains for the streptomycin aptamer.

4.3 Energetic Analysis Discriminates Natural from Random RNA Sequences

With the significantly smaller pool of filtered candidate sequences, we now proceed to evaluate the candidates based on energetic considerations. Because biological

RNA molecules form stable structures at physiological temperatures, specific heat curves and conformational energy landscapes have expected characteristics, as elaborated below.

First, we compute the melting curves and melting temperatures of the candidate sequences and compare them to those of random permutations of the sequence. We plot the melting temperatures of our candidate and its permuted pool against the free energy. A promising candidate should have distinct characteristics from the random pool. In certain cases, we indeed find the candidate sequence to be apart from the bulk of the random sequences; Fig. 3 shows how the sequence of interest falls outside a computed 90% confidence ellipse.

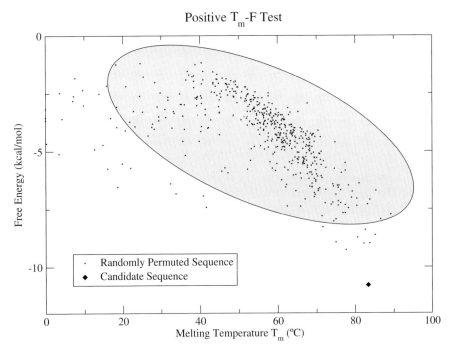

Figure 3. Scatter plot of the melting temperatures (T_m) of the candidate sequence and its 1,000 random permutations versus the free energy of their structure ensembles. Notice that the candidate sequence lies outside the 90% confidence ellipse, signifying an especially stable structure

Secondly, we analyze the conformational energy landscape of each candidate sequence, as shown in Fig. 4. That is, we plot the energy of each suboptimal structure versus the distance of the suboptimal structure to the minimum energy structure. We observe that some candidates possess conformational energy landscapes with multiple low minima, while others have very steep single-minima landscapes. A stable aptamer structure should have a sharp, deep minimum and a funnel-like landscape (Fig. 4, circles). This means that for a given sequence, the more different a fold

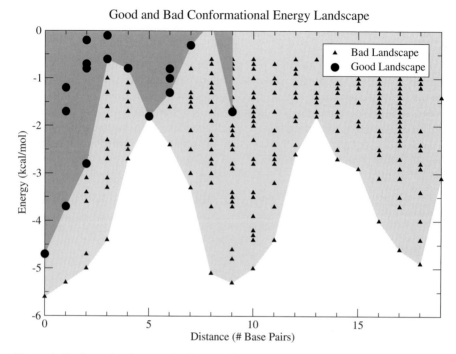

Figure 4. Conformational energy landscapes of two candidate sequences for the neomycin B antibiotic. The "good" candidate, marked with circles, exhibits a sharp, steep slope, while the "bad" candidate, marked with triangles, is more random, with multiple isoenergetic minima

looks from the minimum energy fold, the higher energy it is. A random landscape, or a shallow one with multiple minima (as in Fig. 4, triangles) likely has other different and stable structures for the same sequence that are almost isoenergetic to the minimum energy structure. This information is quantified in the computation of the Valley index. Structures with unfavorable conformational energy landscapes have larger Valley indices. This provides for a quantitative test of the energetic stability of the candidate sequence's minimum energy structure. As described above, each sequence is shuffled 1,000 times and the Valley indices of all the sequences are plotted against their free energy (see Fig. 5). A sequence that falls outside the bulk of the random sequences is probably significantly stable. Using the analyses above, we evaluate the candidate sequences that are most likely to have the physical properties of biological molecules. Our tests are stringent indicators of stability, as only 24% of the candidate sequences pass the melting temperature test, while only 14% pass the Valley index test. Any sequence that passes the Valley index test also passes the melting temperature test. Furthermore, the tests were conducted on several known biological RNAs (5S, U5, U6, U7, and Gln tRNA), and all of them passed the tests with the exception of one sequence which did not pass the Valley index test.

Positive V-F Test

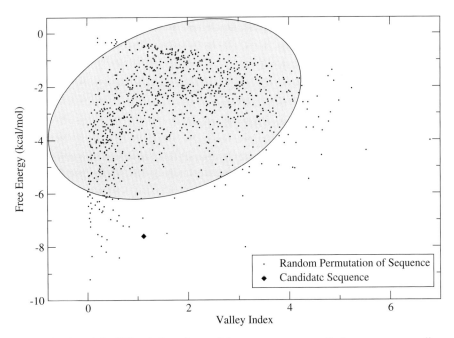

Figure 5. Plot of the Valley index of a candidate sequence versus its free energy, as well as 1,000 random permutations of the sequence. Note how the candidate sequence lies outside of the 90% confidence ellipse, suggesting an especially stable structure

Candidate Sequences

Selected candidate sequences are shown in Table 4, including the results of energetic tests, the computed physical quantities, and their locations in the genome. Some of the sequences occur in non-coding regions, while others occur in genes of known and unknown functions. Many of the sequences occur multiple times in different genomes, and even in the same genome. For example, the two candidates from the ATP pool pass both the melting temperature and the Valley index tests. They occur inside the plasmids of *Halobacterium* in non-coding regions. On the other hand, the streptomycin pool candidate sequence passes neither of the energetic tests, yet shows very positive physical qualities (e.g., high melting temperature) and is also located in a non-coding region. The first two neomycin B sequences pass both energetic tests with the first one in a non-coding sequence while the second is located in a hypothetical protein with a function that is not currently understood. Finally, the chloramphenicol pool sequence passes neither of the tests, and does not exhibit especially stable characteristics. However, it is located in the ECs1492 gene, which encodes for a transcription-repair coupling factor that is responsible for a mutation frequency decline. It is possible that this may explain another mode of action for chloramphenicol, namely that it increases transcription mutation frequency. Finally, based on the data in Tables 1 and 2, it is possible to attribute increased significance to certain matches based on the statistical representation of the motif in a given genome. For example, it is easily seen that *E. coli* accounts for a significantly large number of neomycin B motifs, compared with what is expected. Therefore, the neomycin B matches located in the *E. coli* genomes may have a higher likelihood of existing *in vivo*.

5 Conclusions and Future Directions

The method presented here for searching for artificial aptamer structures in the genomes of various organisms has produced promising RNA sequences that may be functionally important. In response to the assertion of Piganeau and Schroeder [36], it is indeed likely that aptamer structures will be found *in vivo*.

Further work is required to search all genomes comprehensively, investigate other aptamers, develop ways to reliably distinguish biological RNAs from random noise and spurious matches, and most importantly, verify the findings experimentally.

More broadly, the integration of mathematical RNA modeling (such as by graph theory, see Appendix) and experimental methods has the potential to greatly expand our knowledge of RNA's repertoire through the development of new tools for analyzing RNA motifs, genome analysis, and improvement of the *in vitro* selection technology. These tools and technologies will likely broaden the scope of RNA-based methods for biomedical applications including selection of complex synthetic RNAs and identification of ncRNAs associated with various physiological functions. Ultimately, it is prudent to address the challenging problem of connecting and better characterizing the relationships between 2D and 3D RNA folds.

Acknowledgments

We thank Marco Avellaneda, Dinshaw Patel, Renée Schroeder, and Mike Waterman for constructive discussions and many suggestions related to this work. We thank Tom Macke and Ivo Hofacker for technical assistance with RNAMotif and the Vienna RNA package, respectively, and Yanli Wang for preparing Fig. 1. We also thank Dave Scicchitano for supporting this study through the Department of Biology. UL thanks the entire Schlick Lab for many helpful discussions and for helping sort through a myriad of technical problems, and the Howard Hughes Medical Institute for an undergraduate summer research fellowship through the Honors Summer Institute at NYU. We also gratefully acknowledge the support of NSF (DMS-0201160), NIH (GM055164 and ES012692), and the Human Frontier Science Program (RGP0076) for supporting this research.

Table 4. Selected candidate sequences from the candidate pools for streptomycin, chloramphenicol, neomycin B, and ATP (after filtering by looking at predicted 2D structures). Included are the computed energetic data, test results, locations of the start sites of the sequences, and GenBank annotations. The corresponding RNA sequences have T replaced by U

Sequence Genome	Location	Gene	F (kcal/mol)	T_m (°C)	T_m-F Test	V	V-F Test
ATP							
5'-GCTGGTCGAA GACACTGGCT GTCGCTGTCG ACGGCGATCA GC							
Halobacterium sp. NRC1 plasmid	pNRC200 92229	non-coding	−19.43	71.6	+	4.54	+
Halobacterium sp. NRC1 plasmid	pNRC100 92229	non-coding					
STREPTOMYCIN							
5'-GTACCCGGAC GTGCCCTTCC AGGCGTCCAT GGAGGCCTGG CTCGGGGCGG TGC							
S. avermitilis	7323209	non-coding	−28.64	111.4	−	0.84	−
NEOMYCIN B							
5'-TGCGGGCGAA CAGTTTGCA							
E. coli O157:H7 EDL933	5503670	non-coding	−7.61	81.0	+	1.12	+
5'-TTGAGCAGGG GCGTGAAGTT TTTGCTTTG							
E. coli CFT073	3864206	Smf	−10.79	83.4	+	1.44	+
5'-AGTCTGGTGG GCGATATGTT TATTATGAT							
E. coli K12	1324039	yciQ	−6.10	57.8	+	2.24	−
E. coli O157:H7	1826735	ECs1840					
E. coli O157:H7 EDL933	2260441	Z2542					
E. coli CFT073	1569619	yciQ					
CHLORAMPHENICOL							
5'-TCAGAGCTGA AAAACTGGCC CCGAGTGCAG CTAAAAACTG A							
E. coli O157:H7	1532105	ECs1492	−10.58	72.6	−	2.40	−
E. coli O157:H7 EDL933	1617201	Mfd					

Appendix: Use of Graph Theory to analyze RNA 2D Structure and Function

RNA Genomics and Graph Theory

This article focused on using RNA motifs from *in vitro* selection experiments to discover novel functional RNA molecules in genomes. An alternate approach to RNA genomics developed in our group is the use of graph theory for course-grained secondary structure modeling (see, for instance, [11, 14, 26, 35]). Indeed, the utility of the graph theory approach to RNA secondary structure has been known as early as the 1980's with work done on tree edit distances [43], RNA structure comparison [5, 28], and RNA structure statistics [12]. Graph theory analysis of genomes is promising because all RNA structures can be schematically represented as two-dimensional graphs and thus novel graph topologies from graphical enumeration can be used to drive discovery of novel RNA motifs in genomes via methods and analyses similar to those described here. Below, we outline the essentials and advantages of graph theory for describing, cataloguing, and predicting RNA structures in the hope that this will stimulate mathematicians to work in this area.

RNA Structural Motifs and Graph Theory

RNA molecules are hierarchical in nature since their secondary structures are known to be stable independently of their tertiary structures [50]. Thus, many groups approach RNA by focusing on 2D RNA structures [7, 13, 15, 17, 30, 38, 56]. RNA secondary motifs have a network-like topology with stems linking loops, bulges, and junctions (Fig. 6). Such a topological RNA representation allows exploration of RNA

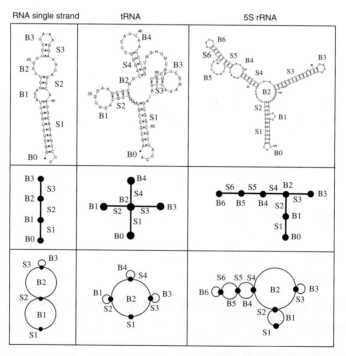

Figure 6. Graphical representations of RNA secondary structures (*top*) as tree (*middle*) and dual (*bottom*) graphs

topologies using graph theory, a field in mathematics widely used for analyzing networks and enumerating structural possibilities, including hydrocarbons, genetic and biochemical networks, ecology, transportation, and the Internet [4, 8, 18].

Figure 6 shows three RNA secondary (tree) motifs represented as tree graphs: the vertices (•) are RNA loops, bulges or junctions, and the edges (lines, —) are RNA stems (precise rules are detailed in [14]). Thus, the schematic tree graphs represent the connectivity between the RNA secondary elements (e.g., stems, loops, bulges, junctions). The tree graphs provide intuitive representations of RNA structures, but they cannot represent other important RNA types, such as pseudoknots. For completeness, we developed another class of RNA graphs called *dual graphs* (third row of Fig. 6; [14]); dual graphs can represent all RNA trees and pseudoknots and can be generalized to represent unusual RNA structures with triple, quadruple, and higher-order helices (e.g., occurring in RNA frameshift signal of HIV-1 and RNAs interacting with antibiotic neomycin [1, 2]).

Since the "RNA graphs" are discrete, they allow us to enumerate all possible 2D RNA motifs using enumeration methods of graphs. Graphical enumeration of RNA topologies can be performed analytically or computationally depending on the complexity of the structures. For example, for unlabeled trees, the number of possible graphs with V vertices is obtained from the coefficients c_i associated with the x_i term of the counting polynomial derived by Harary and Prins [20]:

$$t = \sum_i c_i x^i = x + x^2 + x^3 + 2x^4 + 3x^5 + 6x^6 + 11x^7 + 23x^8 + 47x^9$$
$$+ 106x^{10} + \cdots$$

For example, there is only 1 distinct graph each for $V = 1, 2, 3$ vertices (since $c_1 = c_2 = c_3 = 1$) and 2 distinct 4-vertex graphs ($c_4 = 2$), 3 distinct 5-vertex graphs ($c_5 = 3$), and so on.

These sets of distinct graphs represent libraries of theoretically possible RNA topologies, which include naturally occurring, candidate, and hypothetical RNA motifs theory (see Schlick lab's RNA-As-Graphs (RAG) web resource at monod.-biomath.nyu.edu/rna/ and [11]). Known RNAs in public databases (NDB and others) can thus be matched to the topologies we describe (see Fig. 6). Significantly, because we found that the known 2D RNA motifs represent only a small subset of all possible topologies, we hypothesize that some of the missing motifs may represent undiscovered *naturally occurring* RNAs while others may be designed and then synthesized in the laboratory.

Sequence Space Versus Topology Space

Current theoretical [42] and experimental [53] approaches to RNA structure explore RNA's sequence space. Experimental *in vitro* selection techniques exploit random sequence pools for comprehensive searches for novel RNAs. In the search for RNA genes in genomes, scanning algorithms require sequence and structural motifs as input. In contrast, our RNA analysis focuses on structural motifs rather than sequences

per se. A critical advantage of RNA graph analysis is that the space of topologically distinct structures is vastly smaller than the nucleotide sequence space. In fact,
we estimate, based on Harary-Prins enumeration formula for tree graphs (above)
[19, 20], that the number of distinct RNA tree topologies can be parameterized as
$\sim 2.5^{(N/20)-3}$ for $N > 60$ compared with 4^N for the nucleotide sequence space!
The markedly smaller RNA topology space implies great potential for the search for
novel RNA structures. Once a novel target topology/motif is identified, the corresponding RNA sequences can be found in two ways: for natural RNAs, the selected
motif can be found by scanning the genomes; and, for synthetic RNAs, they can be
designed using modular assembly of existing RNA fragments (i.e., using a library of

V	λ_2	Tree Graph	Secondary Structure
4	0.5858		RNA single strand (PR0021) 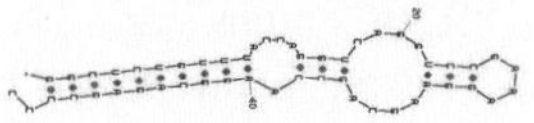
	1.0000	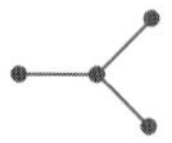	tRNA (PR0019) 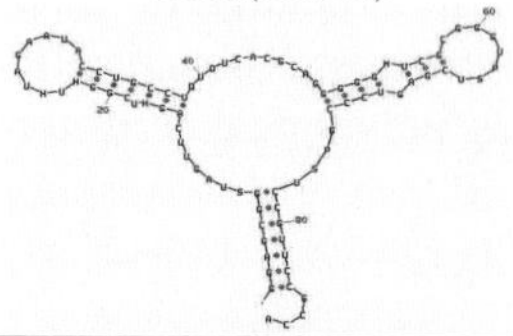
5	0.3820		70S (F) (RR0003) 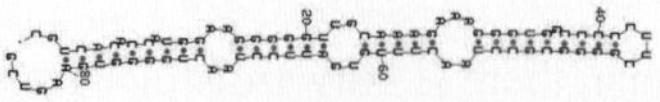
	0.5188		tRNA (TR0001) 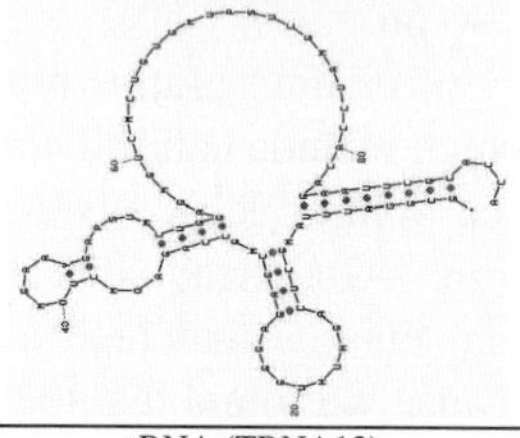
	1.0000		tRNA (TRNA12) 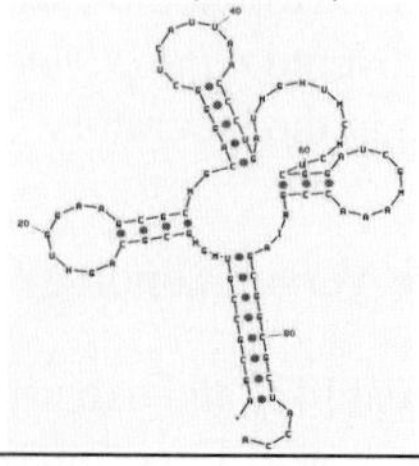

Figure 7. RNA tree motif libraries for $V = 4, 5$. V is vertex number and λ_2 is the second
smallest eigenvalue of the Laplacian matrix

sequence/motif building blocks and application of 2D folding algorithms). Both of these research directions are currently being pursued in our laboratory.

RAG: RNA-As-Graphs Web Resource

Our RNA graphical representations present an opportunity for cataloguing of RNA structures based on their topological properties (Fig. 7). Cataloguing RNA's structural diversity, including hypothetical motifs, is vital for identifying novel RNA structures and for pursuing RNA genomics initiatives. Our RNA-As-Graphs (RAG; monod.biomath.nyu.edu/rna) web resource catalogues and ranks all mathematically possible (including existing and candidate) RNA secondary motifs on the basis of graphical enumeration results. We archive RNA tree motifs as "tree graphs" and other RNAs, including pseudoknots, as general "dual graphs." All RNA motifs are catalogued by graph vertex number (a measure of sequence length) and ranked by topological complexity (second smallest eigenvalue (λ_2) corresponding to the graph's *Laplacian* matrix). RAG's inventory immediately suggests candidates for novel RNA motifs, either naturally occurring or synthetic. Through RAG, we hope to pursue and further stimulate efforts to predict and design novel RNA motifs and thereby contribute to RNA genomics initiatives.

References

[1] D. P. Arya, R. L. Coffee, Jr., and I. Charles. Neomycin-induced hybrid triplex formation. *J. Amer. Chem. Soc.*, 123:11093–11094, 2001.

[2] D. P. Arya, R. L. Coffee, Jr., B. Willis, and A. I. Abramovitch. Aminoglycoside-nucleic acid interactions: Remarkable stabilization of DNA and RNA triple helices by neomycin. *J. Amer. Chem. Soc.*, 123:5385–5395, 2001.

[3] J. P. Bachellerie, J. Cavaille, and A. Huttenhofer. The expanding snoRNA world. *Biochimie*, 84:775–90, 2002.

[4] A. L. Barabási and E. Bonabeau. Scale-free networks. *Sci. Amer.*, 288:60–69, 2003.

[5] G. Benedetti and S. Morosetti. A graph-topological approach to recognition of pattern and similarity in RNA secondary structures. *Biol. Chem.*, 59:179–184, 1996.

[6] D. H. Burke, D. C. Hoffman, A. Brown, M. Hansen, A. Pardi, and L. Gold. RNA aptamers to the peptidyl transferase inhibitor chloramphenicol. *Chem. Biol.*, 4:833–843, 1997.

[7] J. H. Chen, S. Y. Le, and J. V. Maizel. Prediction of common secondary structures of RNAs: a genetic algorithm approach. *Nucl. Acids Res.*, 28:991–999, 2000.

[8] K. J. Devlin. *Mathematics: the New Golden Age*. Penguin, London, 1988.

[9] J. A. Doudna and T. R. Cech. The chemical repertoire of natural ribozymes. *Nature*, 418:222–228, 2002.

[10] A. D. Ellington and J. W. Szostak. In vitro selection of RNA molecules that bind specific ligands. *Nature*, 346:818–822, 1990.

[11] D. Fera, N. Kim, N. Shiffeldrim, J. Zorn, U. Laserson, H. H. Gan, and T. Schlick. RAG: RNA-As-Graphs web resource. *BMC Bioinformatics*, 5:88, 2004.

[12] W. Fontana, D. A. Konings, P. F. Stadler, and P. Schuster. Statistics of RNA secondary structures. *Biopolymers*, 33:1389–1404, 1993.

[13] W. Fontana, P. F. Stadler, E. G. Bornberg-Bauer, T. Griesmacher, I. L. Hofacker, M. Tacker, P. Tarazona, E. D. Weinberger, and P. Schuster. RNA folding and combinatory landscapes. *Phys. Rev. E*, 47:2083–2099, 1993.

[14] H. H. Gan, S. Pasquali, and T. Schlick. Exploring the repertoire of RNA secondary motifs using graph theory; implications for RNA design. *Nucl. Acids Res.*, 31:2926–2943, 2003.

[15] C. Gaspin and E. Westhof. An interactive framework for RNA secondary structure prediction with a dynamical treatment of constraints. *J. Mol. Biol.*, 254:163–174, 1995.

[16] W. W. Gibbs. The unseen genome: Gems among the junk. *Sci. Amer.*, 289:46–53, 2003.

[17] S. Griffiths-Jones, A. Bateman, M. Marshall, A. Khanna, and S. R. Eddy. Rfam: an RNA family database. *Nucl. Acids Res.*, 31:439–441, 2003.

[18] J. Gross and J. Yellen. *Graph Theory and its Applications*. CRC Press, Boca Raton, FL, 1999.

[19] F. Harary. The number of homeomorphically irreducible trees and other species. *Acta Math.*, 101:141–162, 1959.

[20] F. Harary. *Graph Theory*. Addison-Wesley, Reading, MA, 1969.

[21] T. Hermann and D. J. Patel. Adaptive recognition by nucleic acid aptamers. *Science*, 287:820–825, 2000.

[22] I. L. Hofacker, W. Fontana, P. F. Stadler, L. S. Bonhoeffer, M. Tacker, and P. Schuster. Fast folding and comparison of RNA secondary structures. *Monatsh. Chem.*, 125:167–188, 1994. www.tbi.univie.ac.-at/~ivo/RNA/

[23] A. Huttenhofer, J. Brosius, and J. P. Bachellerie. RNomics: Identification and function of small, non-messenger RNAs. *Curr. Opin. Chem. Biol.*, 6:835–843, 2002.

[24] A. Huttenhofer, M. Kiefmann, S. Meier-Ewert, J. O'Brien, H. Lehrach, J. P. Bachellerie, and J. Brosius. RNomics: an experimental approach that identifies 201 candidates for novel, small, non-messenger RNAs in mouse. *EMBO J.*, 20:2943–2953, 2001.

[25] L. Jiang, A. Majumdar, W. Hu, T. J. Jaishree, W. Xu, and D. J. Patel. Saccharide-RNA recognition in a complex formed between neomycin B and an RNA aptamer. *Structure Fold Des.*, 7:817–827, 1999.

[26] N. Kim, N. Shiffeldrim, H. H. Gan, and T. Schlick. Candidates for novel RNA topologies. *J. Mol. Biol.*, 341:1129–1144, 2004.

[27] J. Kitagawa, Y. Futamura, and K. Yamamoto. Analysis of the conformational energy landscape of human snRNA with a metric based on tree representation of RNA structures. *Nucl. Acids Res.*, 31:2006–2013, 2004.

[28] S. Y. Le, R. Nussinov, and J. V. Maizel. Tree graphs of RNA secondary structures and their comparisons. *Comput. Biomed. Res.*, 22:461–473, 1989.

[29] T. J. Macke, D. J. Ecker, R. R. Gutell, D. Gautheret, D. A. Case, and R. Sampath. RNAMotif, an RNA secondary structure definition and search algorithm. *Nucl. Acids Res.*, 29:4724–4735, 2001.

[30] H. Margalit, B. A. Shapiro, A. B. Oppenheim, and J. V. Maizel, Jr. Detection of common motifs in RNA secondary structure. *Nucl. Acids Res.*, 17:4829–4845, 1989.

[31] J. S. Mattick. Non-coding RNAs: the architects of eukaryotic complexity. *EMBO Rep.*, 2:986–991, 2001.

[32] J. S. Mattick and M. J. Gagen. The evolution of controlled multitasked gene networks: the role of introns and other noncoding RNAs in the development of complex organisms. *Mol. Biol. Evol.*, 18:1611–1630, 2001.

[33] A. Nahvi, N. Sudarsan, M. S. Ebert, X. Zou, K. L. Brown, and R. R. Breaker. Genetic control by a metabolite binding mRNA. *Chem. Biol.*, 9:1043–1049, 2002.

[34] Y. Okazaki, M. Furuno, et al. Analysis of the mouse transcriptome based on functional annotation of 60,770 full-length cDNAs. *Nature*, 420:563–573, 2002.

[35] S. Pasquali, H. H. Gan, and T. Schlick. Modular RNA architecture revealed by computational analysis of existing pseudoknots and ribosomal RNAs. *Nucl. Acids Res.*, 2005. In Press.

[36] N. Piganeau and R. Schroeder. Aptamer structures: A preview into regulatory pathways? *Chem. Biol.*, 10:103–104, 2003.

[37] E. Rivas and S. R. Eddy. A dynamic programming algorithm for RNA structure prediction including pseudoknots. *J. Mol. Biol.*, 285:2053–2068, 1999.

[38] E. Rivas and S. R. Eddy. Secondary structure alone is generally not statistically significant for the detection of noncoding RNAs. *Bioinformatics*, 16:583–605, 2000.

[39] K. Salehi-Ashtiani and J. W. Szostak. In vitro evolution suggests multiple origins for the hammerhead ribozyme. *Nature*, 414:82–84, 2001.

[40] M. Sassanfar and J. W. Szostak. An RNA motif that binds ATP. *Nature*, 364:550–553, 1993.

[41] T. Schlick. *Molecular Modeling: An Interdisciplinary Guide.* Springer-Verlag, New York, NY, 2002.

[42] P. Schuster, W. Fontana, P. F. Stadler, and I. L. Hofacker. From sequences to shapes and back: a case study in RNA secondary structures. *Proc. R. Soc. Lond. B. Biol. Sci.*, 255:279–284, 1994.

[43] B. A. Shapiro and K. Z. Zhang. Comparing multiple RNA secondary structures using tree comparisons. *Comput. Appl. Biosci.*, 6:309–318, 1990.

[44] G. A. Soukup and R. R. Breaker. Engineering precision RNA molecular switches. *Proc. Natl. Acad. Sci. USA*, 96:3584–3589, 1999.

[45] G. A. Soukup and R. R. Breaker. Nucleic acid molecular switches. *Trends Biotechnol.*, 17:469–476, 1999.

[46] G. A. Soukup and R. R. Breaker. Allosteric nucleic acid catalysts. *Curr. Opin. Struct. Biol.*, 10:318–325, 2000.

[47] S. Spiegelman. An approach to the experimental analysis of precellular evolution. *Q. Rev. Biophys.*, 4:213–253, 1971.

[48] G. Storz. An expanding universe of noncoding RNAs. *Science*, 296:1260–1263, 2002.

[49] V. Tereshko, E. Skripkin, and D. J. Patel. Encapsulating streptomycin within a small 40-mer RNA. *Chem. Biol.*, 10:175–187, 2003.

[50] I. Tinoco, Jr. and C. Bustamante. How RNA folds. *J. Mol. Biol.*, 293:271–281, 1999.

[51] C. Tuerk and L. Gold. Systematic evolution of ligands by exponential enrichment: RNA ligands to bacteriophage T4 DNA polymerase. *Science*, 249:505–510, 1990.

[52] S. T. Wallace and R. Schroeder. In vitro selection and characterization of streptomycin-binding RNAs: Recognition discrimination between antibiotics. *RNA*, 4:112–123, 1998.

[53] D. S. Wilson and J. W. Szostak. In vitro selection of functional nucleic acids. *Ann. Rev. Biochem.*, 68:611–647, 1999.

[54] W. Winkler, A. Nahvi, and R. R. Breaker. Thiamine derivatives bind messenger RNAs directly to regulate bacterial expression. *Nature*, 419:952–956, 2002.

[55] W. C. Winkler, S. Cohen-Chalamish, and R. R. Breaker. An mRNA structure that controls gene expression by binding FMN. *Proc. Natl. Acad. Sci. USA*, 99:15908–15913, 2002.

[56] M. Zuker, D. H. Mathews, and D. H. Turner. Algorithms and thermodynamics for RNA secondary structure prediction: A practical guide. In J. Barciszewski and B. F. C. Clark, editors, *RNA Biochemistry and Biotechnology*, NATO ASI Series, pages 11–43. Klewer Academic Publishers, Dordrecht, NL, 1999.

Learning to Align Sequences:
A Maximum-Margin Approach

Thorsten Joachims, Tamara Galor, and Ron Elber

Department of Computer Science, Cornell University, Ithaca, New York 14853, USA

Abstract. We propose a discriminative method for learning the parameters of linear sequence alignment models from training examples. Compared to conventional generative approaches, the discriminative method is straightforward to use when operations (e.g. substitutions, deletions, insertions) and sequence elements are described by vectors of attributes. This admits learning flexible and more complex alignment models. While the resulting training problem leads to an optimization problem with an exponential number of constraints, we present a simple algorithm that finds an arbitrarily close approximation after considering only a subset of the constraints that is linear in the number of training examples and polynomial in the length of the sequences. We also evaluate empirically that the method effectively learns good parameter values while being computationally feasible.

1 Introduction

Methods for sequence alignment are common tools for analyzing sequence data ranging from biological applications [3] to natural language processing [11][1]. They can be thought of as measures of similarity between sequences where the similarity score is the result of a discrete optimization problem that is typically solved via dynamic programming. While the dynamic programming algorithm determines the general notion of similarity (e.g. local alignment vs. global alignment), any such similarity measure requires specific parameter values before it is fully specified. Examples of such parameter values are the costs for substituting one sequence elements for another, as well as costs for deletions and insertions. These parameter values determine how well the measure works for a particular task.

In this paper we tackle the problem of inferring the parameter values from training data. Our goal is to find parameter values so that the resulting similarity measure best reflects the desired notion of similarity. Instead of assuming a generative model of sequence alignment (e.g. [11]), we take a discriminative approach to training following the general algorithm described in [14]. A key advantage of discriminative training is that operations can easily be described by features without having to model their dependencies like in generative training. In particular, we aim to find the set of parameter values that corresponds to the best similarity measure a given alignment model can represent. Taking a large-margin approach, we show that we can solve the resulting training problem efficiently for a large class of alignment algorithms that implement a linear scoring function. While the resulting optimization problems have exponentially many constraints, our algorithm finds an arbitrarily good approximation after considering only a subset of constraints that scales polynomially with the length of the sequences and linearly with the number of training examples. We empirically and theoretically analyze the scaling of the algorithm and show that the learned similarity score performs well on test data.

2 Sequence Alignment

Sequence alignment computes a similarity score for two (or more) sequences s_1 and s_2 from an alphabet $\Sigma = \{1, .., \sigma\}$. An alignment a is a sequence of operations that transforms one sequence into the other. In global alignment, the whole sequence is transformed. In local alignment, only an arbitrarily sized subsequence is aligned. Commonly used alignment operations are "match" (m), "substitution" (s), "deletion" (d) and "insertion" (i). An example of a local alignment is given in Fig. 1. In the example, there are 6 matches, 1 substitution, and 1 insertion/deletion. With each operation there is an associated cost/reward. Assuming a reward of 3 for match, a cost of -1 for substitution, and a cost of -2 for insertion/deletion, the total alignment score $D_{\mathbf{w}}(s_1, s_2, a)$ in the example is 15. The optimal alignment a^* is the one that maximizes the score for a given cost model.

$$
\begin{array}{ll}
s_1\text{:} & 5\ 2\ 9\ 2\ \ 1\ 3\ 5\ 5\ \ \ \ 2\ \ 1\ 9\ 2\ 1 \\
s_2\text{:} & \ \ \ \ \ \ 3\ 2\ \ 1\ 2\ 5\ 5\ 3\ 2\ \ 1\ 4\ 7\ 6\ 5\ 2 \\
a\text{:} & \ \ \ \ \ \ \ m\ m\ s\ m\ m\ i\ m\ m
\end{array}
$$

Figure 1. Example of a local sequence alignment

More generally, we consider alignment algorithms that optimize a linear scoring function

$$
D_{\mathbf{w}}(s_1, s_2, a) = \mathbf{w}^T \Psi(s_1, s_2, a) \tag{1}
$$

where $\Psi(s_1, s_2, a)$ is a feature vector describing the alignment a applied to s_1 and s_2. $\mathbf{w}$ is a given cost vector. Instead of assuming a finite alphabet Σ and a finite set of operations, we only require that the reward/cost of each operation a_o on any two characters $c_1, c_2 \in \Sigma$ can be expressed as a linear function over attributes $\phi(c_1, c_2, a_o)$

$$
score(c_1, c_2, a_o) = \mathbf{w}^T \phi(c_1, c_2, a_o) . \tag{2}
$$

$\phi(c_1, c_2, a_o)$ can be thought of as a vector of attributes describing the match of c_1 and c_2 under operation a_o. Note that c_1 and c_2 can take dummy values for insertions, deletions, etc. This representation allows different scores depending on various properties of the characters or the operation. The feature vector for a complete alignment $\Psi(s_1, s_2, a)$ is the sum of the individual feature vectors

$$\Psi(s_1, s_2, a) = \sum_{i=1}^{|a|} \phi(c_1(a_i), c_2(a_i), a_i) \tag{3}$$

where $c_1(a_i)$ and $c_2(a_i)$ indicate the characters that a_i is applied to. Only those operation sequences are valid that transform s_1 into s_2. Note that a special case of this model is the conventional parameterization using a substitution matrix and fixed scores for deletions and insertions. Finding the optimal alignment corresponds to the following optimization problem

$$D_{\mathbf{w}}(s_1, s_2) = max_a \left[D_{\mathbf{w}}(s_1, s_2, a) \right] \tag{4}$$
$$= max_a \left[\mathbf{w}^T \Psi(s_1, s_2, a) \right] \tag{5}$$
$$= max_a \left[\sum_{i=1}^{|a|} score(c_1(a_i), c_2(a_i), a_i) \right]. \tag{6}$$

This type of problem is typically solved via dynamic programming. In the following we consider local alignment via the Smith/Waterman algorithm [12]. However, the results can be extended to any alignment algorithm that optimizes a linear scoring function and that solves (6) globally optimally. This also holds for other structures besides sequences [14].

3 Inverse Sequence Alignment

Inverse sequence alignment is the problem of using training data to learn the parameters w of an alignment model and algorithm so that the resulting similarity measure $D_{\mathbf{w}}(s_1, s_2)$ best represents the desired notion of similarity on new data. While previous approaches to this problem exist [5, 13], they are limited to special cases and small numbers of parameters. We will present an algorithm that applies to any linear alignment model with no restriction on the function or number of parameters, instantiating the general algorithm we described in [14]. An interesting related approach is outlined in [9, 10], but it is not clear in how far it leads to practical algorithms.

We assume the following two scenarios, for which the notation is inspired by protein alignment.

3.1 Alignment Prediction

In the first scenario, the goal is to predict the optimal sequence of alignment operations a for a given pair of sequences s^N and s^H, which we call native and homolog sequence. We assume that examples are generated i.i.d. according to a distribution

$P(s^N, s^H, a)$. We approach this prediction problem using the following linear pre-diction rule which is parameterized by $\mathbf{w}$.

$$\hat{a} = argmax_a \left[D_{\mathbf{w}}(s^N, s^H, a) \right] \tag{7}$$

This rule predicts the alignment sequence $\hat{a}$ which scores highest according to the linear model. By changing the cost of alignment operations via $\mathbf{w}$, the behavior of the prediction rule can be modified. The error of a prediction $\hat{a}$ compared to the true alignment a is measured using a loss function $L(a, \hat{a})$. The goal of learning is to find a $\mathbf{w}$ that minimizes the expected loss (i.e. risk).

$$R_P^L(\mathbf{w}) = \int L\left(a, argmax_a \left[D_{\mathbf{w}}(s^N, s^H, a) \right] \right) dP(s^N, s^H, a) \tag{8}$$

One reasonable loss function $L(.,.)$ to use is the number of alignment operations that are different in a and $\hat{a}$. For simplicity, however, we will only consider the 0/1-loss $L_\Delta(.,.)$ in the following. It return the value 0 if both arguments are equal, and value 1 otherwise.

3.2 Homology Prediction

In the second scenario the goal is to predict whether two proteins are homol-ogous. We assume that examples are generated i.i.d. according to a distribution $P(s^N, s^H, S^D)$. s^N is the native sequence, s^H the homologous sequence, and S^D is a set of decoy sequences $s^{D_1}, ..., s^{D_d}$. The goal is a similarity measure $D_{\mathbf{w}}(.,.)$ so that native sequence s^N and homolog s^H are more similar than the native sequence s^N and any decoy s^{D_j}, i.e.

$$D_{\mathbf{w}}(s^N, s^H) > D_{\mathbf{w}}(s^N, s^{D_j}). \tag{9}$$

The goal of learning is to find the cost parameters $\mathbf{w}$ that minimize the probability $Err_P^\Delta(\mathbf{w})$ that the similarity with any decoy sequence $D_{\mathbf{w}}(s^N, s^{D_j})$ is higher than the similarity with the homolog sequence $D_{\mathbf{w}}(s^N, s^H)$.

$$Err_P^L(\mathbf{w}) = \int L_\Delta \left(s^H, \arg\max_{s \in S^D \cup \{s^H\}} D_{\mathbf{w}}(s^N, s) \right) dP(s^N, s^H, S^D) \tag{10}$$

Again, we assume a 0/1-loss $L_\Delta(.,.)$.

4 A Maximum-Margin Approach to Learning the Cost Parameters

In both scenarios, the data generating distributions $P(s^D, s^H, a)$ and $P(s^D, s^H, S^D)$ are unknown. However, we have a training sample S drawn i.i.d from $P(.)$. This training sample will be used to learn the parameters $\mathbf{w}$. We will first consider the case of Alignment Prediction, and then extend the algorithm to the problem of Homology Prediction.

4.1 Alignment Predictions

Given is a training sample $S = ((s_1^D, s_1^H, a_1), ..., (s_n^D, s_n^H, a_n))$ of n sequence pairs with their desired alignment. In the following, we will design a discriminative training algorithm that finds a parameter vector $\mathbf{w}$ for rules of type (7) by minimizing the loss on the training data S.

$$R_S^{L_\Delta}(\mathbf{w}) = \frac{1}{n} \sum_{i=1}^{n} L_\Delta \left(a_i, argmax_a \left[D_\mathbf{w}(s_i^N, s_i^H, a)\right]\right) \tag{11}$$

First, consider the case where there exists a $\mathbf{w}$ so that the training loss $R_S^{L_\Delta}(\mathbf{w})$ is zero. Since we assume a scoring function that is linear in the parameters

$$D_\mathbf{w}(s_1, s_2, a) = \mathbf{w}^T \Psi(s_1, s_2, a), \tag{12}$$

the condition of zero training error can be written as a set of linear inequality constraints. For each native/homolog pair s_i^N / s_i^H, we need to introduce one linear constraint for each possible alignment a of s_i^N into s_i^H.

$$\forall a \neq a_1 : \ D_\mathbf{w}(s_1^N, s_1^H, a) < D_\mathbf{w}(s_1^N, s_1^H, a_1)$$
$$\cdots \tag{13}$$
$$\forall a \neq a_n : \ D_\mathbf{w}(s_n^N, s_n^H, a) < D_\mathbf{w}(s_n^N, s_n^H, a_n)$$

Any parameter vector $\mathbf{w}^*$ that fulfills this set of constraints has a training loss $R_S^{L_\Delta}(\mathbf{w}^*)$ of zero. This approach of writing the training problem as a linear system follows the method in [8] proposed for the special case of global alignment without free insertions/deletions. However, for the general case in (13) the number of constraints is exponential, since the number of alignments a between s_i^N and s_i^H can be exponential in the length of s_i^N and s_i^H. Unlike the restricted case in [8], standard optimization algorithms cannot handle this size of problem. To overcome this limitation, in Sect. 5 we will propose an algorithm that exploits the special structure of (13) so that it needs to examine only a subset that is polynomial in the length of s_i^N and s_i^H.

 If the set of inequalities in (13) is feasible, there will typically be more than one solution $\mathbf{w}^*$. To specify a unique solution, we select the $\mathbf{w}^*$ for which each score $D_\mathbf{w}(s_i^N, s_i^H, a_i)$ is uniformly most different from $\max_{a \neq a_i} D_\mathbf{w}(s_i^N, s_i^H, a)$ for all i. This corresponds to the maximum-margin principle employed in Support Vector Machines (SVMs) [15]. Denoting the margin by δ and restricting the L_2 norm of $\mathbf{w}$ to make the problem well-posed, this leads to the following optimization problem.

$$max_\mathbf{w} \ \delta \tag{14}$$
$$\forall a \neq a_1 : \ D_\mathbf{w}(s_1^N, s_1^H, a) \leq D_\mathbf{w}(s_1^N, s_1^H, a_1) - \delta$$
$$\cdots \tag{15}$$
$$\forall a \neq a_n : \ D_\mathbf{w}(s_n^N, s_n^H, a) \leq D_\mathbf{w}(s_n^N, s_n^H, a_n) - \delta$$
$$||\mathbf{w}|| = 1 \tag{16}$$

Due to the linearity of the similarity function (12), the length of $\mathbf{w}$ is a free variable and we can fix it to $1/\delta$. Substituting for δ and rearranging leads to the equivalent optimization problem

$$min_{\mathbf{w}} \quad \frac{1}{2}\mathbf{w}^T\mathbf{w} \tag{17}$$

$$\forall a \neq a_1 : \ \left(\Psi(s_1^N, s_1^H, a_1) - \Psi(s_1^N, s_1^H, a)\right)\mathbf{w} \geq 1 \tag{18}$$

$$... \tag{19}$$

$$\forall a \neq a_n : \ \left(\Psi(s_n^N, s_n^H, a_n) - \Psi(s_n^N, s_n^H, a)\right)\mathbf{w} \geq 1 \tag{20}$$

Since this quadratic program (QP) has a positive-definite objective function and (feasible) linear constraints, it is strictly convex. This means it has a unique global minimum and no local minima [4]. The constraints are similar to the ordinal regression approach in [6] and it has a structure similar to the Ranking SVM described in [7] for information retrieval. However, the number of constraints is much larger.

To allow errors in the training set, we introduce slack variables ξ_i [2]. Corresponding to the error measure $R_P^{L\Delta}(\mathbf{w})$ we have one slack variable for each native sequence. This is different from a normal classification or regression SVM, where there is a different slack variable for each constraint. The slacks enter the objective function according to a trade-off parameter C. For simplicity, we consider only the case where the slacks enter the objective function squared.

$$min_{\mathbf{w},\boldsymbol{\xi}} \quad \frac{1}{2}\mathbf{w}^T\mathbf{w} + C\sum_{i=1}^{n}\xi_i^2 \tag{21}$$

$$\forall a \neq a_1 : \ \left(\Psi(s_1^N, s_1^H, a_1) - \Psi(s_1^N, s_1^H, a)\right)\mathbf{w} \geq 1 - \xi_1 \tag{22}$$

$$... \tag{23}$$

$$\forall a \neq a_n : \ \left(\Psi(s_n^N, s_n^H, a_n) - \Psi(s_n^N, s_n^H, a)\right)\mathbf{w} \geq 1 - \xi_n \tag{24}$$

Analogous to classification and regression SVMs [15], this formulation minimizes a regularized upper bound on the training loss $R_S^{L\Delta}(\mathbf{w})$.

Proposition 1. *For any feasible point $(\mathbf{w}, \boldsymbol{\xi})$ of (21)-(24), $\frac{1}{n}\sum_{i=1}^{n}\xi_i^2$ is an upper bound on the training loss $R_S^{L\Delta}(\mathbf{w})$ for the 0/1-loss $L_\Delta(.,.)$.*

The proof is given in [14]. The quadratic program and the proposition can be extended to any non-negative loss function.

4.2 Homology Prediction

For the problem of Homology Prediction, we can derive a similar training problem. Here, the training sample S consists of native sequences $s_1^N, ..., s_n^N$, homolog sequences $s_1^H, ..., s_n^H$, and a set of decoy sequences $S_1^D, ..., S_n^D$ for each native $s_1^N, ..., s_n^N$. As a simplifying assumption, we assume that between native and homolog sequences the alignment $a_1^{NH}, ..., a_n^{NH}$ of maximum score is known[1]. The

---

[1]For protein alignment, for example, this could be generated via structural alignment.

goal is to find an optimal $\mathbf{w}$ so that the error rate $Err_P^{LA}(\mathbf{w})$ is low. Again, finding a $\mathbf{w}$ such that the error on the training set

$$Err_P^{LA}(\mathbf{w}) = \frac{1}{n} \sum_{i=1}^{n} L_{\Delta}\left(s_i^H, \underset{s \in S_i^D \cup \{s_i^H\}}{\arg\max} D_{\mathbf{w}}(s_i^N, s)\right) \tag{25}$$

is zero can be written as a set of linear inequality constraints. There is one constraint for each combination of native sequence s_i^N, decoy sequence $s_i^{D_j}$, and possible alignment a of s_i^N into $s_i^{D_j}$.

$$\forall s_1^{D_j} \in S_1^D \forall a : \ D_{\mathbf{w}}(s_1^N, s_1^{D_j}, a) < D_{\mathbf{w}}(s_1^N, s_1^H, a_1^{NH})$$

$$\cdots \tag{26}$$

$$\forall s_n^{D_j} \in S_n^D \forall a : \ D_{\mathbf{w}}(s_n^N, s_n^{D_j}, a) < D_{\mathbf{w}}(s_n^N, s_n^H, a_n^{NH})$$

Similar to the case of Alignment Prediction, one can add a margin criterion and slacks and arrives at the following convex quadratic program.

$$min_{\mathbf{w}, \boldsymbol{\xi}} \ \frac{1}{2}\mathbf{w}^T\mathbf{w} + C \sum_{i=1}^{n} \xi_i^2 \tag{27}$$

$$\forall s_1^{D_j} \in S_1^D \forall a : \ \left(\Psi(s_1^N, s_1^H, a_1^{NH}) - \Psi(s_1^N, s_1^{D_j}, a)\right)\mathbf{w} \geq 1 - \xi_1 \tag{28}$$

$$\cdots \tag{29}$$

$$\forall s_n^{D_j} \in S_n^D \forall a : \ \left(\Psi(s_n^N, s_n^H, a_n^{NH}) - \Psi(s_n^N, s_n^{D_j}, a)\right)\mathbf{w} \geq 1 - \xi_n \tag{30}$$

Again, $\frac{1}{n}\sum_{i=1}^{n} \xi_i^2$ is an upper bound on the training error $Err_S^{LA}(\mathbf{w})$.

5 Training Algorithm

Due to the exponential number of constraints in both the optimization problem for Alignment Prediction and Homology Prediction, naive use of off-the-shelf tools for their solution is computationally intractable for problems of interesting size. However, by exploiting the special structure of the problem, we propose the algorithms shown in Figs. 2 and 3 that find the solutions of (21)-(24) and (27)-(30) after examining only a small number of constraints. The algorithms proceeds by greedily adding constraints from (21)-(24) or (28)-(30) to a working set K. The algorithms stop, when all constraints in (21)-(24) or (28)-(30) are fulfilled up to a precision of ϵ.

The following two theorems show that the algorithms return a solutions of (21)-(24) and (27)-(30) that are accurate with a precision of some predefined ϵ, and that they stop after a polynomially bounded number of iterations through the repeat-loop.

Theorem 1. (CORRECTNESS)
The algorithms return an approximation that has an objective value not higher than the solution of (21)-(24) and (27)-(30), and that fulfills all constraints up to a precision of ϵ. For $\epsilon = 0$, the algorithm returns the exact solution $(\mathbf{w}^*, \boldsymbol{\xi}^*)$.

Input: native sequences $s_1^N, ..., s_n^N$, homolog sequences $s_1^H, ..., s_n^H$, alignments $a_1, ..., a_n$, tolerated approximation error $\epsilon \geq 0$.

$K = \emptyset$, $\mathbf{w} = 0$, $\boldsymbol{\xi} = 0$

repeat

- $K_{org} = K$
- for i from 1 to n
  - find $\hat{a} = argmax_a \left[\mathbf{w}^T \Psi(s_i^N, s_i^H, a) \right]$ via dynamic programming
  - if $\mathbf{w}^T (\Psi(s_i^N, s_i^H, a_i) - \Psi(s_i^N, s_i^H, \hat{a})) < 1 - \xi_i - \epsilon$
    - $K = K \cup \left\{ \mathbf{w}^T (\Psi(s_i^N, s_i^H, a_i) - \Psi(s_i^N, s_i^H, \hat{a})) \geq 1 - \xi_i \right\}$
    - solve QP $(\mathbf{w}, \boldsymbol{\xi}) = argmin_{\mathbf{w}, \boldsymbol{\xi}} \frac{1}{2} \mathbf{w}^T \mathbf{w} + C \sum_{i=1}^n \xi_i^2$ subject to K.

until$(K = K_{org})$

Output: $\mathbf{w}$

Figure 2. Sparse Approximation Algorithm for the Alignment Prediction task

Input: native sequences $s_1^N, ..., s_n^N$, homolog sequences $s_1^H, ..., s_n^H$, alignments $a_1^{NH}, ..., a_n^{NH}$, sets of decoy sequences $S_1^D, ..., S_n^D$, tolerated approximation error $\epsilon \geq 0$.

$K = \emptyset$, $\mathbf{w} = 0$, $\boldsymbol{\xi} = 0$

repeat

- $K_{org} = K$
- for i from 1 to n
  - for j from 1 to $|S_i^D|$
    - find $\hat{a} = argmax_{a \neq a_i} \left[\mathbf{w}^T \Psi(s_i^N, s_i^{D_j}, a) \right]$ via dynamic programming
    - if $\mathbf{w}^T (\Psi(s_i^N, s_i^H, a_i^{NH}) - \Psi(s_i^N, s_i^{D_j}, \hat{a})) < 1 - \xi_i - \epsilon$
    - $K = K \cup \left\{ \mathbf{w}^T (\Psi(s_i^N, s_i^H, a_i^{NH}) - \Psi(s_i^N, s_i^{D_j}, \hat{a})) \geq 1 - \xi_i \right\}$
    - solve QP $(\mathbf{w}, \boldsymbol{\xi}) = argmin_{\mathbf{w}, \boldsymbol{\xi}} \frac{1}{2} \mathbf{w}^T \mathbf{w} + C \sum_{i=1}^n \xi_i^2$ subject to K.

until$(K = K_{org})$

Output: $\mathbf{w}$

Figure 3. Sparse Approximation Algorithm for the Homology Prediction task

Proof. Let $\mathbf{w}^*$ and ξ_i^* be the solution of (21)-(24) or (27)-(30) respectively. Since the algorithm solves the QP on a subset of the constraints in each iteration, it returns a solution $\mathbf{w}$ with $\frac{1}{2} \mathbf{w}^T \mathbf{w} + C \sum_{i=1}^n \xi_i^2 \leq \frac{1}{2} \mathbf{w}^{*T} \mathbf{w}^* + C \sum_{i=1}^n \xi_i^{*2}$. This follows from the fact that restricting the feasible region cannot lead to a lower minimum.

It is left to show that the algorithm does not terminate before all constraints (21)-(24) or (28)-(30) are fulfilled up to precision ϵ. In the final iteration, the algorithms find the most violated constraint. For Alignment Prediction, this is the constraint $\mathbf{w}^T (\Psi(s_i^N, s_i^H, a_i) - \Psi(s_i^N, s_i^H, \hat{a})) < 1 - \xi_i$ corresponding to the highest scoring alignment $\hat{a}$. For Homology Prediction, it is the constraint $\mathbf{w}^T (\Psi(s_i^N, s_i^H, a_i^{NH}) - \Psi(s_i^N, s_i^{D_j}, \hat{a})) < 1 - \xi_i$ corresponding to the highest scoring alignment $\hat{a}$ for each decoy. Three cases can occur: First, the constraint can be violated by more than ϵ and the algorithm will not terminate yet. Second, it is already in K and is fulfilled by construction. Third, the constraint is not in K but fulfilled anyway and it is not added. If the constraint is fulfilled, the constraints for all other alignments into this decoy are fulfilled as well, since we checked the constraint for which the margin was smallest for the given $\mathbf{w}$. It follows that the algorithm terminates only if all constraints are fulfilled up to precision ϵ.

It is left to show that the algorithms terminates after a number of iterations that is smaller than the set of constraints. The following theorem shows that the algorithm stops after a polynomial number of iterations.

Theorem 2. (TERMINATION)
The algorithms stops after adding at most

$$\frac{2VR^2}{\epsilon^2} \tag{31}$$

constraints to the set K. V is the minimum of (21)-(24) or (27)-(30) respectively. R^2 is a constant bounded by the maximum of $(\Psi(s_i^N, s_i^H, a_i) - \Psi(s_i^N, s_i^H, a))^2 + \frac{1}{2C}$ or $(\Psi(s_i^N, s_i^H, a_i^{NH}) - \Psi(s_i^N, s_i^{D_j}, a))^2 + \frac{1}{2C}$ respectively.

Proof. In the following, we focuses on the Homology Prediction task. The proof for the Alignment Prediction task is analogous. The first part of the proof is to show that the objective value increases by some constant with every constraint that is added to K. Denote with V_k the solution $V_k = P(\mathbf{w}_k^*, \boldsymbol{\xi}_k^*) = min_{\mathbf{w}, \boldsymbol{\xi}} \frac{1}{2}\mathbf{w}^T\mathbf{w} + C\sum_{i=1}^n \xi_i^2$ subject to K_k after adding k constraints. This primal optimization problem can be transformed into an equivalent problem of the form $V_k = P(\mathbf{w}_k') = min_{\mathbf{w}'} \frac{1}{2}\mathbf{w}'^T\mathbf{w}'$ subject to K_k', where each constraint has the form $\mathbf{w}'^T\mathbf{x} \geq 1$ with $\mathbf{x} = (\Psi(s_i^N, s_i^H, a_i^{NH}) - \Psi(s_i^N, s_i^{D_j}, \hat{a}); 0; ...; 0; 1/\sqrt{2C}; 0; ...; 0)$. Its corresponding Wolfe dual is $D(\boldsymbol{\alpha}_k^*) = max_{\boldsymbol{\alpha} \geq 0} \sum_{i=1}^k \alpha_i - \frac{1}{2}\sum_{i=1}^k \sum_{j=1}^k \alpha_i \alpha_j \mathbf{x}_i \mathbf{x}_j$. At the solution $D(\boldsymbol{\alpha}_k^*) = P(\mathbf{w}_k'^*) = P(\mathbf{w}_k^*, \boldsymbol{\xi}_k^*) = V_k$ and for every feasible point $D(\boldsymbol{\alpha}) \leq P(\mathbf{w}, \boldsymbol{\xi})$. Primal and dual are connected via $\mathbf{w}'^* = \sum_{i=1}^k \alpha_i^* \mathbf{x}_i$. Adding a constraint to the dual with $\mathbf{w}'^{*T}\mathbf{x}_{k+1} = \sum_{i=1}^k \alpha_i^* \mathbf{x}_i \mathbf{x}_{k+1} \leq 1 - \epsilon$ means extending the dual to

$$D_{k+1}(\boldsymbol{\alpha}_{k+1}^*) = \max_{\alpha_{k+1} \geq 0} \sum_{i=1}^k \alpha_i - \frac{1}{2}\sum_{i=1}^k \sum_{j=1}^k \alpha_i \alpha_j \mathbf{x}_i \mathbf{x}_j + \alpha_{k+1} - \alpha_{k+1} \sum_{i=1}^k \alpha_i \mathbf{x}_i \mathbf{x}_{k+1} - \frac{1}{2}\alpha_{k+1}^2 \mathbf{x}_{k+1}^2$$

$$\geq D_k(\boldsymbol{\alpha}_k^*) + \max_{\alpha_{k+1} \geq 0} \alpha_{k+1} - \alpha_{k+1} \sum_{i=1}^k \alpha_i^* \mathbf{x}_i \mathbf{x}_{k+1} - \frac{1}{2}\alpha_{k+1}^2 \mathbf{x}_{k+1}^2$$

$$\geq D_k(\boldsymbol{\alpha}_k^*) + \max_{\alpha_{k+1} \geq 0} \alpha_{k+1} - \alpha_{k+1}(1 - \epsilon) - \frac{1}{2}\alpha_{k+1}^2 \mathbf{x}_{k+1}^2$$

Solving the remaining scalar optimization problem over α_{k+1} shows that $\alpha_{k+1}^* \geq 0$ and that $V_{k+1} \geq V_k + \frac{\epsilon^2}{2R^2}$.

Since the algorithm only adds constraints that are violated by the current solution by more than ϵ, after adding $k_{max} = \frac{2VR^2}{\epsilon^2}$ constraints the solution $V_{k_{max}}$ over the subset $K_{k_{max}}$ is at least $V_{k_{max}} \geq V_0 + \frac{2VR^2}{\epsilon^2} \frac{\epsilon^2}{2R^2} = 0 + V$. Any additional constraint that is violated by more than ϵ would lead to a minimum that is larger than V. Since the minimum over a subset of constraints can only be smaller than the minimum over all constraints, there cannot be any more constraints violated by more than ϵ and the algorithm stops.

Since V can be upper bounded as $V \leq C.n$ using the feasible point $\mathbf{w} = 0$ and $\boldsymbol{\xi} = 1$ in (21)-(24) or (27)-(30), the theorem directly leads to the conclusion that the maximum number of constraints in K scales linearly with the number of training examples n. Furthermore, it scales only polynomially with the length of the sequences, since R is polynomial in the length of the sequences.

While the number of constraints can potentially explode for small values of ϵ, experience with Support Vector Machines for classification showed that relatively large values of ϵ are sufficient without loss of generalization performance. We will verify the efficiency and the prediction performance of the algorithm empirically in the following.

6 Experiments

To analyze the behavior of the algorithm under varying conditions, we constructed a synthetic dataset according to the following sequence and alignment model. While this simple model does not exploit the flexibility of the parameterized linear model $D_{\mathbf{w}}(s_1, s_2, a) = \mathbf{w}^T \Psi(s_1, s_2, a)$, it does serve as a feasibility check of the learning algorithm. The native sequence and the decoys are generated by drawing randomly from a 20 letter alphabet $\Sigma = \{1, .., 20\}$ so that letter $c \in \Sigma$ has probability $c/210$. Each sequence has length 50, and there are 10 decoys per native. To generate the homolog, we generate an alignment string of length 30 consisting of 4 characters "match", "substitute", "insert", "delete". For simplicity of illustration, substitutions are always $c \rightarrow (c \bmod 20) + 1$. While we experiment with several alignment models, we only report typical results here where matches occur with probability 0.2, substitutions with 0.4, insertion with 0.2, deletion with 0.2. The homolog is created by applying the alignment string to a randomly selected substring of the native. The shortening of the sequences through insertions and deletions is padded by additional random characters.

In the following experiments, we focus on the problem of Homology Prediction. Fig. 4 shows training and test error rates for two models depending on the number of training examples averaged over 10 trials. The first model has only 3 parameters ("match", "substitute", "insert/delete") and uses a uniform substitution matrix. This makes the feature vectors $\phi(c_1, c_2, a_i)$ three-dimensional with a 1 indicating the appropriate operation. The second model also learns the 400 parameters of the substitution matrix, resulting in a total of 403 parameters. Here, $\phi(c_1, c_2, a_i)$ indicates the element of the substitution matrix in addition to the operation type. We chose $C = 0.01$ and $\epsilon = 0.1$. The left-hand graph of Fig. 4 shows that for the 403-parameter model, the generalization error is high for small numbers of training

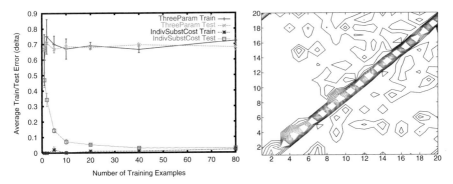

Figure 4. *Left*: Train and test error rates for the 3 and the 403 parameter model depending on the number of training examples. *Right*: Typical learned substitution matrix after 40 training examples for the 403-parameter model

examples, but quickly drops as the number of examples increases. The 3-parameter model cannot fit the data as well. Its training error starts out much higher and training and test error essentially converge after only a few examples. The right-hand graph of Fig. 4 shows the learned matrix of substitution costs for the 403-parameter model. As desired, the elements of the matrix are close to zero except for the off-diagonal. This captures the substitution model $c \to (c \bmod 20) + 1$.

Figure 5 analyzes the efficiency of the algorithm via the number of constraints that are added to K before convergence. The left-hand graph shows the scaling with the number of training examples. As predicted by Theorem 2, the number of constraints grows (sub-)linearly with the number of examples. Furthermore, the actual number of constraints encountered during any iteration of the algorithm is small enough to be handled by standard quadratic optimization software. The right-hand graph shows how the number of constraints in the final K changes with $\log(\epsilon)$. The observed scaling appears to be better than suggested by the upper bound in

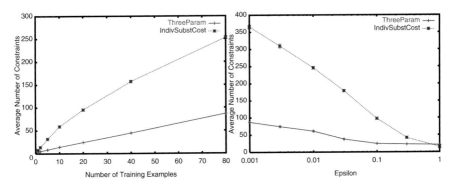

Figure 5. Number of constraints added to K depending on the number of training examples (*left*) and the value of ϵ (*right*). If not stated otherwise, $\epsilon = 0.1$, $C = 0.01$, and $n = 20$

Theorem 2. A good value for ϵ is 0.1. We observed that larger values lead to worse prediction accuracy, while smaller values decrease efficiency while not providing further benefit.

7 Conclusions

The paper presented a discriminative learning approach to inferring the cost parameters of a linear sequence alignment model from training data. We proposed an algorithm for solving the resulting training problem and showed that it is computationally efficient. Experiments show that the algorithm can effectively learn the alignment parameters on a synthetic task. We are currently applying the algorithm to learning alignment models for protein homology detection and protein alignment prediction.

References

[1] R. Barzilay and L. Lee. Bootstrapping lexical choice via multiple-sequence alignment. In *Conference on Empirical Methods in Natural Language Processing (EMNLP)*, 2002.
[2] Corinna Cortes and Vladimir N. Vapnik. Support–vector networks. *Machine Learning Journal*, 20:273–297, 1995.
[3] R. Durbin, S. Eddy, A. Krogh, and G. Mitchison. *Biological Sequence Analysis*. Cambridge University Press, 1998.
[4] P. E. Gill, W. Murray, and M. H. Wright. *Practical Optimization*. Academic Press, 1981.
[5] D. Gusfield and P. Stelling. Parametric and inverse-parametric sequence alignment with xparal. *Methods in Enzymology*, 266:481–494, 1996.
[6] R. Herbrich, T. Graepel, and K. Obermayer. Large margin rank boundaries for ordinal regression. In *Advances in Large Margin Classifiers*, pages 115–132. MIT Press, Cambridge, MA, 2000.
[7] T. Joachims. Optimizing search engines using clickthrough data. In *Proceedings of the ACM Conference on Knowledge Discovery and Data Mining (KDD)*, 2002.
[8] J. Meller and R. Elber. Linear programming optimization and a double statistical filter for protein threading protocols. *Proteins Structure, Function, and Genetics*, 45:241–261, 2001.
[9] L. Pachter and B. Sturmfelds. Parametric inference for biological sequence analysis. In *Proceedings of the National Academy of Sciences*, volume 101, pages 16138–16143, 2004.
[10] L. Pachter and B. Sturmfelds. Tropical geometry of statistical models. In *Proceedings of the National Academy of Sciences*, volume 101, pages 16132–16137, 2004.
[11] S. E Ristad and P. N. Yianilos. Learning string edit distance. *IEEE Transactions on Pattern Recognition and Machine Intelligence*, Vol. 20(5):522–532, 1998.

[12] T. Smith and M. Waterman. Identification of common molecular subsequences. *Journal of Molecular Biology*, 147:195–197, 1981.

[13] Fangting Sun, D. Fernandez-Baca, and Wei Yu. Inverse parametric sequence alignment. In *International Computing and Combinatorics Conference (CO-COON)*, 2002.

[14] I. Tsochantaridis, T. Hofmann, T. Joachims, and Y. Altun. Support vector machine learning for interdependent and structured output spaces. In *International Conference on Machine Learning (ICML)*, 2004.

[15] V. Vapnik. *Statistical Learning Theory*. Wiley, Chichester, GB, 1998.

Part II

Minimization of Complex Molecular Landscapes

Overcoming Energetic and Time Scale Barriers Using the Potential Energy Surface

David J. Wales, Joanne M. Carr and Tim James

Department of Chemistry, Lensfield Road, Cambridge CB2 1EW, UK

Abstract. Sampling stationary points of the potential energy surface provides an intuitive way to coarse-grain calculations of both thermodynamic and dynamic properties. Functions such as internal energy, entropy, free energy and the heat capacity can be obtained from the superposition approximation, where the total partition function is written as a sum of contributions from a database of local minima. Rates can be calculated if the database is augmented to include transition states that connect the minima, and the discrete path sampling method provides a systematic approach to this problem. Transforming the potential energy surface into the basins of attraction of local minima also provides a powerful global optimisation algorithm via the basin-hopping approach.

Key words: Energy landscapes. Discrete path sampling. Rare events. Global optimisation. Basin-hopping.

1 Introduction

Standard simulation techniques, namely the molecular dynamics and Monte Carlo approaches, often encounter serious difficulties when applied to systems that exhibit large free energy barriers or long relaxation time scales between the states of interest. An alternative approach to such problems is provided by methods based upon direct sampling of stationary points of the underlying potential energy surface [75]. Stationary points for an N-atom system are configurations where the $3N$-dimensional gradient of the potential energy vanishes. At a local minimum all the normal mode frequencies are positive (or zero). A formally exact representation of the complete partition function can be obtained using a sum over the contributions of local minima, i.e.

$$\Omega(E) = \sum_\alpha \Omega_\alpha(E) \quad \text{and} \quad Z(T) = \sum_\alpha Z_\alpha(T), \tag{1}$$

for the microcanonical density of states and the canonical partition function, respectively. Here, the term corresponding to local minimum α can be defined from the 'catchment basin' [51] or 'basin of attraction', which is the volume of configuration space from which steepest-descent paths converge to that minimum [75]. The super-position method is an old idea [9, 35, 40, 50, 68, 69, 73, 75, 79], which has become increasingly popular in recent years with improvements in geometry optimisation algorithms and increased computer power. It is usually applied using the classical harmonic approximation for the vibrational density of states, although anharmonic [12, 23] and quantum [13] effects can also be included. Full details can be found in reference [75].

To treat kinetics within the same framework we must locate transition states in configuration space, which are here defined as stationary points of the potential energy that possess a single imaginary force constant [52]. The Murrell-Laidler theorem states that if two local minima are connected by a path involving a saddle point with two or more imaginary normal mode frequencies, then a lower energy path exists involving only true transition states [52]. Hence, by finding transition states we can characterise pathways between local minima that involve the lowest barriers. We can also employ statistical rate theories to calculate rate constants [28, 29, 34, 60, 85], and the most common approximation again involves harmonic densities of states [75].

Finding transition states is significantly more difficult than locating local minima. However, a number of algorithms have been developed for treating large systems, including biomolecules modelled by empirical force fields. All the results in the following sections employ a hybrid eigenvector-following approach [56, 75]. Once a connected database of stationary points has been constructed partition functions are calculated for the individual local minima, together with rate constants for the minimum-to-minimum transitions. The global dynamics can then be treated using master equation [47, 70] or kinetic Monte Carlo [7, 36–39] techniques.

The total number of stationary points generally grows exponentially with system size [25, 69, 78], and it is therefore necessary to develop schemes to provide an appropriate sampling for either thermodynamic or dynamic properties. Sect. 2 provides an outline of the discrete path sampling (DPS) approach [74–76], which corresponds to a coarse-grained version of transition path sampling [4, 20] based on stationary points of the potential energy surface (PES). This method has now been applied to a number of different systems [18, 32, 33, 74, 76].

Coarse-graining the PES into catchment basins for local minima is also the starting point for the basin-hopping approach to global optimisation [77], which provides a generalisation of the 'Monte Carlo plus energy minimisation' procedure of Li and Scheraga [49]. This algorithm is outlined in Sect. 3, together with some results for protonated water clusters.

2 Discrete Path Sampling

Various alternative strategies have been suggested for overcoming the time scale problem in computer simulations [4, 14, 17, 19, 20, 26, 27, 30, 31, 41, 42, 46, 55, 58, 59, 61–64, 67, 71, 72]. The DPS approach focuses upon 'discrete paths', defined as connected sequences of minima and the intervening transition state(s) [74]. The

simplest discrete path comprises a single transition state and the two minima it links
by (approximate) steepest-descent paths commencing parallel and antiparallel to the
Hessian eigenvector corresponding to the unique negative eigenvalue; paths between
minima well separated in configuration space are likely to contain many transition
states and intervening minima.

The DPS approach [74] provides a means of constructing databases of minima
and the connecting transition states on, and in the vicinity of, discrete paths between
two states, A and B. Within this framework, the algorithm also enables rate constants
to be calculated by summing over contributions from all the possible paths in the
database.

The A and B states are groups of local potential energy minima defined according to an order parameter of some form. In order for single-exponential decay from
an initial non-equilibrium distribution to be observed, the minima within each of the
A and B states must be in local equilibrium:

$$p_a(t) = \frac{p_a^{eq} P_A(t)}{P_A^{eq}} \quad \text{and} \quad p_b(t) = \frac{p_b^{eq} P_B(t)}{P_B^{eq}} . \tag{2}$$

$p_a(t)$ is the occupation probability of minimum a at time t, $P_A(t) = \sum_{a\in A} p_a(t)$
etc., a lower-case subscript denotes a member of the corresponding upper-case state,
and the superscript 'eq' stands for 'equilibrium'.

The theory begins from the linear master equation: the set of coupled first-order
differential equations determining the time evolution of the components of the vector
of occupation probabilities for the local minima, i.e.

$$\frac{dP_\alpha(t)}{dt} = \sum_{\alpha \neq \beta} [k_{\alpha\beta} P_\beta(t) - k_{\beta\alpha} P_\alpha(t)] , \tag{3}$$

where $k_{\beta\alpha}$ is the rate constant from minimum α to minimum β and the sum is over
all geometrically distinct minima directly connected to α. The linearity of equation
(3) results from the assumption that the system has time to equilibrate sufficiently
between transitions to lose its memory of how it reached the current location, i.e. the
dynamics are Markovian.

When all the minima belong to either the A or the B state, the master equation
becomes

$$\frac{dP_A(t)}{dt} = k_{AB} P_B(t) - k_{BA} P_A(t)$$
$$\text{and} \quad \frac{dP_B(t)}{dt} = k_{BA} P_A(t) - k_{AB} P_B(t) , \tag{4}$$

where

$$k_{AB} = \frac{1}{P_B^{eq}} \sum_{a\in A}\sum_{b\in B} k_{ab} p_b^{eq} \quad \text{and} \quad k_{BA} = \frac{1}{P_A^{eq}} \sum_{a\in A}\sum_{b\in B} k_{ba} p_a^{eq} . \tag{5}$$

Thus, the phenomenological rate constants can be expressed as a weighted sum over
all the minimum-to-minimum transitions involving a single transition state across the

boundary between the two regions. The weighting factor is the conditional equilibrium probability of finding the system in the starting minimum given that the system is in the appropriate A or B state, $p_a^{\text{eq}}/P_A^{\text{eq}}$ or $p_b^{\text{eq}}/P_B^{\text{eq}}$.

In the more usual case, where not all of the minima can be classified as having either A or B character, we can admit a third, intervening (I) set. Effective two-state dynamics is recovered within the steady-state approximation for the occupation probability of each intervening minimum, i:

$$\frac{\mathrm{d}p_i(t)}{\mathrm{d}t} = \sum_{\alpha \neq i} k_{i\alpha} p_\alpha(t) - p_i(t) \sum_{\alpha \neq i} k_{\alpha i} \approx 0 \,. \tag{6}$$

This assumption gives expressions for the (low) occupation probabilities of the intervening minima which, when inserted into the appropriate master equation, yield the phenomenological rate constants as

$$k_{AB} = \frac{1}{P_B^{\text{eq}}} \sum_{a \leftarrow b} \frac{k_{ai_1} k_{i_1 i_2} \dots k_{i_n b}\, p_b^{\text{eq}}}{\sum_{\alpha_1} k_{\alpha_1 i_1} \sum_{\alpha_2} k_{\alpha_2 i_2} \dots \sum_{\alpha_n} k_{\alpha_n i_n}} \,,$$

$$k_{BA} = \frac{1}{P_A^{\text{eq}}} \sum_{b \leftarrow a} \frac{k_{bi_1} k_{i_1 i_2} \dots k_{i_n a}\, p_a^{\text{eq}}}{\sum_{\alpha_1} k_{\alpha_1 i_1} \sum_{\alpha_2} k_{\alpha_2 i_2} \dots \sum_{\alpha_n} k_{\alpha_n i_n}} \,. \tag{7}$$

Again, we are summing over all paths that start and finish on the A and B boundaries, but now we include paths that start on the A/I or B/I boundary and pass through only intervening minima, as well as direct $A \leftrightarrow B$ transitions. The sums over α_j in the denominator include all geometrically distinct minima directly connected to intervening minimum i_j.

The number of steps in a given discrete path is the number of transition states in the sequence, which is one less than the total number of minima. For each discrete path containing three or more steps, an infinite number of paths can be generated from the same stationary points by allowing an arbitrary number of recrossings of any transition state involving only intervening minima. As the contribution to the total rate constant diminishes with each recrossing, the overall rate constant for a particular sequence of unique stationary points can be obtained as a convergent sum using the weighted adjacency matrix with elements

$$A_{i_\alpha i_\beta} = \begin{cases} k_{i_\alpha i_\beta} / \sum_\gamma k_{\gamma i_\alpha}, & |i_\alpha - i_\beta| = 1 \,, \\ 0, & \text{otherwise,} \end{cases} \tag{8}$$

where γ is a minimum directly connected to i_α. The dimension of $\mathbf{A}$ is the number of intervening minima. The total contribution from a particular $a \leftarrow b$ or $b \leftarrow a$ discrete path is then

$$k_{ab} = \frac{p_b^{eq}}{P_B^{eq}} k_{ai_1} \frac{k_{i_n b}}{\sum_\gamma k_{\gamma i_n}} \sum_{p=n-1}^{\infty} [\mathbf{A}^p]_{1n}$$

or $\quad k_{ba} = \dfrac{p_a^{eq}}{P_A^{eq}} k_{bi_n} \dfrac{k_{i_1 a}}{\sum_\gamma k_{\gamma i_1}} \displaystyle\sum_{p=n-1}^{\infty} [\mathbf{A}^p]_{n1} \,.$ \hfill (9)

The sums are over p odd or even for n even or odd, respectively, and may in practice be terminated after some convergence criterion is met.

The total DPS $A \leftarrow B$ and $B \leftarrow A$ rate constants are then sums over discrete paths of the rate constants from (9), which includes the recrossings correction. A similar analysis can be performed for all the discrete paths that can be constructed from the known set of stationary points, using the weighted adjacency matrix for all the connected minima.

It should be noted that recrossings, where the system moves from minimum i to minimum $i + 1$ via transition state j and then returns to minimum i via the same transition state at some later time along the path, are different from the dynamical corrections for recrossing of the dividing surface in the course of a single minimum-to-minimum transition. The latter corrections could be admitted through reactive flux calculations for the elementary minimum-to-minimum rate constants [1, 5, 15, 19].

The second aspect of the DPS method is concerned with the sampling of discrete paths that are kinetically relevant in the appropriate temperature range. Starting from the current path, a minimum on the path is 'perturbed' by replacing it with an off-pathway minimum that is usually connected to the original minimum by a single transition state. Attempts to locate a connected path between this replacement minimum and the two minima a chosen number of steps along the path in either direction are then made. When the algorithm commences, the current path is the initial path. If the perturbation-connection procedure produces a faster path (in terms of the rate constant including recrossings), then that path becomes the current one and the process is repeated. Otherwise, the old path is perturbed again until a specified number of replacements have been tried, after which point the fastest known path for which fewer than the maximum number of perturbations have been performed becomes the current one for the next cycle. It is only necessary to sample paths in one sense ($A \leftarrow B$ or $B \leftarrow A$) if the sets of A and B minima are fixed, as detailed balance enables us to calculate the other rate constant from $k_{AB}/k_{BA} = P_A^{eq}/P_B^{eq}$. It is noteworthy that the contributions to the overall rate constant include a conditional probability weighting; fast paths that would not contribute because they are highly improbable need not be sampled.

The resulting databases of stationary points can be used to visualise the PES using a disconnectivity graph approach [6, 75, 81] and analysed using any appropriate dynamical method (which need not invoke the steady-state approximation for the intervening minima), such as kinetic Monte Carlo [7, 36–39], or a master equation formulation [47, 70]. Recent results for biomolecules include the calculation of pathways and rate constants for the neurotransmitter peptide met-enkephalin [32] and the 16-amino acid β hairpin-forming sequence from residues 41-56 of the B1 domain of protein G [33]. Both studies employed the CHARMM19 force field [3]

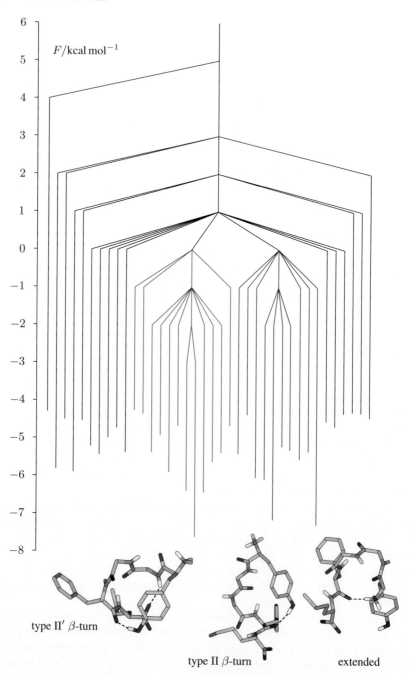

Figure 1. Free energy disconnectivity graph for met-enkephalin at 298 K. Each node represents a group of minima constructed as described in reference [32]. The lowest 38 groups are shown, as these are calculated to contain 90 % of the population. The energy is in units of kcal mol^{-1}. The low energy region of the graph contains two funnels, highlighted in red and blue

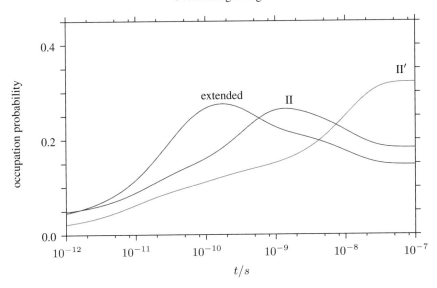

Figure 2. Results of master equation dynamics calculations for met-enkephalin at 298 K starting from a high temperature distribution (800 K) [32]

and the EEF1 implicit solvation potential [48]. Some results for met-enkephalin are shown in Figs. 1 and 2. The calculated rate constants indicate that the peptide can fold from extended conformations to a compact, low-energy type II′ β-turn structure in around 0.1 μs [32]. A type II β-turn structure was also found to represent a significant kinetic trap [32].

3 Basin-Hopping Global Optimisation

Basin-hopping [77], also known as Monte Carlo plus minimisation [49], is a hypersurface deformation global optimisation technique. As the name implies, such methods attempt to improve the efficiency of the search for the global minimum by manipulating the high-dimensional space in which the search progresses. In the case of basin-hopping, the transformation applied at each point of the surface is a local minimisation:

$$\widetilde{E}(\mathbf{X}) = \min\{E(\mathbf{X})\}\,, \tag{10}$$

where E represents the original surface as a function of the nuclear degrees of freedom, $\mathbf{X}$, $\widetilde{E}$ represents the transformed surface, and min indicates local minimisation. In our recent work we have employed a modified version of the limited-memory quasi-Newton routine (LBFGS) by Nocedal [8] as the minimiser, but in principle any method could be used. The effect of the basin-hopping transformation is to collapse the energy landscape onto a series of plateaux, each of which corresponds to the basin of attraction for a particular local minimum on the untransformed surface. This process is illustrated for a one-dimensional landscape in Fig. 3. As can be seen

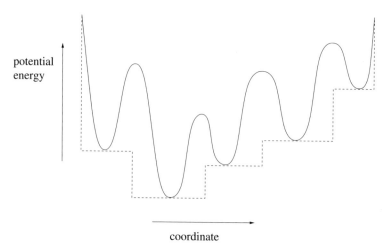

potential energy

coordinate

Figure 3. Illustration of the basin-hopping landscape transformation. The solid line shows the original potential and the deformed surface is indicated as a dashed line

from the diagram, the transformation removes the downhill barriers and can significantly reduce the uphill ones. Importantly, unlike some other hypersurface deformation methods [65], the basin-hopping transformation does not change the position of the global minimum. This feature ensures that the global minimum on the deformed surface, once located, maps directly to the global minimum on the original surface without the need for a reverse transformation.

The basin-hopping method has been applied to investigate clusters of various types [43, 77, 80], and has been further combined with multi-canonical techniques [87]. An equivalent landscape transformation is also implicitly contained within certain evolutionary global optimisation algorithms [10, 21, 57, 82]. Kinetically, basin-hopping improves the rate of interbasin motion because transitions between basins on the transformed hypersurface can occur at any point along their shared boundaries. This is different from the original landscape, where transitions are typically constrained to occur along certain low-energy pathways between minima. Crucially, a change in the thermodynamics of the system is also observed: on the transformed landscape the occupation probability distributions for different structures are broadened [24]. Transitions to the funnel containing the global minimum structure from funnels corresponding to competing morphologies are therefore enhanced.

3.1 Application to Protonated Water Clusters

Recently, we have used the basin-hopping method to investigate the low-energy structures of singly protonated water clusters, $(H_2O)_n H^+$. For this purpose, we employed the multistate empirical valence bond potential (MSEVB) of Schmitt and Voth [66]. Empirical valence bond (EVB) methods [2, 11, 54, 83] are one way in which chemically reactive species, such as protonated water systems, may be modelled. In the EVB approach, the electronic wavefunction, Ψ, is expanded in a basis set of valence bond states, $|\phi_i>$:

$$\Psi = \sum_i c_i |\phi_i> \ . \tag{11}$$

The computational complexity of the problem is reduced by limiting the expansion to a set of intuitively selected, chemically reasonable states. The EVB wavefunction is thus an approximation to the complete valence bond description [45]. The ground state of the system is obtained by solving the equation:

$$\mathbf{Hc} = E\mathbf{Sc} \ , \tag{12}$$

where $\mathbf{c}$ is the vector of coefficients, $\mathbf{H}$ and $\mathbf{S}$ are the Hamiltonian and overlap matrices, and E is the energy. Orthonormality amongst the basis functions ($S_{ij} = \delta_{ij}$) is normally assumed. However, the key approximation of the method is the calculation of the elements of the EVB Hamiltonian using empirical terms rather than electronic structure methods. The diagonal elements of the Hamiltonian represent the interactions within each valence bond state, and are usually given by conventional force field expressions. The off-diagonal elements represent the interactions between different states, and are used to correct the potential in regions of configuration space where the assumptions implicit in the force fields break down, specifically in regions characterised by bond breaking and formation. Functional forms for the various Hamiltonian elements are proposed and then parameterised by fitting to experimental or high-level *ab initio* data [2] for the system of interest. In the MSEVB model, each valence bond state is constructed by assigning a single hydronium ion and $(n-1)$ water molecules from amongst the atoms present in the system. The expansion of the wavefunction begins with an assignment for the current nuclear geometry of a 'pivot' state, which we presume will be the most favourable (lowest energy). Further valence bond states are added in hydration shells around the hydronium ion from the pivot state, up to a maximum of three shells. This process is illustrated in Fig. 4 for an $(H_2O)_4H^+$ system. The diagonal elements of the MSEVB Hamiltonian are given by standard force field expressions representing the various intra and intermolecular interactions. For example, the intermolecular water interactions are given by a flexible TIP3P potential [22, 44], which has been linked by some authors to various deficiencies of the model when compared to *ab initio* calculations on small protonated clusters [16]. The off-diagonal MSEVB Hamiltonian

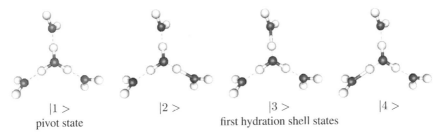

$$|1> \qquad\qquad |2> \qquad\qquad |3> \qquad\qquad |4>$$
pivot state   first hydration shell states

Figure 4. Valence bond states for the $(H_2O)_4H^+$ system

elements are assigned during the hydration shell expansion. For example, in the expansion depicted in Fig. 4, state 2 is generated from the pivot state by reassigning one of the hydrogen atoms, hence the term H_{12} represents the interaction between these two states. Details of the expressions for the interactions are given in reference [66]. The MSEVB potential allows for a flexible description of protonated water clusters. The identity of the excess proton is not fixed, hence its local environment can vary between Zundel-like ($H_5O_2^+$) [86] and Eigen-like ($H_9O_4^+$) [84] forms. It has been suggested by recent quantum simulations [53] that these ions represent the limiting forms for an excess proton in bulk solution.

Combinations of rigid-body translations and rotations of the H_2O and H_3O^+ species identified by the pivot state assignment were used to generate trial moves in the basin-hopping runs for the MSEVB potential. It is unlikely that randomly translating individual atoms would be efficient for this model, because only the excess proton has a reasonable probability of being moved a significant distance without producing a completely unrelated structure. Translating individual atoms would also create major problems for the valence bond assignment procedure. Even employing a hydrogen-bond angle cutoff to limit the hydration shell expansion, the potential is prone to produce cold fusion events if small oxygen-oxygen separations are allowed in the configurations that are passed to the local minimisation routine. Therefore, any trial rigid-body move that generated clashes of this sort was undone and another of the same type (rotation or translation) applied to the same molecule.

We have applied our procedure to investigate clusters of sizes n=2-4, 8 and 20-22. Following the work of Wales and Hodges on neutral water clusters [80], we have found it more efficient for the larger systems to separate the trial basin-hopping steps into blocks of purely translational and purely rotational moves. It may also be more efficient to do this for smaller clusters, but we did not find it necessary. The improvement for larger clusters probably results from reducing the interference between changes of the hydrogen-bonding pattern (angular moves) and the cluster framework (translational moves). In particular, the energy range spanned by clusters with the same basic morphology, but different hydrogen-bonding patterns, can be very large [80]. It is therefore helpful to search the possible hydrogen-bonding schemes more thoroughly before a change in morphology is considered.

Putative global minima for selected clusters, and their MSEVB energies, are shown in Fig. 5. For comparison, the mean first-encounter 'times' of the global minima for n=8 and 21 were 630±480 and 150,000±170,000 steps, respectively. These statistics are based on ten independent basin-hopping runs for each system size, the starting structures for which were selected at regular energy intervals from amongst the minima generated in an initial high-temperature run. The values represent the mean ± the standard deviation, and the observation that the standard deviation is similar to the mean is typical for the basin-hopping approach. This result is consistent with a decaying exponential probability distribution for the first-encounter time, which results if the conditional encounter probability is time independent [75]. The purpose of this sampling procedure was not to exhaustively cover all possible starting points but rather to decrease the probability that in our studies we only investigate structures belonging to a particular morphology.

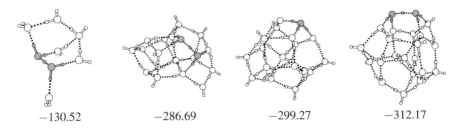

$$-130.52 \qquad -286.69 \qquad -299.27 \qquad -312.17$$

Figure 5. Putative global minima for protonated water clusters, $(H_2O)_n H^+$, $n=8$, 20-22, as described by the MSEVB potential. Energies are in $kcal\, mol^{-1}$. The excess proton unit is shaded in each case for clarity

References

[1] Anderson J.B.: Statistical theories of chemical reactions. Distributions in the transition region. J. Chem. Phys, **58**, 4684–4692 (1973)

[2] Åqvist J., Warshel A.: Simulation of enzyme reactions using valence bond force fields and other hybrid quantum/classical approaches. Chem. Rev., **93**, 2523–2544 (1993)

[3] Brooks B.R., Bruccoleri R.E., Olafson B.D., States D.J., Swaminathan S., Karplus M.: CHARMM: A program for macromolecular energy, minimization, and dynamics calculations. J. Comp. Chem., **4**, 187–217 (1983)

[4] Bolhuis P.G., Chandler D., Dellago C., Geissler P.L.: Transition path sampling: Throwing ropes over rough mountain passes, in the dark. Annu. Rev. Phys. Chem., **53**, 291–318 (2002)

[5] Bennett C.H.: Molecular dynamics and transition state theory: the simulation of infrequent events. In: Christofferson R.E. (ed) Algorithms for Chemical Computations, 63–97, American Chemical Society, Washington, D.C. (1977)

[6] Becker O.M., Karplus M.: The topology of multidimensional potential energy surfaces: Theory and application to peptide structure and kinetics. J. Chem. Phys., **106**, 1495–1517 (1997)

[7] Bortz A.B., Kalos M.H., Leibowitz J.L.: A new algorithm for Monte Carlo simulation of Ising spin systems. J. Comput. Phys., **17**, 10–18 (1975)

[8] Byrd R.H., Lu P., Nocedal J.: A limited memory algorithm for bound constrained optimization. SIAM Journal on Scientific and Statistical Computing, **16**, 1190–1208 (1995)

[9] Burton J.J.: Vibrational frequencies and entropies of small clusters of atoms. J. Chem. Phys., **56**, 3133–3138 (1972)

[10] Cheng L., Cai W., Shao X.: A connectivity table for cluster similarity checking in the evolutionary optimization method. Chem. Phys. Lett., **389**, 309–314 (2004)

[11] Coulson C.A., Danielsson U.: Ionic and covalent contributions to the hydrogen bond. Ark. Fys., **8**, 239–255 (1954)

[12] Calvo F., Doye J.P.K., Wales D.J.: Characterization of anharmonicities on complex potential energy surfaces: Perturbation theory and simulation. J. Chem. Phys., **115**, 9627–9636 (2001)

[13] Calvo F., Doye J.P.K., Wales D.J.: Quantum partition functions from classical distributions. Application to rare gas clusters. J. Chem. Phys., **114**, 7312–7329 (2001)

[14] Czerminski R., Elber R.: Reaction-path study of conformational transitions in flexible systems—applications to peptides. J. Chem. Phys., **92**, 5580–5601 (1990)

[15] Chandler D.: Statistical mechanics of isomerization dynamics in liquids and the transition state approximation. J. Chem. Phys., **68**, 2959–2970 (1978)

[16] Christie R.A., Jordan K.D.: Theoretical investigation of the $H_3O^+(H_2O)_4$ cluster. J. Phys. Chem. A, **105**, 7551–7558 (2001)

[17] Chekmarev S.F., Krivov S.V.: Confinement of the molecular dynamics trajectory to a specified catchment area on the potential surface. Chem. Phys. Lett., **287**, 719–724 (1998)

[18] Calvo F., Spiegelman F., Wales D.J.: Theoretical study of finite-temperature spectroscopy in $CaAr_n$ clusters. II. time-dependent absorption spectra. J. Chem. Phys., **118**, 8754–8762 (2003)

[19] Dellago C., Bolhuis P.G., Chandler D.: On the calculation of reaction rate constants in the transition path ensemble. J. Chem. Phys., **110**, 6617–6625 (1999)

[20] Dellago C., Bolhuis P., Geissler P.L.: Transition path sampling. Adv. Chem. Phys., **123**, 1–78 (2002)

[21] Deaven D.M., Ho K.M.: Molecular geometry optimization with a genetic algorithm. Phys. Rev. Lett., **75**, 288–291 (1995)

[22] Dang L.X., Pettitt B.M.: Simple intramolecular model potentials for water. J. Phys. Chem., **91**, 3349–3354 (1987)

[23] Doye J.P.K., Wales D.J.: Calculation of thermodynamic properties of small Lennard-Jones clusters incorporating anharmonicity. J. Chem. Phys., **102**, 9659–9672 (1995)

[24] Doye J.P.K., Wales D.J.: Thermodynamics of global optimization. Phys. Rev. Lett., **80**, 1357–1360 (1998)

[25] Doye J.P.K., Wales D.J.: Saddle points and dynamics of Lennard-Jones clusters, solids, and supercooled liquids. J. Chem. Phys., **116**, 3777–3788 (2002)

[26] Elber R., Ghosh A., Cardenas A.: Long time dynamics of complex systems. Accounts Chem. Res., **35**, 396–403 (2002)

[27] Eastman P., Grønbech-Jensen N., Doniach S.: Simulation of protein folding by reaction path annealing. J. Chem. Phys., **114**, 3823–3841 (2001)

[28] Evans M.G., Polanyi M.: Some applications of the transition state method to the calculation of reaction velocities, especially in solution. Trans. Faraday Soc., **31**, 875–894 (1935)

[29] Evans M.G., Polanyi M.: On the introduction of thermodynamical variables into reaction kinetics. Trans. Faraday Soc., **33**, 448–452 (1937)

[30] Elber R., Meller, J., Olender, R.: Stochastic path approach to compute atomically detailed trajectories: Application to the folding of C peptide. J. Phys. Chem. B, **103**, 899–911 (1999)

[31] Elber R., Shalloway D.: Temperature dependent reaction coordinates. J. Chem. Phys., **112**, 5539–5545 (2000)

[32] Evans D.A., Wales D.J.: The free energy landscape and dynamics of met-enkephalin. J. Chem. Phys., **119**, 9947–9955 (2003)

[33] Evans D.A., Wales D.J.: Folding of the GB1 hairpin peptide from discrete path sampling. J. Chem. Phys., **121**, 1080–1090 (2004)

[34] Eyring H.: The activated complex and the absolute rate of chemical reactions. Chem. Rev., **17**, 65–77 (1935)

[35] Franke G., Hilf E.R., Borrmann P.: The structure of small clusters: Multiple normal-modes model. J. Chem. Phys., **98**, 3496–3502 (1993)

[36] Fichthorn K.A., Weinberg W.H.: Theoretical foundations of dynamical Monte Carlo simulations. J. Chem. Phys, **95**, 1090–1096 (1991)

[37] Gillespie D.T.: A general method for numerically simulating the stochastic time evolution of coupled chemical reactions. J. Comput. Phys., **22**, 403–434 (1976)

[38] Gillespie D.T.: Exact stochastic simulation of coupled chemical reactions. J. Phys. Chem., **81**, 2340–2361 (1977)

[39] Gilmer G.H.: Computer models of crystal growth. Science, **208**, 355–363 (1980)

[40] Hoare M.R.: Structure and dynamics of simple microclusters. Adv. Chem. Phys., **40**, 49–135 (1979)

[41] Huo S., Straub J.E.: The MaxFlux algorithm for calculating variationally optimized reaction paths for conformational transitions in many body systems at finite temperature. J. Chem. Phys., **107**, 5000–5006 (1997)

[42] Huo S., Straub J.E.: Direct computation of long time dynamical processes in peptides and proteins: reaction path study of the coil to helix transition in polyalanine. Proteins: Structure, Function, Genetics, **36**, 249–261 (1999)

[43] Hodges M.P., Wales D.J.: Global minima of protonated water clusters. Chem. Phys. Lett., **324**, 279–288 (2000)

[44] Jorgensen W.L., Chandrasekhar J., Madura J.D.: Comparison of simple potential functions for simulating liquid water. J. Chem. Phys., **79**, 926–935 (1983)

[45] Jensen F.: Introduction to Computational Chemistry, chapter 7, p. 195. Wiley (1999)

[46] Krivov S.V., Chekmarev S.F., Karplus M.: Potential energy surfaces and conformational transitions in biomolecules: A successive confinement approach applied to a solvated tetrapeptide. Phys. Rev. Lett., **88**, 038101 (4 pages) (2002)

[47] Kunz R.E.: Dynamics of first-order phase transitions. Deutsch, Thun (1995)

[48] Lazaridis T., Karplus M.: Effective energy function for proteins in solution. Proteins: Structure, Function, Genetics, **35**, 133–152 (1999)

[49] Li Z., Scheraga H.A.: Monte Carlo-minimisation approach to the multiple-minima problem in protein folding. Proc. Natl. Acad. Sci, USA, **84**, 6611–6615 (1987)

[50] McGinty D.J.: Vapor phase homogenous nucleation and the thermodynamic properties of small clusters of argon atoms. J. Chem. Phys., **55**, 580–588 (1971)

[51] Mezey P.G.: Catchment region partitioning of energy hypersurfaces, I. Theo. Chim. Acta, **58**, 309–330 (1981)

[52] Murrell J.N., Laidler K.J.: Symmetries of activated complexes. J. Chem. Soc. Faraday Trans., **64**, 371–377 (1968)

[53] Marx D., Tuckerman M.E., Hutter J., Parrinello M.: The nature of the hydrated excess proton in water. Nature, **397**, 601–604 (1999)

[54] Mulliken R.S.: The interaction of electron donors and acceptors. J. Chim. Phys., **61**, 20–38 (1964)

[55] Montalenti F., Voter A.F.: Exploiting past visits or minimum-barrier knowledge to gain further boost in the temperature-accelerated dynamics method. J. Chem. Phys., **116**, 4819–4828 (2002)

[56] Munro L.J., Wales D.J.: Defect migration in crystalline silicon. Phys. Rev. B, **59**, 3969–3980 (1999)

[57] Niesse J.A., Mayne H.R.: Global optimization of atomic and molecular clusters using the space-fixed modified genetic algorithm method. J. Comp. Chem., **18**, 1233–1244 (1997)

[58] Olender R., Elber R.: Calculation of classical trajectories with a very large time step: Formalism and numerical examples. J. Chem. Phys., **105**, 9299–9315 (1996)

[59] Passerone D., Parrinello M.: Action-derived molecular dynamics in the study of rare events. Phys. Rev. Lett., **87**, 108302 (4 pages) (2001)

[60] Pelzer H., Wigner E.: Über die Geschwindigkeitskonstante von Austauschreaktionen. Z. Phys. Chem., **B15**, 445–463 (1932)

[61] Rahman J.A., Tully J.C.: Puddle-skimming: An efficient sampling of multidimensional configuration space. J. Chem. Phys., **116**, 8750–8760 (2002)

[62] Rahman J.A., Tully J.C.: Puddle-jumping: a flexible sampling algorithm for rare event systems. Chem. Phys., **285**, 277–287 (2002)

[63] Sevick E.M., Bell A.T., Theodorou D.N.: A chain of states method for investigating infrequent event processes occurring in multistate, multidimensional systems. J. Chem. Phys., **98**, 3196–3212 (1993)

[64] Straub J.E., Guevara J., Huo S., Lee J.P.: Long time dynamic simulations: Exploring the folding pathways of an Alzheimer's amyloid abeta-peptide. Accounts Chem. Res., **35**, 473–481 (2002)

[65] Stillinger F.H., Stillinger D.K.: Cluster optimization simplified by interaction modification. J. Chem. Phys., **93**, 6106–6107 (1990)

[66] Schmitt U.W., Voth G.A.: The computer simulation of proton transport in water. J. Chem. Phys., **111**, 9361–9381 (1999)

[67] Sørensen M.R., Voter A.F.: Temperature-accelerated dynamics for simulation of infrequent events. J. Chem. Phys., **112**, 9599–9606 (2000)

[68] Stillinger F.H., Weber T.A.: Hidden structure in liquids. Phys. Rev. A, **25**, 978–989 (1982)

[69] Stillinger F.H., Weber T.A.: Packing structures and transitions in liquids and solids. Science, **225**, 983–989 (1984)

[70] van Kampen N.G.: Stochastic processes in physics and chemistry. North-Holland, Amsterdam (1981)

[71] Voter A.F.: Hyperdynamics: accelerated molecular dynamics of infrequent events. Phys. Rev. Lett., **78**, 3908–3911 (1997)

[72] Voter A.F.: A method for accelerating the molecular dynamics simulation of infrequent events. J. Chem. Phys., **106**, 4665–4677 (1997)

[73] Wales D.J.: Coexistence in small inert-gas clusters. Mol. Phys., **78**, 151–171 (1993)

[74] Wales D.J.: Discrete path sampling. Mol. Phys., **100**, 3285–3306 (2002)

[75] Wales D.J.: Energy landscapes. Cambridge University Press, Cambridge (2003)

[76] Wales D.J.: Some further applications of discrete path sampling to cluster isomerization. Mol. Phys., **102**, 891–908 (2004)

[77] Wales D.J., Doye J.P.K.: Global optimization by basin-hopping and the lowest energy structures of Lennard-Jones clusters containing up to 110 atoms. J. Phys. Chem. A, **101**, 5111–5116 (1997)

[78] Wales D.J., Doye J.P.K.: Stationary points and dynamics in high-dimensional systems. J. Chem. Phys., **119**, 12409–12416 (2003)

[79] Wales D.J., Doye J.P.K., Miller M.A., Mortenson P.N., Walsh T.R.: Energy landscapes: from clusters to biomolecules. Adv. Chem. Phys., **115**, 1–111 (2000)

[80] Wales D.J., Hodges M.P.: Global minima of water clusters $(H_2O)_n$, $n \leq 21$, described by an empirical potential. Chem. Phys. Lett., **286**, 65–72 (1998)

[81] Wales D.J., Miller M.A., Walsh T.R.: Archetypal energy landscapes. Nature, **394**, 758–760 (1998)

[82] Wales, D.J., Scheraga, H.A.: Global optimization of clusters, crystals and biomolecules. Science, **285**, 1368–1372 (1999)

[83] Warshel A., Weiss R.M.: An empirical valence bond approach for comparing reactions in solutions and in enzymes. J. Am. Chem. Soc., **102**, 6218–6226 (1980)

[84] Wicke E., Eigen M., Ackermann T.: Über den Zustand des Protons (Hydroniumions) in wäßriger Lösung. Z. Physik. Chem. (N. F.), **1**, 340–364 (1954)

[85] Wynne-Jones W.F.K., Eyring H.: The absolute rate of reactions in condensed phases. J. Chem. Phys., **3**, 492–502 (1935)

[86] Zundel G., Metzger H.: Energiebänder der tunnelnden Überschuß-Protonen in flüssigen Säuren. Eine IR-spektroskopische Untersuchung der Natur der Gruppierungen $H_5O_2^+$. Z. Physik. Chem. (N. F.), **58**, 225–245 (1968)

[87] Zhan L., Piwowar B., Liu W.K., Hsu P.J., Lai S.K., Chen J.Z.Y.: Multicanonical basin hopping: A new global optimization method for complex systems. J. Chem. Phys., **120**, 5536–5542 (2004)

The Protein Folding Problem

H. A. Scheraga, A. Liwo, S. Oldziej, C. Czaplewski, J. Pillardy, J. Lee, D.R. Ripoll, J.A. Vila, R. Kazmierkiewicz, J.A. Saunders, Y.A. Arnautova, K.D. Gibson, A. Jagielska, M. Khalili, M. Chinchio, M. Nanias, Y.K. Kang, H. Schafroth, A. Ghosh, R. Elber and M. Makowski

Baker Laboratory of Chemistry and Chemical Biology, Cornell University, Ithaca, New York 14583-1301, USA

Abstract. The two types of protein folding problems (theoretical predictions of protein structure and folding pathways) are discussed. Most of the effort in our laboratory has been devoted to computing three-dimensional structures of proteins from amino acid sequence, using either an all-atom or a united-residue force field. More recently, folding pathways have been computed with a stochastic difference equation method, and folding trajectories have been simulated by molecular dynamics with the united-residue force field.

Key words: Protein folding, computer simulations, force fields

1 Introduction

Anfinsen [1] demonstrated that the amino acid sequence of a polypeptide chain contains all the information necessary to fold it into the three-dimensional structure of a native protein. Based on this observation, he proposed that the native structure is the thermodynamically most stable one in a given environment. There are consequently two protein folding problems of interest to theoretical chemists. The first is to compute the native structure from a knowledge of the amino acid sequence, and the second is to compute the structural pathways and rates from the unfolded to the folded form. This article is concerned primarily with the first problem, but also with initial efforts to consider the second problem. It is a summary of our efforts to deal with these two problems. Our approach has been an *ab initio* physics-based one in order to identify the inter-residue interactions that control the folding process and the final, native structure.

2 Early Approaches to Structure Prediction

Many methods have been used for global optimization of an empirical all-atom force field to predict the three-dimensional structure of a native protein. A variety of such force fields are available in the literature. The all-atom force field that was developed in our laboratory (ECEPP, Empirical Conformational Energy Program for Peptides) keeps the bond lengths and bond angles fixed, and varies the dihedral angles for rotation about the bonds of the backbone and side chains [2]. This empirical force field has Lennard-Jones, electrostatic, hydrogen-bond, and intrinsic torsional components.

The global optimization procedures include a chain build-up [3], Monte Carlo-with-minimization (MCM) [4, 5], a Self-Consistent Electric Field (SCEF) method [6], a combination of MCM and SCEF denoted as EDMC (Electrostatically-Driven Monte Carlo) [7], a diffusion equation method (DEM)[8] and, with a simplified force field, a Self-Consistent Basin-to-Deformed-Basin Mapping (SCBDBM) method [9], and a Conformational Family Monte Carlo (CFMC) method [10]. These procedures, which are computationally intensive, were applied to the linear pentapeptide methionine enkephalin, the cyclic decapeptide gramicidin S, polyalanine chains containing up to 70 residues, polytripeptide models of collagen, and protein A, and are summarized in detail in reference [11].

3 Global Optimization of Crystal Structures

Motivated by the need to circumvent problems that appeared in the DEM, the SCB-DBM and CFMC methods were developed, and also applied to the prediction of crystal structures. This is another of many applications of global optimization in physics, and this section is a diversion to illustrate how our methodology can treat another problem of physical interest. As with our approach to protein-structure calculations, we use an *ab initio* approach, without using knowledge of the space group, i.e., to predict both the arrangement of the molecules in the crystal and also the lattice constants. The first application was made to a nonpolar crystal of S_6 molecules [12] in which only a Lennard-Jones potential is involved in the intermolecular interactions. In the computed structure, the molecular positions, as well as the lattice vectors, agree quite well with the experimental ones.

Application to a crystal of a nonpolar molecule, benzene, but one with a partial-charge representation without a permanent dipole moment, required use of the Ewald summation to treat the electrostatic contribution [12], and led to a structure in which there is the familiar edge-to-face arrangement of the molecules. Further calculations were carried out for crystals of polar molecules with permanent dipole moments [13], with results that showed good agreement between the computed and experimental molecular positions and lattice vectors.

Participation in two recent blind tests on crystal structure prediction [14, 15], organized by the Cambridge Crystallographic Data Centre, provided a good test for our global search method and showed that it is efficient enough to predict crystals of rigid and flexible molecules if accurate potentials are used. When applied to crystals, the global search methods provide information about their potential energy surface, and therefore, can be used as a tool for evaluating potentials [13]. The information they provide enabled us to develop a new global-optimization-based method for parameter optimization [16, 17].

While all the results of crystal calculations showed small deviations between experimental and calculated structures, global optimization of the potential energy can reduce these deviations by improving the parameters of the potential function to force it to lead to better agreement between computed and experimental structures. This global optimization-based approach is now being used to obtain an improved all-atom potential function [17–19].

4 All-atom Treatment of Protein A and the Villin Headpiece

It is now possible to extend the all-atom global optimization procedures to larger systems than those mentioned in Sect. 2 because of the availability of a Beowulf-type cluster. In particular, the EDMC method has thus far been applied to protein A, consisting of 46 residues [20], and to the villin headpiece, consisting of 36 residues [21].

Calculations on protein A[20] were carried out with ECEPP/3 [2] and the EDMC procedure [7], together with two implicit hydration models, OONS [22] and SR-FOPT [23], starting from four different random conformations. Three of the four runs converged within an RMSD of 3.8-4.6Å to the native-like fold, while the fourth converged to the mirror-image conformation. The EDMC method was subsequently applied to the villin headpiece [21] with results of similar quality.

5 Hierarchical Approach to Predict Structures of Large Protein Molecules

It is not yet clear whether the all-atom approach of Sect. 4 can be applied, within a reasonable amount of computing time, to larger protein molecules containing of the order of 100-200 residues. Therefore, a hierarchical approach was developed to treat this problem [24–26].

In this hierarchical approach, global optimization is carried out by using a Conformational Space annealing (CSA) method [27] with a united-residue (UNRES) representation of the protein chain [24]. This is the key stage of the hierarchical algorithm. It is designed to locate the *region* of the global minimum rapidly and efficiently. The lowest-energy structures obtained from the UNRES representation in this stage are then converted to the all-atom representation [28, 29], and a local search is carried out in the restricted region located with the UNRES/CSA approach. This is accomplished with the EDMC method and the ECEPP/3 force field [2], together with the SRFOPT hydration model [23]. Initially, the backbone of the chain is constrained to the structures obtained by UNRES and CSA, but the constraints are gradually reduced as the calculations proceed.

The UNRES model [24] consists of a virtual-bond chain, i.e., a sequence of α-carbons, united peptide groups, and united side chains (whose size depends on the nature of the amino acid residue). The α-carbons are not centers of interaction, but merely serve to locate the backbone. The centers of interaction are the united peptide groups and united side chains, with a united-residue potential given by (1),

$$
\begin{aligned}
U = &\sum_j \sum_{i<j} U_{SC_i SC_j} + w_{SCp} \sum_j \sum_{i \neq j} U_{SC_i p_j} + w_{el} \sum_j \sum_{i<j-1} U_{p_i p_j} \\
&+ w_{tor} \sum_i U_{tor}(\gamma_i) + w_{tord} \sum_i U_{tord}(\gamma_i, \gamma_{i+1}) \\
&+ w_b \sum_i U_b(\theta_i) + w_{rot} \sum_i U_{rot}(\alpha_{sc_i}, \beta_{sc_i}) + \sum_{m=2}^{N_{corrr}} w_{corr}^{(m)} U_{corr}^{(m)}
\end{aligned}
\tag{1}
$$

where the successive terms represent side chain-side chain, side chain-peptide, peptide-peptide, torsional, double torsional, bond-angle bending, side-chain angles α and β, and multi-body (correlation) interactions, respectively. The w's are the relative weights of each term. The correlation terms arise from a cumulant expansion [24] of the Restricted Free Energy (RFE) function (or potential of mean force) of the simplified chain obtained from the all-atom (e.g., ECEPP) energy surface by integrating out the secondary degrees of freedom. The variables to change conformation are the angles (θ_i) between virtual bonds, the torsional angle (γ_i) for rotation about the virtual bonds, and the position angle (α_i) and rotational angle (β_i) of the side chains.

The CSA method [27] starts with a widely-separated set of 50-100 local energy minima in the multi-dimensional conformational energy space. The method is based essentially on a build-up and a genetic algorithm to force the local minima to coalesce to the *region* of the global minimum. All of these UNRES minimum-energy conformations in the coalesced clusters are converted to the all atom representation [28, 29], and the global optimization search is continued from these starting conformations with the EDMC procedure, as indicated above.

After carrying out initial tests of this hierarchical procedure on proteins of known structure, we participated in successive blind tests (CASP, Critical Assessment of Protein Structure Prediction), beginning with CASP3 in 1998. Various improvements of the procedure were implemented in successive tests.

6 Performance in CASP Tests

According to Reva et al. [30], a prediction of a protein structure at an RMSD of 6Å should be considered a successful one. Thus, the RMSD from an experimental structure should be regarded as a reasonable scoring function to assess the quality of a structure prediction.

In CASP3, we predicted the structure of residues 25-85 of HDEA (80% of the whole protein) with an RMSD of 4.2Å, and the segment from residues 16-42 (36% of the whole protein) with an RMSD of 2.9Å. The structure of Mar A, a DNA-binding protein consisting of two domains, was also predicted; one of its (61-residue) domains, which is 53% of the whole protein, was obtained with a 6.0Å RMSD. The predictions in CASP3 were successful only for α-type proteins.

To be able to predict β, as well as α, structures, extra terms in the cumulant expansion (represented by the last Σ term in eq. (1)) that reflect β-type structure were added. In the CASP4 test, a 68-residue portion of an α protein was predicted with an RMSD of 5.9Å, and a 57-residue fragment of a double-hairpin β portion of another protein was predicted with an RMSD of 6.5Å.

In CASP3 and CASP4, the double torsional term (U_{tord}) in eq. (1) had not yet been included. Inclusion of this term in CASP5 and CASP6, and separation of the non-native structures into several levels during Z-score optimization [25], improved the predictability. In addition, the UNRES/CSA procedure was extended from single to multiple chains [31]. With imposition of rotational symmetry, the known structure of the retro-GCN4 leucine zipper tetramer was computed with an RMSD of 2.34 Å, and that of a synthetic domain-swapped dimer was predicted with an RMSD of 5.65 Å[31]. For an α-type protein, residues 93-179 were predicted with an RMSD of 6.0 Å, and residues 222-291 of a β-type protein were predicted with an RMSD of 5.3 Å, in CASP5.

In CASP5 and CASP6, we used secondary-structure predictions to initialize the starting CSA procedure; however, runs started from random structures were also carried out. Also, the final models were selected based on energy alone. In CASP5, we also used secondary-structure information to constrain the search, because our force field was not yet mature. For a large α-helical protein in CASP6, a large bank of structures was produced by using a very simplified interaction potential [32], and a novel approach was then used to select the best model according to its UNRES energy (unpublished work).

In preparation for CASP6, the parameters of the cumulant terms were modified by an improved Z-score approach trained on four proteins [$(\alpha+\beta)$, β-type, α-type, and $(\alpha+\beta)$-type] [26].

The following results were obtained for CASP6 targets. Residues 9-212 of a six-α-helical protein were obtained with an RMSD of 8Å (and residues 36-111, viz. two of these helices, were obtained with a 6Å RMSD with the all-atom, instead of the UNRES, force field). Residues 24-76 of a three-helix α-type protein were obtained with an RMSD of 3.5Å. Residues 1-70 of an $(\alpha+\beta)$ protein were obtained with an RMSD of 5.5Å. Residues 127-225 were obtained with an RMSD of 6Å. Residues 116-207 and residues 142-207 of an $(\alpha+\beta)$ protein were obtained with RMSDs of 8Å and 6Å, respectively. Finally, residues 1-104 of an $(\alpha+\beta)$ protein were obtained with an RMSD of 7Å.

7 Computation of Folding Pathways

A second type of protein folding problem is to compute the folding pathways from a given initial unfolded state i (with no native contacts or native hydrogen bonds) to the known final folded structure f, obtained by x-ray or NMR experiments. One approach makes use of the stochastic difference equation method of Elber et al. [33], and has been applied with a full-atom treatment to protein A. This method provides the sequence of formation of intermediate structures between i and f, but not the kinetics, there being no computation of the time scale for folding. This boundary value problem may be contrasted with an initial-value formulation in which one starts with the initial unfolded protein and uses molecular dynamics to compute the folding trajectory. The stochastic difference equation method avoids the problem that molecular dynamics requires femtosecond steps, and would consume an enormous amount of computer time to reach, say, a microsecond level, whereas most proteins (except for some very fast folders) fold in the millisecond-to-second time scale.

The method requires that the action S, as in eqs. (2) and (3),

$$S = \int_{Y_u}^{Y_f} \sqrt{2\,(E - U)}\,dl \, , \tag{2}$$

which is approximated by

$$S \cong \sum_i \sqrt{2\,(E - U)}\Delta l_{i,i+1} \tag{3}$$

remain stationary along the length l of the pathway from i to f, where E is the total energy, U is the potential energy, and Y_u and Y_f are coordinates of the initial unfolded protein and of the final folded protein, respectively. To achieve a pathway with a stationary action, the function T of (4) must be optimized.

$$T = \sum_i \left(\frac{\partial S/\partial Y_i}{\Delta l_{i,j+1}}\right)^2 \Delta l_{i,i+1} + \lambda \sum_i (\Delta l_{i,i+1} - <\Delta l>)^2 \, , \tag{4}$$

with

$$<\Delta l> = (1/N) \sum_i \Delta_{i,i+1} \, , \tag{5}$$

where the parameter λ is the strength of a penalty function that keeps all of the length elements, $\Delta l_{i,i+1}$, equal to the average length given by (5), and equal to each other.

In treating the three-helix protein A, an initial ensemble of 130 representatives of the unfolded protein were generated, and 130 separate trajectories to reach the final folded structure were computed, and an average was evaluated over all those trajectories [34]. None of the 130 initial conformations had any native contacts or native hydrogen bonds.

The progress along the trajectories shows that helix 3 appears to fold first, followed by helix 1 and then by helix 2. A more detailed view of the folding pathways is obtained by dividing the trajectory into five equal segments. It is found that, contrary to the current view that there is an initial hydrophobic collapse, followed by a slow rearrangement to form the native structure, there is a wide distribution of radii of gyration with very few native hydrogen bonds or native contacts formed in the initial stage. In the second 20% of the trajectory, only one or two additional (native) hydrogen bonds form but there is still the wide distribution of radii of gyration. Only in the third 20% is there a concomitant drop in the distribution of radii of gyration and a slight increase in the formation of native hydrogen bonds and native contacts. This behavior continues into the fourth and fifth stages in which the final radius of gyration and full complement of native hydrogen bonds and native contacts appears. Future planned calculations of this type on proteins, for which experimental pathways are known, will provide a direct experimental test of theoretically-calculated folding pathways.

8 Molecular Dynamics with the UNRES Model

As pointed out in Sect. 7, all-atom molecular dynamics (MD) is too slow to reach the millisecond or second time scales in which most proteins fold. In order to extend the time scale, the UNRES model has been used instead of the all-atom model [35–37]. For this purpose, eq. (1) has been modified by addition of the term

$$w_{vib} \sum_i U_{vib}(d_i), \tag{6}$$

where this term takes account of variation in the length of the ith virtual bond, with d_i being the length of this bond, with the energy expressed as

$$U_{vib}(d_i) = \frac{1}{2}k_{d_i}(d - d_i^o)^2, \tag{7}$$

with k_{d_i} being the force constant of the ith virtual bond, and d_i^o being its average length corresponding to that used in the fixed-bond UNRES representation.

Langevin dynamics has been applied to the UNRES model as

$$\mathbf{G}\ddot{\mathbf{q}} = -\nabla_\mathbf{q} U(\mathbf{q}) + \mathbf{f}^{\text{fric}} + \mathbf{f}^{\text{rand}} \tag{8}$$

with the matrix $\mathbf{G}$ defined as

$$\mathbf{G} = \mathbf{A}^\mathsf{T}\mathbf{M}\mathbf{A} + \mathbf{H} \tag{9}$$

where U is the UNRES potential energy, $\mathbf{f}^{\text{fric}}$ and $\mathbf{f}^{\text{rand}}$ are the friction and random forces, respectively; $\mathbf{A}$ is the matrix of a linear transformation from the space of generalized coordinates and velocities ($\mathbf{q}$ and $\dot{\mathbf{q}}$) to the space of Cartesian coordinates and velocities of the interacting sites, $\mathbf{M}$ is the diagonal matrix of the masses of the interacting sites, $\mathbf{H}$ (a diagonal matrix) is the part of the inertia matrix that corresponds to the internal (stretching) motions of the virtual bonds, and ∇_q is a vector of the first derivatives with respect to $\mathbf{q}$'s. Stokes law is used to compute the friction forces, $\mathbf{f}^{\text{fric}}$ and the random forces $\mathbf{f}^{\text{rand}}$ are sampled from a normal distribution with zero mean and variance related to temperature and friction coefficients [37, 38]. Together, the stochastic and friction forces constitute a thermostat that maintains the average temperature at the pre-set value.

We use a simplified version [37] of the stochastic velocity Verlet algorithm [38] to solve the Langevin equations of motion. Taking the Ala_{10} polypeptide in water as an example, we found that UNRES MD offers a 4000- and 60-fold speed-up relative to all-atom MD simulations with explicit and implicit water, respectively [36, 37]. This speed-up is a result of (i) substantial reduction of computational cost and (ii) averaging out the secondary (fast-moving) degrees of freedom when passing from the all-atom representation to UNRES [36]. Compared to all-atom molecular dynamics, the UNRES event-based time scale is 4-7 times wider [36, 37]. We also found [36, 37] that, with UNRES, Ala_{10} folds in 0.4 ns on average, while the experimental times of α-helix formation are of the order of 0.5 μs[39], and that the average folding time of protein A (a 46-residue three-helix bundle) with UNRES is 4.2 ns, while even the fastest-folding mutants of this protein fold in microseconds [39]. This means that the event-based time scale for UNRES is larger by three orders of magnitude than the experimental time scale. This is caused by averaging out the secondary degrees of freedom, and strongly suggests that UNRES MD can be used in *ab initio* studies of protein folding in real time.

To test the capability of the UNRES MD approach to fold proteins, we carried out test MD simulations[35] on a number of proteins with lengths from 28 to 75 amino-acid residues for which the native-like structures were global minima as found by the CSA method. We used our latest force field determined by hierarchical optimization[26]. Most of the test proteins folded to native-like structures, although the force field was optimized using the CSA-generated and not MD-generated decoy sets. The average folding time was only 2.3 ns even for 1CLB, which was the largest protein considered (75 residues); for this protein the folding required only about 5 wall-clock hours with a single AMD Athlon(tm) MP 2800+ processor on average, this wall-clock time being similar to that required for global optimization of this protein with the CSA method, which requires using about 100 processors. This means that UNRES MD (i) is a practical approach to study folding pathways and (ii) is a very efficient method for searching the conformational space of united-residue chains. However, UNRES MD with the present force field generally failed to fold most of the β- and $(\alpha+\beta)$- proteins; α-helical structures were obtained instead of native folds [35], even though the native folds of these proteins are global minima of the potential-energy surface as found by CSA [26]. The reason for this is neglect of the entropy factor in the present parameterization of UNRES. The decoys for optimization are local energy minima generated with CSA and they form only a very tiny fraction of conformational space, while MD at the folding temperature is not restricted to exploring local minima; we noted that the lowest potential energy achieved in MD simulations for a 46-residue protein is about 100 kcal/mol higher than that of the global minimum because of thermal motion [35]. Therefore, a great part of our effort in the near future will be directed at optimizing the force field for MD simulations.

9 Conclusions

The results presented here demonstrate that the available physics is capable of predicting protein structure and protein-folding pathways. Current efforts to refine both the all-atom and UNRES potential functions and also the search procedures (especially UNRES/MD) should, hopefully, lead to computed structures (for proteins containing 100-200 residues) with RMSDs closer to the native ones. Further details of this work are available in reference [11].

Acknowledgements

Support was received from the National Institutes of Health (GM-14312) and the National Science Foundation (MCB00-03722). The research was conducted by using the resources of the Cornell Theory Center, and the National Science Foundation Terascale Computing System at the Pittsburgh Supercomputer Center.

References

[1] C.B. Anfinsen. Principles that govern the folding of protein chains, *Science*, 181:223–230, 1973.

[2] G. Némethy, K.D. Gibson, K.A. Palmer, C.N. Yoon, G. Paterlini, A. Zagari, S. Rumsey, and H.A. Scheraga. Energy parameters in polypeptides. 10. improved geometrical parameters and nonbonded interactions for use in the ECEPP/3 algorithm, with application to proline containing peptides. *J. Phys. Chem.*, 96:6472–6484, 1992.

[3] K. D. Gibson and H. A. Scheraga. Revised algorithms for the build up procedure for predicting protein conformations by energy minimization. *J. Comput. Chem.*, 8:826–834, 1987.

[4] Z. Li and H. A. Scheraga. Monte Carlo minimization approach to the multiple minima problem in protein folding. *Proc. Natl. Acad. Sci., U.S.A*, 84:6611–6615, 1987.

[5] Z. Li and H. A. Scheraga. Structure and free energy of complex thermodynamic systems. *J. Molec. Str. (Theochem).* 179:333–352, 1988.

[6] L. Piela and H. A. Scheraga. On the multiple minima problem in the conformational analysis of polypeptides. I. Backbone degrees of freedom for a perturbed α helix, *Biopolymers*, 26:S33–S58, 1987.

[7] D.R. Ripoll, A. Liwo, and H. A. Scheraga. New developments of the electrostatically driven Monte Carlo method: Test on the membrane-bound portion of melittin. *Biopolymers*, 46:117–126, 1998.

[8] L. Piela, J. Kostrowicki, and H. A. Scheraga. The multiple minima problem in the conformational analysis of molecules. Deformation of the potential energy hypersurface by the diffusion equation method. *J. Phys. Chem.*, 93:3339–3346, 1989.

[9] J. Pillardy, A. Liwo, M. Groth, and H.A. Scheraga. An efficient deformation-based global optimization method for off-lattice polymer chains; self-consistent basin-to-deformed-basin mapping (SCBDBM). Application to united-residue polypeptide chains. *J. Phys. Chem., B*, 103:7353–7366, 1999.

[10] J. Pillardy, C. Czaplewski, W.J. Wedemeyer, and H.A. Scheraga. Conformation-family Monte Carlo (CFMC): An efficient computational method for identifying the low-energy states of a macromolecule. *Helv. Chim. Acta*, 83:2214–2230, 2000.

[11] H.A. Scheraga, A. Liwo, S. Oldziej, C Czaplewski, J. Pillardy, D.R. Ripoll, J.A. Vila, R. Kazmierkiewicz, J.A. Saunders, Y.A. Arnautova, A. Jagielska,

M. Chinchio, and M. Nanias. The protein folding problem: Global optimization of force fields. *Frontiers in Bioscience*, 9:3296–3323, 2004.

[12] R. J. Wawak, J. Pillardy, A. Liwo, K.D. Gibson, and H. A. Scheraga. The protein folding problem: Global optimization of force fields. *J. Phys. Chem.*, 102:2904–2918, 1998.

[13] J. Pillardy, R.J. Wawak, Y.A. Arnautova, C. Czaplewski, and H.A. Scheraga. Crystal structure prediction by global optimization as a tool for evaluating potentials: Role of the dipole moment correction term in successful predictions. *J. Am. Chem. Soc.*, 122:907–921, 2000.

[14] W.D.S. Motherwell, H.L. Ammon, J.D. Dunitz, A. Dzyabchenko, P. Erk, A. Gavezzotti, D.W.M. Hofmann, F.J.J. Leusen, J.P.M. Lommerse, W.T.M. Mooij, S.L. Price, H. Scheraga, B.Schweizer, M.U. Schmidt, B.P. van Eijck, P. Verwer, and D.E. Williams. Crystal structure prediction of small organic molecules: a second blind test. *Acta Cryst. B*, 58:647–661, 2002.

[15] G.M. Day, W.D.S. Motherwell, H. Ammon, S.X.M. Boerrigter, R.G. Della Valle, E. Venuti, A. Dzyabchenko, J. Dunitz, B. Schweizer, B.P. van Eijck, P. Erk, J.C. Facelli, V.E. Bazterra, M.B. Ferraro, D.W.M. Hofmann, F.J.J. Leusen, C. Liang, C.C. Pantelides, P.G. Karamertzanis, S.L. Price, T.C. Lewis, H. Nowell, A. Torrisi, H.A. Scheraga, Y.A. Arnautova, M.U. Schmidt, and P. Verwer. A third blind test of crystal structure prediction. *Acta Cryst. B*, 61:511–527, 2005.

[16] J. Pillardy, Y.A. Arnautova, C. Czaplewski, K.D. Gibson, and H.A. Scheraga. Conformation-family Monte Carlo: A new method for crystal structure prediction. *Proc. Natl. Acad. Sci., U.S.A.*, 98:12351–12356, 2001.

[17] Y.A. Arnautova, J. Pillardy, C. Czaplewski, and H.A. Scheraga. Global optimization-based method for deriving intermolecular potential parameters for crystals. *J. Phys. Chem. B*, 107:712–723, 2003.

[18] Y.A. Arnautova, A. Jagielska, J. Pillardy, and H.A. Scheraga. Derivation of a new force field for crystal-structure prediction using global optimization: nonbonded potential parameters for hydrocarbons and alcohols. *J. Phys. Chem. B*, 107:7143–7154, 2003.

[19] A. Jagielska, Y.A. Arnautova, and H. A. Scheraga. Derivation of a new force field for crystal - structure prediction using global optimization: nonbonded potential parameters for amines, imidazoles, amides and carboxylic acids. *J. Phys. Chem. B*, 108:12181–12196, 2004.

[20] J.A. Vila, D.R. Ripoll, and H.A. Scheraga. Atomically detailed folding simulation of the B domain of staphylococcal protein A from random structures. *Proc. Natl. Acad. Sci., U.S.A.*, 100:14812–14816, 2003.

[21] D.R. Ripoll, J.A. Vila, and H.A. Scheraga. Folding of the villin headpiece subdomain from random structures. Analysis of the charge distribution as a function of pH. *J. Mol. Biol.*, 339:915–925, 2004.

[22] T. Ooi, M. Oobatake, G. Némethy, and H. A. Scheraga. Acessible surface areas as a measure of the thermodynamic parameters of hydration of peptides. *Proc. Natl. Acad. Sci., U.S.A.*, 84:3086–3090, 1987 (Erratum: ibid., 84, 6015 (1987))

[23] J. Vila, R. L. Williams, M. Vasquez, and H. A. Scheraga. Empirical solvation models can be used to differentiate native from near native conformations of bovine pancreatic trypsin inhibitor. *Proteins: Structure, Function, and Genetics*, 10:199–218, 1991.

[24] A. Liwo, C. Czaplewski, J. Pillardy, and H. A. Scheraga. Cumulant-based expressions for the multibody terms for the correlation between local and electrostatic interactions in the united-residue force field. *J. Chem. Phys.*, 115:2323–2347, 2001.

[25] A. Liwo, P. Arlukowicz, C. Czaplewski, S. Oldziej, J. Pillardy, and H.A. Scheraga. A method for optimizing potential-energy functions by a hierarchical design of the potential-energy landscape: Application to the UNRES force field. *Proc. Natl. Acad. Sci., U.S.A.*, 99:1937–1942, 2002.

[26] S. Oldziej, J. Lagiewka, A. Liwo, C Czaplewski, M. Chinchio, M. Nanias, and H.A. Scheraga. Optimization of the UNRES force field by hierarchical design of the potential-energy landscape. 3. Use of many proteins in optimization. *J. Phys. Chem. B*, 108:16950–16959, 2004.

[27] J. Lee, H. A. Scheraga, and S. Rackovsky. New optimization method for conformational energy calculations on polypeptides: Conformational space annealing. *J. Comput. Chem.*, 18:1222–1232, 1997.

[28] R. Kazmierkiewicz, A. Liwo, and H.A. Scheraga. Energy-based reconstruction of a protein backbone from its α-carbon trace by a Monte-Carlo method. *J. Comput. Chem.*, 23:715–723, 2002.

[29] R. Kazmierkiewicz, A. Liwo, and H.A. Scheraga. Addition of side chains to a known backbone with defined side-chain centroids. *Biophys. Chem.*, 100:261–280, 2003. (Erratum: *Biophys. Chem.*, 106, 91 (2003).)

[30] B.A. Reva, A.V. Finklestein, and J. Skolnick. What is the probability of a chance prediction of a protein structure with an RMSD of 6Å ? *Folding & Design*, 3:141–147, 1998.

[31] J.A. Saunders and H.A. Scheraga. *Ab initio* structure prediction of two α-helical oligomers with a multiple-chain united residue force field and global search. *Biopolymers, 68, 300–317 (2003).*, 68:300–317, 2003.

[32] M. Nanias, M. Chinchio, J. Pillardy, D.R. Ripoll, and H.A. Scheraga. Packing helices in proteins by global optimization of a potential energy function. *Proc. Natl. Acad. Sci., U.S.A.*, 100:1706–1710, 2003.

[33] R. Elber, A. Ghosh, and A. Cárdena. Long time dynamics of complex systems. *Accts. of Chem. Res.*, 35:396–403, 2002.

[34] A. Ghosh, R. Elber, and H.A. Scheraga. An atomically detailed study of the folding pathways of protein A with the stochastic difference equation. *Proc. Natl. Acad. Sci., U.S.A.*, 99:10394–10398, 2002.

[35] A. Liwo, M. Khalili, and H.A. Scheraga. *Ab initio* simulations of protein-folding pathways by molecular dynamics with the united-residue model of polypeptide chains. *Proc. Natl. Acad. Sci, U.S.A.*, 102:2362–2367, 2005.

[36] M. Khalili, A. Liwo, and H.A. Scheraga. Molecular dynamics with the united-residue (UNRES) model of polypeptide chains. I. Lagrange equations

of motion and tests of numerical stability in the microcanonical mode. *J. Phys. Chem. B,* 109:13785–13797, 2005.

[37] M. Khalili, A. Liwo, A. Jagielska, and H.A. Scheraga. Molecular dynamics with the united-residue (UNRES) model of polypeptide chains. II. Langevin and Berendsen-bath dynamics and tests on model alpha-helical systems. *J. Phys. Chem. B,* 109:13798–13810, 2005.

[38] F. Guarnieri and W.C. Still. A rapidly convergent simulation method; mixed Monte Carlo-Stochastic Dynamics. *J. Comp. Chem.*, 15:1302–1310, 1994.

[39] J. Kubelka, J. Hofrichter, and W.A. Eaton. The protein folding 'speed limit'. *Curr. Opinion Struct. Biol.*, 14:76–88, 2004.

Part III

**Dynamical and Stochastic-Dynamical Foundations
for Macromolecular Modelling**

Biomolecular Sampling: Algorithms, Test Molecules, and Metrics

Scott S. Hampton, Paul Brenner, Aaron Wenger, Santanu Chatterjee, and Jesús A. Izaguirre

Department of Computer Science and Engineering, University of Notre Dame, Notre Dame, Indiana 46556, USA

Abstract. We compare the effectiveness of different simulation sampling techniques by illustrating their application to united atom butane, alanine dipeptide, and a small solvated protein, BPTI. We introduce an optimization of the Shadow Hybrid Monte Carlo algorithm, a rigorous method that removes the bias of molecular dynamics. We also evaluate the ability of constant-temperature MD methods (based on Langevin and Nosé-Hoover dynamics) to achieve uniform thermal equilibrium. Our results show the superiority of Langevin dynamics over Nosé-Hoover dynamics to achieve thermal equilibrium. They also illustrate the inherent limitation of protocols that rely on sampling the microcanonical and canonical ensemble at only one temperature, and the importance of generalized ensemble approaches such as replica exchange that sample using different temperatures. Finally, we show how SHMC is able to remove bias and scale with timestep and system size. Presented, herein, are a set of sampling algorithms, test molecules, and metrics. We make the sampling methods discussed available via the open source and freely distributed PROTOMOL molecular simulation framework.

Key words: Monte Carlo, Molecular Dynamics, Replica Exchange, Shadow Hybrid Monte Carlo, Langevin Impulse, Nosé-Hoover dynamics.

1 Introduction

For a system of N atoms, define configuration space as a dN dimensional vector of positions, where d is the number of degrees of freedom per atom. Similarly, phase space is defined by both positions and momenta and is $2dN$ dimensional. The problem of sampling the configuration space of biomolecules, such as proteins, is an inherently difficult problem. The difficulties are due to the high dimensionality of phase space $(6N)$, the ruggedness of the energy functions, the presence of multiple time scales, high energy barriers, and the introduction of systematic error (bias) by sampling methods.

We compare the effectiveness of different simulation sampling techniques by illustrating their application to two simple molecules, united atom butane and alanine dipeptide, and in a few cases to a small solvated protein, BPTI. The methods we test are based on molecular dynamics (MD) or Monte Carlo (MC) sampling techniques. We introduce an optimization of our Shadow Hybrid Monte Carlo algorithm (SHMC), a rigorous method that removes the bias of MD. We also evaluate the ability of constant-temperature MD methods (based on Langevin and Nosé-Hoover dynamics) to achieve uniform thermal equilibrium. Our results show the inherent limitation of protocols that rely on sampling the microcanonical and canonical ensemble at only one temperature, and the importance of generalized ensemble approaches such as the replica exchange method (REM). REM is based on making several simulations at different temperatures, and combining the results to get better sampling at the target temperature. In REM, a random walk is performed in temperature space, which in turn induces a random walk in configuration space.

At the same time, we present a set of sampling algorithms, test molecules, and metrics. We make the sampling methods tested available via the open source and freely distributed PROTOMOL[26, 27] molecular simulation framework.

The problem of sampling can be thought of as estimating expectation values for a function $A(\Gamma)$ with respect to a probability density function (p.d.f.) $\rho(\Gamma)$, where $\Gamma = \begin{pmatrix} X \\ P \end{pmatrix}$, and X and P are the vectors of collective positions and momenta. The expected value of $A(\Gamma)$ is

$$\langle A(\Gamma) \rangle_\rho = \int A(\Gamma)\rho(\Gamma)\mathrm{d}\Gamma . \tag{1}$$

Examples of observables A are potential energy, pressure, free energy, and distribution of solvent molecules in vacancies [11, 21, 34]. For the sampling of configuration space of biological molecules, ρ typically corresponds to a constant system size N, constant temperature T, and constant volume V ensemble (canonical ensemble),

$$\rho_{\mathrm{NVT}}(\Gamma) \propto \frac{\exp(-\mathcal{H}(\Gamma)/(k_\mathrm{B}T))}{\int \exp(-\mathcal{H}(\Gamma)/(k_\mathrm{B}T))\mathrm{d}\Gamma} , \tag{2}$$

such that $\mathcal{H}$ is the Hamiltonian or total energy of the system, k_B is Boltzmann's constant, and T is temperature. Integration is performed over all of phase space. Another common ensemble is the microcanonical, with constant N, V, and E; where V is the volume and E is the target energy of the system. The p.d.f. then becomes [1]:

$$\rho_{\mathrm{NVE}}(\Gamma) \propto \frac{\delta(\mathcal{H}(\Gamma) - E)}{\int \delta(\mathcal{H}(\Gamma) - E)\mathrm{d}\Gamma} , \tag{3}$$

where $\delta()$ chooses only those energy states equal to E.

Biomolecular sampling is a general term which should be qualified based on the sampling quantity of interest. We subgroup the quantities into geometric conformations and thermodynamic properties (free energy, entropy, chemical potential, kinetic and potential energy, temperature, and pressure). We consider both conformational sampling and computation of observables in our tests. In Sect. 1.1, we refer the reader to multiple papers which further describe the fundamental hurdles of sampling.

We first present the sampling algorithms tested. Second, we introduce the test molecules, simulation parameters, and applicable sampling metrics. Third, we present the results from applying the sampling algorithms to the test molecules. We conclude with a discussion of results and overview of promising methods.

1.1 Related Work

There are several reviews of classical biomolecular sampling methods and newer methods [2, 5, 9, 22, 29, 39]. Leach [22] provides a good review of conformational analysis techniques but simulation details are unavailable. Berne and Straub [5] provide a fairly thorough discussion of sampling methods and introduce several sampling metrics. Nardi and Wade [29] reviewed the application of several sampling techniques to simple biomolecules, showing extensive results. These authors do not include a discussion of SHMC or REM.

Our presentation of molecular simulation parameters and test metrics is based on the work by Barth, Leimkuhler, and Reich [2]. We have extended the metrics to consider the problem of sampling and quantification of bias in more depth.

2 Sampling Algorithms

We study simulation-based methods for conformational sampling. The key technique for sampling the configuration space of a biomolecule is constant-temperature MD, which is widely applicable and flexible. We compare Nosé dynamics to Langevin dynamics in their ability to achieve thermal equilibrium (needed to do sampling). MD methods have a timestep dependent bias, which we quantify for sample metrics in Sect. 4. MC methods, in contrast, are rigorous sampling methods. A rigorous method removes the effect of systematic errors such as discretization error, although the effect of rounding errors is neglected. We study two MC methods: the Hybrid Monte Carlo (HMC) method, that uses MD as a global MC move, and SHMC, which improves the scaling of HMC with system size and MD timestep. Finally, we consider replica exchange, which can be used with either MD or MC methods, and which greatly improves their sampling ability.

2.1 Molecular Dynamics

Molecular dynamics solves Newton's equations of motion to sample from the microcanonical ensemble. In the thermodynamic limit of infinite system size, the microcanonical and canonical ensembles are equivalent for ergodic systems. In practice, it is more convenient to sample from the canonical ensemble.

Propagators for MD, such as Leapfrog, can be written as

$$\Gamma^{n+1} = \Psi(\Gamma^n) , \tag{4}$$

where Ψ is the MD integrator, and the superscript denotes time. For example, the velocity-form of the Verlet method can be expressed as:

$$X^{n+1} = X^n + \Delta t M^{-1} P^n - \frac{1}{2} \Delta t^2 M^{-1} F(X^n) \,, \tag{5}$$

$$P^{n+1} = P^n - \frac{1}{2} \Delta t \left(F(X^n) + F(X^{n+1}) \right) \,, \tag{6}$$

where M is the matrix of atomic masses, and $F(X) = -\nabla U(X)$ is the force, assumed to be the negative gradient of a potential energy function.

The temperature of the system is related to the time-average of the kinetic energy of the system. A popular way of controlling temperature can be achieved by coupling the system to an external heat bath, such as in Nosé-Hoover dynamics [14, 30]. Nosé proposed a Hamiltonian of the form:

$$\mathcal{H}_{\mathrm{Nose}} = \mathcal{H}(x, p/s) + \frac{p_s^2}{2Q} + gkT \ln s \,, \tag{7}$$

where Q is the Nosé mass, s is the thermostatting variable, p_s is its conjugate momentum and x and p are positions and momenta respectively.

Langevin dynamics is also a way of generating the canonical ensemble. Here, the equations of motion are modified to include a damping factor γ in the velocity and a random forcing term from a Gaussian distribution with zero mean and variance $\sqrt{2\gamma k_{\mathrm{B}} T} M^{-1/2}$, cf. [7].

2.2 Hybrid Monte Carlo

HMC combines an MD trajectory with an MC rejection step. It takes advantage of the long steps in phase space that can be achieved through MD while eliminating the inaccuracies due to a finite timestep and other numerical artifacts through the MC step. One step of HMC is shown in Algorithm 1, where $\beta = (k_{\mathrm{B}} T)^{-1}$. The output of Algorithm 1 is either a new $\Gamma' = \Psi(\Gamma)$, or simply the original input point in phase space, $\Gamma' = \Gamma$.

Algorithm 1 Hybrid Monte Carlo

Hybrid Monte Carlo:

---

1. Given $X := x$
2. Generate P with Gaussian p.d.f.
3. $\Gamma := \begin{pmatrix} X \\ P \end{pmatrix}$
4. $\Gamma' := \Psi(\Gamma)$ (run NVE MD for a small number of steps using, for example, the Verlet MD integrator Ψ given by Eqs. (5) and (6))
5. Accept Γ' with probability
   $\min \left\{ 1, \frac{\exp(-\beta(\mathcal{H}(\Gamma')))}{\exp(-\beta(\mathcal{H}(\Gamma)))} \right\}$
6. If rejected, $\Gamma' := \Gamma$

2.3 Shadow Hybrid Monte Carlo

SHMC is a new method recently published by Izaguirre and Hampton [17] for sampling the phase space of large molecules, particularly biological molecules. It improves sampling of HMC by allowing larger timesteps and system sizes in the MD step. The acceptance rate of HMC decreases exponentially with increasing system size N or timestep Δt. For a proof see [17, p. 588] and [8, 13, 28]. This poor performance is due to the effect of discretization errors introduced by the numerical integrator on the acceptance function.

SHMC achieves an asymptotic $O(N^{1/4})$ speedup over HMC by sampling from all of phase space using high order approximations to a shadow or modified Hamiltonian exactly integrated by a symplectic MD integrator. SHMC satisfies microscopic reversibility and is a rigorous sampling method. SHMC requires extra storage, modest computational overhead, and a reweighting step to obtain averages from the canonical ensemble. This method has been validated by numerical experiments that compute observables for different molecules, ranging from a small n-alkane butane with 4 united atoms to a larger solvated protein with 14,281 atoms. In these experiments, SHMC achieves an order of magnitude speedup over HMC in sampling efficiency for medium sized proteins.

Description of SHMC:

SHMC samples the target density $\tilde{\rho}(\mathbf{x}, \mathbf{p})$, where

$$\tilde{\rho}(\mathbf{x}, \mathbf{p}) \propto \exp\left(-\beta \tilde{\mathcal{H}}(\mathbf{x}, \mathbf{p})\right), \tag{8}$$

$$\tilde{\mathcal{H}}(\mathbf{x}, \mathbf{p}) = \max\left\{\mathcal{H}(\mathbf{x}, \mathbf{p}), \mathcal{H}_{[2k]}(\mathbf{x}, \mathbf{p}) - c\right\}. \tag{9}$$

Here, $\mathcal{H}_{[2k]}(\mathbf{x}, \mathbf{p})$ is the more accurate shadow Hamiltonian, defined by Skeel and Hardy (2001), and c is an arbitrary constant that limits the amount by which $\mathcal{H}_{[2k]}$ is allowed to depart from $\mathcal{H}(\mathbf{x}, \mathbf{p})$. In Fig. 1, we plot the 4th an 8th order shadow Hamiltonians, as well as the true Hamiltonian for a typical system. As can be seen in the figure, the shadow Hamiltonian has much smaller fluctuations.

Algorithm 2 lists one iteration of SHMC. The first step is to generate a set of momenta, P, often chosen via a Gaussian distribution. P is accepted based on a Metropolis criterion step proportional to the difference of the total and shadow energies. This step is repeated until a set of momenta are accepted. Next, the system is integrated using MD and accepted with probability proportional to (8).

After all iterations of SHMC have been completed, in order to calculate unbiased values, the observables are reweighted using the following formula:

$$\langle A \rangle_{\rho_{\mathrm{NVT}}} = \frac{\sum_{i=1}^{m} w_i A_i}{\sum_{i=1}^{m} w_i}, \tag{10}$$

where

$$w_i = \frac{\exp\left(-\beta \mathcal{H}(\Gamma_i)\right)}{\exp\left(-\beta \tilde{\mathcal{H}}(\Gamma_i)\right)} \tag{11}$$

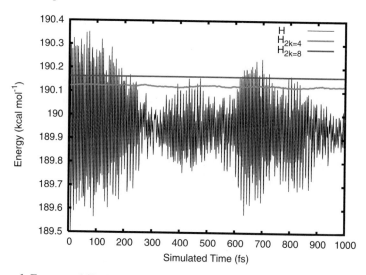

Figure 1. Energy and Shadow energy (4th and 8th order) for a Decalanine molecule

---

Algorithm 2 Shadow Hybrid Monte Carlo .

---

1. **MC Step:** Given $X := x$, generate P with p.d.f. $\tilde{\rho}(X, p)$, using the acceptance-rejection method:
   a) Generate P having Gaussian p.d.f.
   b) Accept with probability $\min \left\{ 1, \frac{\exp(-\beta(\mathcal{H}_{[2k]}(X,P)-c))}{\exp(-\beta\mathcal{H}(X,P))} \right\}$
   c) Repeat (1a) - (1b) until P is accepted.

2. **MD Step:** Given $\Gamma := \begin{pmatrix} X \\ P \end{pmatrix}$, and $R := \begin{pmatrix} I & 0 \\ 0 & -I \end{pmatrix}$,
   a) $\Gamma' := R\Psi(\Gamma)$, (Run NVE MD for a small number of steps, where the MD integrator Ψ nearly conserves $\mathcal{H}_{[2k]}$)
   b) Accept Γ' with probability $\min \left\{ 1, \frac{\tilde{\rho}(\Gamma')}{\tilde{\rho}(\Gamma)} \right\}$
   c) If rejected, $\Gamma' = \Gamma$.

There are two criteria in the optimization of c. The first is to minimize the difference in the shadow and total energies so that the reweighted observables of $\mathcal{H}_{[2k]}$ have a reasonable variance. If the difference is left unbounded, the mean of the observable will be consistent with the correct ensemble average, but the standard deviation grows with system size and timestep. In particular, without introducing c in SHMC, it is not possible to define the ensemble from which the method samples and thus prove microscopic reversibility, see [17]. The second criterion is to maximize the computational efficiency of SHMC. Figs. 2a and 2b demonstrate how the choice of c affects the runtime of a simulation. First, we plot the number of momenta rejections per SHMC step, followed by the number of position rejections per SHMC step. It is important to note that rejecting new positions is much more costly than rejecting the momenta as the forces must be calculated for an MD trajectory to produce new

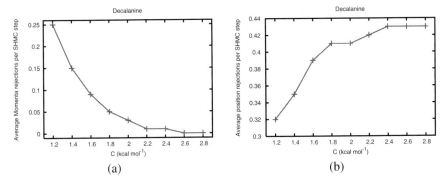

Figure 2. (a) Average number of rejected momenta per step; and **(b)** Average number of position rejections per step as a function of the choice of c for a 66-atom decalanine. Simulations run for 100 ps with 25 MD steps per SHMC step

positions but not for new momenta. For large c, our algorithm reduces to HMC. Thus, the HMC-regime has many rejections of positions since c increases with system size and timestep, whereas for large negative c, SHMC has a high rejection of momenta instead.

In general, we are not able to predict c *a priori*, but we have devised a heuristic that can choose an optimal c at runtime. In order to automatically choose c, we first run initialization steps (20) of our algorithm using a default value for c. During each warm-up step, we record the difference between the total and shadow energies two times: at the beginning of the MD run, immediately after the momenta are accepted, and at the end of the MD run. In a well equilibrated system, the energies at these two points should be fluctuating about some average value. This gives us forty samples and the mean value of this difference is chosen as an initial guess for c, denoted c_0.

Up to this point, this heuristic tends to choose large c. Therefore, we do additional relaxation steps (20) where the value of c is allowed to change slightly. Even when the initial value of c_0 is large, we noticed that big changes during relaxation are not effective. For this reason, each relaxation step is allowed to change c by no more than $\pm \log_{10} c_0$. We also want c to converge during the relaxation instead of jumping between predetermined values. In order to accomplish this, we multiply any suggested change by a decreasing factor, $1 - \frac{i}{N_R}$, where N_R is the number of relaxation steps and i is the current step, $0 \leq i < N_R$. During this phase, we may or may not change c depending on a predetermined criteria. In [17], we suggest that rejecting a few momenta during each step of SHMC is beneficial to the algorithm. We use an additional rejection factor to influence c to give us the desired number of momenta rejections. Based on our tests, we change c at each step of the relaxation according to Table 1. Thus, if after the third step of the relaxation there were 17 momenta rejections, the new value of c would be:

$$c_4 = c_3 + (1 - \frac{3}{20}) \times 1.5 \times \log_{10} c_0 .$$

Table 1. Change in c for next relaxation step based on the number of rejected momenta for the current step.

Rejected momenta	0	1–5	6–10	11–15	16–20	21–25	> 25
Rejection factor	−1	0	0.5	1	1.5	2	3

In this way, we lower the value of c if no rejections occur and raise the value if we reject too many. We have found that this method will, on average, give us the desired 5–10 momenta rejections per timestep for most values of the timestep Δt. As Δt grows so does the discretization error and the number of rejections of the momenta will also grow.

2.4 Replica Exchange Method

The method of replica exchange or "exchange Monte Carlo method", as introduced by Hukushima and Nemoto [15], uses a number of non-interacting MD or MC simulations running in parallel each at a different temperature. Replica coordinates are swapped between temperature neighbors according to a probability function, with an intent to enable a replica at low temperature to surpass barriers between its local energy minima. The method has found favor because the probability weighting factors are known *a priori*, unlike other generalized-ensemble algorithms which require detailed determination. The method is shown in Algorithm 3 following Okamoto [31].

Algorithm 3 Replica Exchange

1. Each replica in the canonical ensemble of fixed temperature is simulated simultaneously and independently for a certain number of MD or MC steps
2. A pair of replicas at neighboring temperatures, say $\gamma_m^{[i]}$ and $\gamma_{m+1}^{[j]}$, are exchanged with the probability

$$w(\gamma_m^{[i]} | \gamma_{m+1}^{[j]}) = \min(1, \exp(-\beta_m - \beta_{m+1})(U(x^{[j]}) - U(x^{[i]})) , \qquad (12)$$

where $U(x)$ is the potential energy.

Quantities of interest such as conformational distributions and thermodynamic properties can be derived from the individual replica output data sets through utilization of single histogram or multiple-histogram reweighting techniques such as the weighted histogram analysis method WHAM [20].

3 Test Systems, Methods, and Metrics

3.1 United Atom Butane

The 4-atom, united atom model butane, recently summarized by Barth *et al.* [2], is useful for elementary evaluation of sampling algorithms. Analytical calculations, relative to the dihedral energy, are easily generated for verification and validation purposes. Our model uses a two component term for the dihedral energy equation which closely approximates the original model [33]. Baseline analytical values are reported for subsequent comparison with the sampling algorithms.

Molecule Parameters:

Parameters for the united atom butane which we utilized in our testing are listed in Table 2. The non-bonded Lennard-Jones forces were evaluated using all-pairs with no cutoffs, and excluding scaled1-4. Coulombic forces were not used for this system.

Table 2. Parameters for United Atom Butane (Based on CHARMM 19 [24, 25])

Bonds	
CH2-CH2	$\kappa_b = 225.0\,\text{kcal/mol}, b_{eq} = 1.52\,\text{Å}$
CH2-CH3	$\kappa_b = 225.0\,\text{kcal/mol}, b_{eq} = 1.54\,\text{Å}$

Angle	
CH3-CH2-CH2	$\kappa_\theta = 45.0\,\text{kcal/mol}, \theta_{eq} = 110°$

Dihedral	
CH3-CH2-CH2-CH3	$\kappa_\phi = 1.6\,\text{kcal/mol}, \phi_{eq} = 0°, n_\phi = 3$
	$\kappa_\phi = 0.6\,\text{kcal/mol}, \phi_{eq} = 0°, n_\phi = 1$

Lennard-Jones	
CH2	$\epsilon = 0.1142\,\text{kcal/mol}, \sigma = 2.235\,\text{Å}$
CH3	$\epsilon = 0.1811\,\text{kcal/mol}, \sigma = 2.165\,\text{Å}$

Metric - Average Dihedral Energy:

The average dihedral energy for united-atom butane can be computed analytically according to the following equation reported by Fischer *et al.* [10]:

$$\langle U_D \rangle = \frac{\int U_D(\omega) \exp[-\beta U_D(\omega)]d\omega}{\int \exp[-\beta U_D(\omega)]d\omega}, \tag{13}$$

where U_D is the potential energy for the dihedral angle and ω is over the interval $[-\pi, \pi]$. The value at 300 K is given in Table 3.

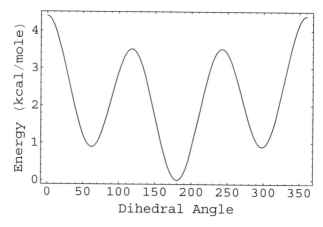

Figure 3. United atom butane energy for dihedral angle

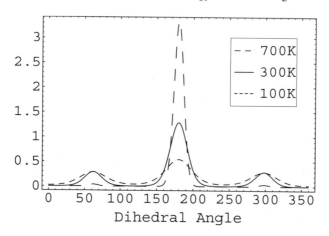

Figure 4. United atom butane dihedral distribution

Metric - Dihedral Distribution:

The dihedral energy of butane is shown in Fig. 3. The probability density of dihedral values for united atom butane at 300 K is plotted in Fig. 4 as calculated by

$$\frac{\exp[-\beta U_D(\omega)]d\omega}{\int \exp[-\beta U_D(\omega)]d\omega} \ . \tag{14}$$

3.2 Alanine Dipeptide

Alanine dipeptide is a frequently used test problem in MD simulations. It has only 22 atoms and it exhibits conformational flexibility in the two-dimensional $\phi-\psi$ plane. Also, it has multiple CO and NH units for H-bonding and a pendant side chain (methyl group) off of the main backbone (Fig. 5).

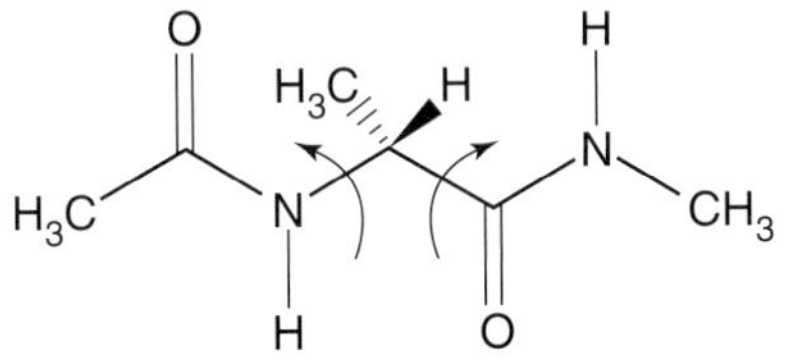

Figure 5. Diagram of Alanine Dipeptide (source: http://www.chem.umass.edu/nermmw/alanine.html)

The change in conformation seen through the dynamics of the backbone dihedral and its size make it a suitable molecule for studying conformational sampling. In alanine dipeptide, using CHARMM22 force field, the lowest free energy structure is the extended β conformation, $(\phi, \psi) = (-81, 81)$, while in solution, the extended β, $(\phi, \psi) = (-81, 153)$ and right-handed α-helical, $(\phi, \psi) = (-81, 63)$, conformations are nearly isoenergetic. In solution, a secondary minimum at $(\phi, \psi) = (63, -81)$ occurs at approximately 2.3 kcal/mol, a global free-energy minimum [32]. The ψ dihedral angle corresponds to N-C_α-C-N and the ϕ to C-N-C_α-C.

3.3 Bovine Pancreatic Trypsin Inhibitor (BPTI)

BPTI, protein data bank entry 6PTI [4], is commonly studied because of its small size. We use a solvated version with 73 additional water molecules for a total of 1,101 atoms. We also solvate BPTI for a larger system size of 14,281 atoms. The systems were prepared for simulation using NAMD 2.5 [18]. First, we ran minimization for 1,000 steps. Next, 20,000 steps of constant temperature, constant pressure (NPT) MD were run to equilibrate the system at a temperature of 300 K and pressure of 1 atm.

Metric - Conformation Strings

In order to identify unique conformations, we use a "dihedral string" method similar to that reported in [17] and [19]. A set of key dihedrals from the molecule are observed and their values resolved into an alphanumeric tag representing which energetic well the dihedral currently resides in. The energetic maximas for each dihedral in the input evaluation set are determined during initialization using a modified Brent's method and dihedral parameter knowledge from the topology (non zero phase shifts are normalized). By concatenating the tags together, we form a conformation string for each timestep. Comparing strings, we can classify and analyze the number of unique conformations.

4 Simulation Results

4.1 United Atom Butane

In order to compare and contrast the ability of conformational sampling algorithms to surpass energy barriers we ran two simulation sets. The first simulation set included the HMC and SHMC algorithms. We also ran the same test set at longer times steps (4 and 8 fs). No significant differentiation from the 1 fs data was observed as one is able to take longer steps with united atom butane than would be possible for larger all-atom proteins. Simulations were run using 10 MD steps per MC step, periodic boundary conditions, the Leapfrog integrator, all bonded forces, and full Lennard-Jones nonbonded forces. Performance at three temperatures (100, 300, and 700 K) was evaluated over 1 ns of simulation time. Average dihedral energies and the dihedral distributions were calculated. The performance at 300 K can be found in Table 3. For this simple molecule, HMC and SHMC performed similarly so SHMC data is omitted. The results relative to BPTI more appropriately contrast SHMC performance over HMC.

For the second simulation set, we performed three 1 ns tests using NVE Leapfrog, NVT Langevin Impulse, and a replica exchange with NVT Langevin Impulse. All tests utilized a 1 fs timestep, periodic boundary conditions, and the same force field used in the HMC simulations. Langevin simulations were run with a γ value of $5000\,\mathrm{ps}^{-1}$. The replica exchange simulation consisted of eight replicas at temperatures: 200 K, 239 K, 286 K, 342 K, 409 K, 489 K, 585 K, and 700 K [37]. Each replica was simulated using the Langevin impulse NVT method for 125 ps with a timestep of 1 fs. A replica exchange was attempted every 400 fs. It should be noted that all three tests represent the same amount of simulation time (a total of 1 ns), but the REM approach offers the additional benefit of fully parallel execution.

The results of the REM were analyzed using the "temperature" variant of the weighted histogram analysis method [12]. As is evident from Fig. 6, 7, and 8, the REM simulation achieved wider and more accurate sampling of the dihedral space than the standard MD simulations, suggesting that REM is superior for investigating the conformation space. In turn, NVT Langevin Impulse was far superior to NVE Leapfrog. We include these results in Table 3.

Table 3. Analytical, NVE Leapfrog, Hybrid Monte Carlo (HMC), Langevin Impulse, and Replica Exchange using Langevin Impulse (REM) values for united atom butane at 300 K

	$\langle U_D \rangle$ in kcal/mol	gauche- in %	trans in %	gauche+ in %
Analytical	0.628	16.0	68.0	16.0
Leapfrog	0.096	0.0	100.0	0.0
Langevin	0.71	5.8	55.3	38.9
HMC	0.604	14.6	72.3	13.1
REM	0.619	18.4	66.5	15.1

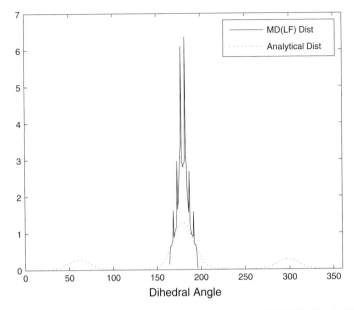

Figure 6. United atom butane dihedral distributions: Leapfrog NVE MD (1 ns). The dashed lines show the correct analytical distributions. Unitless normalized distribution versus dihedral angle in degrees

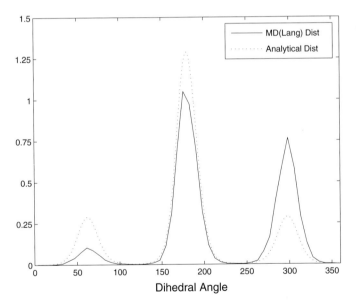

Figure 7. United atom butane dihedral distributions: Langevin Impulse NVT MD (1 ns). The dashed lines show the correct analytical distributions. Unitless normalized distribution versus dihedral angle in degrees

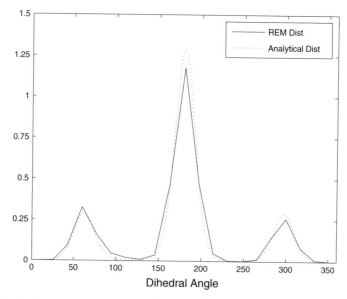

Figure 8. United atom butane dihedral distributions: Replica Exchange Langevin NVT (1 ns total simulation time for 8 replicas - each replica 125 ps only). The dashed lines show the correct analytical distributions. Unitless normalized distribution versus dihedral angle in degrees

4.2 Alanine Dipeptide

A brief analysis of two different MD implementations of the NVT ensemble, the Nosé-Hoover method using single thermostat and the Langevin impulse method, was performed to determine which achieved thermal equilibration more quickly. Langevin Impulse is exact for constant forces and has been shown to be a good discretization of the Langevin equation [16, 35]. Thermal equilibrium is important in the context of sampling because one typically starts sampling only when the system has achieved equilibrium. In addition, some methods such as umbrella sampling [6, 38] use biasing potentials to explore different conformations, and quick equilibration is necessary for efficient sampling.

We equilibrated alanine dipeptide for 500 ps using a 1 fs timestep in the NPT ensemble. Additionally, we examined the temperature of subsystems within the larger system. We expect an equipartition of the kinetic energy. Thus, we use the following criteria to judge the effectiveness of the NVT ensembles: 1) Does the ensemble achieve the desired system temperature? In other words, is the average system temperature equal to the target and is the standard deviation small? 2) Does the ensemble maintain a consistent temperature among system components?

An alanine dipeptide molecule solvated with 607 water molecules was used in the simulations. For each method, a 10 ps simulation was performed with a timestep of 0.1 fs. All bonded forces were utilized, including bond, bond angle, dihedral, and improper. For non-bonded forces, both Lennard-Jones and Coulombic forces were calculated using an all-pairs algorithm. Additionally, an artificial harmonic weighting potential was added to the system to force the ϕ dihedral to a desired value. The weighting potential was a function of the dihedral angle: $U(\phi) = k(\phi_{cur} - \phi_{goal})^2$ where ϕ_{cur} is the current value of the dihedral, ϕ_{goal} is the desired value, and k is a force constant. We want to simulate using MD techniques around the point in phase space represented by the reaction coordinate ϕ_{goal}. This is similar to umbrella sampling where a weighting function is used to bias a reaction coordinate to a particular configuration. The weighting potential was utilized to create biasing in the system to adequately test the NVT methods. We used a γ value of 5000 ps^{-1} for the simulation using the Langevin Impulse method. The ϕ_{goal} has been set to $60°$. The k was set to 100 kcal/mol.

As is clear from Figs. 9a and 9b and Table 4, the Langevin impulse method was superior in our test simulations. It maintained all subsystems near the target temperature throughout the simulation. Conversely, the Nosé-Hoover method, despite maintaining the total system temperature near the target value, produced a significant temperature differential between the alanine dipeptide molecule and the water. Even after 10 ps, the Nosé-Hoover method did not achieve the desired subsystem temperatures. These results are in agreement with the contribution of Barth, Leimkuhler, and Sweet in this volume.

Table 4. Subsystem temperatures for NVT ensembles with a target temperature of 300 K

	Nosé-Hoover NVT		Langevin impulse NVT	
Subsystem	Average (K)	Std Dev	Average (K)	Std Dev
Alanine	1215.50	596.55	302.10	84.27
Water	288.88	9.63	300.05	5.79
Total System	300.11	7.38	300.23	5.78

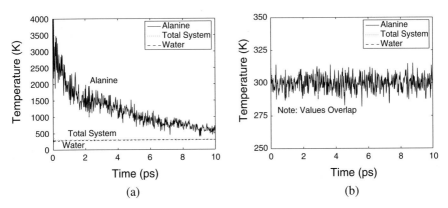

Figure 9. Thermal equilibration for water and solute of solvated alanine dipeptide (**a**) Nosé-Hoover NVT at 300 K; (**b**) Langevin impulse NVT at 300 K

4.3 BPTI

Figures 10 and 11 show the cost for HMC and SHMC of producing a different conformation (using the conformation string) as a function of the timestep for 1,101-
atom and 14,281-atom solvated BPTI respectively. These charts were selected from
our recently published SHMC work [17] to demonstrate how SHMC allows much
longer timesteps than HMC. The 1,101 atom uses an MD trajectory of 42 fs for each
HMC or SHMC step, whereas the 14,281 uses an MD trajectory of 15 fs for each
HMC or SHMC step. The speedup in sampling efficiency, given by the ratio of computational cost of HMC to SHMC, ranges from 2 to 10 as the timestep is increased.

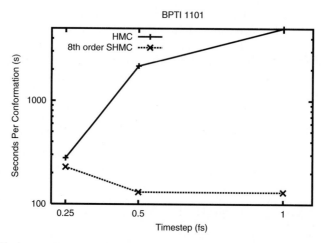

Figure 10. Average computer time per discovered conformation for 1,101 atom BPTI

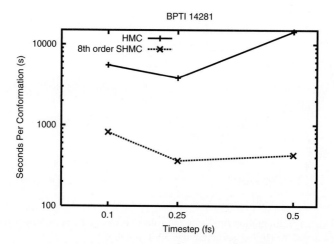

Figure 11. Average computer time per discovered conformation for the 14,281 atom BPTI

4.4 Bias

In Fig. 12, we show the effect of bias on an MD simulation, as well as how MC methods remove the bias. We simulated BPTI with 1,101 atoms for 20 ps, using Leapfrog, HMC, and SHMC for 6 different timesteps. HMC and SHMC use 25 MD steps per MC step. Error bars are 1 standard deviation and are averaged over 3 runs. From the figure, it can be seen that as the timestep increases, there is an obvious upward drift in the potential energy that is greater than the statistical error. A least squares fit is plotted through the points. SHMC, on the other hand, is fluctuating about an average value that falls within the statistical error. Thus, the bias is removed. For the larger timesteps, SHMC is beginning to show more deviation due to a decrease in accepted moves. HMC does not show removal of bias because its poor scalability does not allow it to accept enough moves beyond a timestep of 0.5 fs in this test.

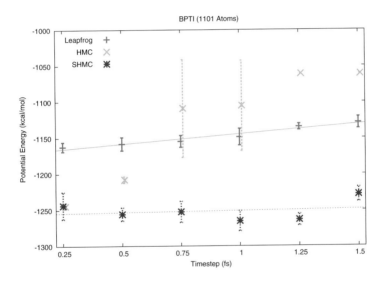

Figure 12. Bias in the average potential energy of MD, HMC, and SHMC as a function of timestep

5 Discussion

From our results with butane, it is evident that REM could be considered one of the best methods for conformational sampling of biomolecules. For optimal performance, one needs to choose the right distribution of the temperatures. REM has the advantage that the weighting factor is known in advance and is a suitable candidate for parallel implementation. Indeed, this method allows us to sample around all energy wells under low temperature. REM can work with either MD or MC methods.

Thus, if one needs to be exact, one can use HMC or SHMC for each replica simulation.

There are several different optimizations proposed based on general REM. One of them is multidimensional replica exchange method (MREM). In this method, a pair of replicas with different temperature or different parameters of the potential energy are interchanged [36]. This method provides greater flexibility over the general REM and it is based on the following observation: For the M noninteracting replicas, the Hamiltonian $\mathcal{H}$ of the system could be different for different replicas depending on some parameter(s). One particularly interesting instance of MREM is replica-exchange umbrella-sampling.

SHMC is exact since it eliminates the bias associated with MD sampling of molecular systems. SHMC can use much higher timesteps than classical HMC, particularly as the system size increases. In practice, one needs to choose an extra parameter c. We have presented an efficient method for automatically determining this additional parameter and shown its effectiveness in computing potential energy of solvated BPTI. We show that SHMC is greater than ten times more efficient than HMC for a medium size solvated BPTI with 14,281 atoms.

Nosé dynamics is a popular scheme to simulate molecular systems under constant temperature, but it is based on the ergodic hypotheses, which does not necessarily hold for small or stiff systems. In our experiment with alanine dipeptide, we have found a significant gap in subsystem temperature from the target value. Several extensions of the original Nosé-Hoover method exist which address this issue. There are extensions which use generalized heat baths to control the temperature of the subsystem. Auxiliary heat baths are coupled into thermostatting variables. But the design of auxiliary heat bath could slow down the simulation significantly. One of the interesting extensions along this line is due to Leimkuhler and Sweet. Their work is based on the idea of adding multiple thermostats to a Hamiltonian which has been modified by Nosé's method, retaining sampling from the canonical ensemble. They employ a regularizing term in the Hamiltonian to bound the integrals over the thermostats. A detailed discussion on this topic and related methods can be found in [23]. Recursive multiple thermostat (RMT) is a recent development by which equilibration is obtained from a more complicated interaction of thermostat variables with the physical variables. The work by Barth *et al.* [3] have shown that the RMT is most promising of all methods based on Nosé-Hoover chains for correct thermostatting.

From our experiment with NVT integrators on solvated alanine dipeptide it is evident that Langevin Impulse can be used to achieve thermal equilibration quickly and accurately compared to the Nosé-Hoover method. The former method kept the temperature of the biomolecule and the solvent close to the target temperature for the entire course of the MD run. On the other hand, the Nosé-Hoover formulation is more flexible and can be used to derive other equations of motion.

A complex but powerful solution to the problems associated with potential barriers is due to Rosso et al. [32]. The idea is to separate out a reactive subspace using a coordinate transformation. Then an adiabatic separation is imposed over this subspace and from the adiabatic probability distribution free energy surface is computed. Roughness of the free-energy manifolds have been explored using a higher temperature for a certain subset of reaction coordinates. This technique, known as adiabatic free-Energy dynamics (AFED), can also be applied to efficiently sample over the entire conformational space. However, this method has a high overhead for larger molecules since the coordinate transformations have to be derived on-the-fly. This is an instance of a broader class of algorithms based on smoothing the potential energy surface. A user of sampling methods will generally need to have access to a whole range of tools to solve their problems. Our hope is that this paper serves as an introduction to some of the very useful and new methods.

References

[1] M. P. Allen and D. J. Tildesley. *Computer Simulation of Liquids*. Clarendon Press, Oxford, New York, 1987. Reprinted in paperback in 1989 with corrections.

[2] E. Barth, B. Leimkuhler, and S. Reich. A test set for molecular dynamics algorithms. In T. Barth, M. Griebel, D. Keyes, R. Nieminen, D. Roose, and T. Schlick, editors, *Computational Methods for Macromolecules: Challenges and Applications*, volume 24 of *Lecture Notes in Computational Science and Engineering*, pages 73–103. Springer, 2002.

[3] E. Barth, B. Leimkuhler, and C. Sweet. Approach to thermal equilibrium in biomolecular simulation. 2005. To appear in *Proc. of Algorithms for Macromolecular Modeling IV*. Lecture Notes in Computational Science and Engineering (LNCSE). Springer Verlag. New York and Berlin.

[4] H. M. Berman, J. Westbrook, Z. Feng, G. Gilliland, T. N. Bhat, H. Weissig, I. N. Shindyalov, and P. E. Bourne. The protein data bank. *Nucleic Acids Research*, pages 235–242, 2000.

[5] B. J. Berne and J. E. Straub. Novel methods of sampling phase space in the simulation of biological systems. *Curr. Opin. Struct. Biol.*, 7:181–189, 1997.

[6] D. L. Beveridge and F. M. DiCapua. Free energy via molecular simulation: Applications to chemical and biological systems. *Ann. Rev. Biophys. Biophys. Chem.*, 18:431–492, 1989.

[7] A. Brünger, C. B. Brooks, and M. Karplus. Stochastic boundary conditions for molecular dynamics simulations of ST2 water. *Chem. Phys. Lett.*, 105:495–500, 1982.

[8] M. Creutz. Global Monte Carlo algorithms for many-fermion systems. *Phys. Rev. D*, 38(4):1228–1238, 1988.

[9] R. Elber. Reaction path studies of biological molecules. In R. Elber, editor, *Recent Developments in Theoretical Studies of Proteins (Advanced Series in Physical Chemistry, Vol. 7)*, Singapore, 1996. World Scientific.

[10] A. Fischer, F. Cordes, and C. Schütte. Hybrid Monte Carlo with adaptive temperature in mixed-canonical ensemble: Efficient conformational analysis of RNA. *J. Comp. Chem.*, 19(15):1689–1697, 1998.

[11] D. Frenkel and B. Smit. *Understanding Molecular Simulation*. Academic Press, San Diego, 2nd edition, 2002.

[12] E. Gallicchio, M. Andrec, A. K. Felts, and R. M. Levy. Temperature weighted histogram analysis method, replica exchange, and transition. *J. Chem. Phys.*, 109:6722–6731, 2005.

[13] S. Gupta, A. Irbäck, F. Karsch, and B. Petersson. The acceptance probability in the hybrid Monte-Carlo method. *Phys. Lett. B*, 242:437–443, 1990.

[14] W. G. Hoover. Canonical dynamics: Equilibrium phase-space distribution. *Phys. Rev. A*, 31(3):1695–1697, 1985.

[15] K. Hukushima and K. Nemoto. Exchange Monte Carlo method and application to spin glass simulations. *J. Phys. Soc. of Japan*, 65(6):1604–1608, 1996.

[16] J. A. Izaguirre, D. P. Catarello, J. M. Wozniak, and R. D. Skeel. Langevin stabilization of molecular dynamics. *J. Chem. Phys.*, 114(5):2090–2098, 2001.

[17] J. A. Izaguirre and S. S. Hampton. Shadow hybrid Monte Carlo: An efficient propagator in phase space of macromolecules. *J. Comput. Phys.*, 200(2):581–604, 2004.

[18] L. Kalé, R. Skeel, M. Bhandarkar, R. Brunner, A. Gursoy, N. Krawetz, J. Phillips, A. Shinozaki, K. Varadarajan, and K. Schulten. NAMD2: Greater scalability for parallel molecular dynamics. *J. Comput. Phys.*, 151:283–312, 1999.

[19] P. D. Kirchhoff, M. B. Bass, B. A. Hanks, J. Briggs, A. Collet, and J. A. Mc-Cammon. Structural fluctuations of a cryptophane host: A molecular dynamics simulation. *J. Am. Chem. Soc.*, 118:3237–3246, 1996.

[20] S. Kumar, D. Bouzida, R. H. Swendsen, P. A. Kollman, and J. M. Rosenberg. The weighted histogram analysis method for free-energy calculations on biomolecules. i: The method. *J. Comp. Chem.*, 13(8):1011–1021, 1992.

[21] A. R. Leach. *Molecular Modelling: Principles and Applications.* Addison-Wesley, Reading, Massachusetts, July 1996.

[22] A. R. Leach. *Molecular Modelling: Principles and Applications.* Prentice Hall, 2nd edition, Mar. 2001.

[23] B. J. Leimkuhler and C. R. Sweet. The canonical ensemble via symplectic integrators using Nosé and Nosé-Poincaré chains. *J. Chem. Phys.*, 121(1):108–116, 2004.

[24] A. D. MacKerell Jr., D. Bashford, M. Bellott, R. L. Dunbrack Jr., J. Evanseck, M. J. Field, S. Fischer, J. Gao, H. Guo, S. Ha, D. Joseph, L. Kuchnir, K. Kuczera, F. T. K. Lau, C. Mattos, S. Michnick, T. Ngo, D. T. Nguyen, B. Prodhom, I. W. E. Reiher, B. Roux, M. Schlenkrich, J. Smith, R. Stote, J. Straub, M. Watanabe, J. Wiorkiewicz-Kuczera, D. Yin, and M. Karplus. All-hydrogen empirical potential for molecular modeling and dynamics studies of proteins using the CHARMM22 force field. *J. Phys. Chem. B*, 102:3586–3616, 1998.

[25] A. D. MacKerell Jr., D. Bashford, M. Bellott, R. L. Dunbrack Jr., J. Evanseck, M. J. Field, S. Fischer, J. Gao, H. Guo, S. Ha, D. Joseph, L. Kuchnir, K. Kuczera, F. T. K. Lau, C. Mattos, S. Michnick, T. Ngo, D. T. Nguyen, B. Prodhom, B. Roux, M. Schlenkrich, J. Smith, R. Stote, J. Straub, M. Watanabe, J. Wiorkiewicz-Kuczera, D. Yin, and M. Karplus. Self-consistent parameterization of biomolecules for molecular modeling and condensed phase simulations. *FASEB J.*, A143:6, 1992.

[26] T. Matthey. *Framework Design, Parallelization and Force Computation in Molecular Dynamics*. PhD thesis, University of Bergen, Bergen, Norway, 2002.

[27] T. Matthey, T. Cickovski, S. S. Hampton, A. Ko, Q. Ma, M. Nyerges, T. Raeder, T. Slabach, and J. A. Izaguirre. PROTOMOL: An object-oriented framework for prototyping novel algorithms for molecular dynamics. *ACM Trans. Math. Softw.*, 30(3):237–265, 2004.

[28] B. Mehlig, D. W. Heermann, and B. M. Forrest. Hybrid Monte Carlo method for condensed-matter systems. *Phys. Rev. B*, 45(2):679–685, 1992.

[29] F. Nardi and R. Wade. *Molecular Dynamics. From Classical to Quantum Methods*, chapter 21, pages 859–898. Elsevier Science B.V., first edition, 1999.

[30] S. Nosé. A unified formulation of the constant temperature molecular dynamics methods. *J. Chem. Phys.*, 81(1):511–519, 1984.

[31] Y. Okamoto. Generalized-ensemble algorithms: enhanced sampling techniques for Monte Carlo and molecular dynamics simulations. *J. Molecular Graphics and Modelling*, 22:425–439, 2004.

[32] L. Rosso, J. B. Abrams, and M. E. Tuckerman. Mapping the backbone dihedral free-energy surfaces in small peptides in solution using adiabatic free-energy dynamics. *J. Phys. Chem. B*, 109:4162–4167, 2005.

[33] J.-P. Ryckaert and A. Bellemans. Molecular dynamics of liquid alkanes. *Faraday Discussions*, 66:95–106, 1978.

[34] T. Schlick. *Molecular Modeling and Simulation - An Interdisciplinary Guide*. Springer-Verlag, New York, NY, 2002.

[35] R. D. Skeel and J. A. Izaguirre. An impulse integrator for Langevin dynamics. *Mol. Phys.*, 100(24):3885–3891, 2002.

[36] Y. Sugita, A. Kitao, and Y. Okamoto. Multidimensional replica-exchange method for free-energy calculations. *J. Chem. Phys.*, 113(15):6042–6051, 2000.

[37] Y. Sugita and Y. Okamoto. Replica-exchange molecular dynamics method for protein folding. *Chem. Phys. Lett.*, 314:141–151, 1999.

[38] Torrie and Valleau. Nonphysical sampling distributions in Monte Carlo free-energy estimation: Umbrella sampling. *J. Comput. Phys.*, 23:187–199, 1977.

[39] W. F. van Gunsteren, T. Huber, and A. E. Torda. Biomolecular modelling: Overview of types of methods to search and sample conformational space. volume 330, pages 253–268. AIP, 1995.

Approach to Thermal Equilibrium
in Biomolecular Simulation

Eric Barth[1], Ben Leimkuhler[2], and Chris Sweet[2]

[1] Department of Mathematics, Kalamazoo College, Kalamazoo, Michigan, USA 49006
[2] Centre for Mathematical Modelling, University of Leicester, University Road,
Leicester LE1 7RH, UK

Abstract. The evaluation of molecular dynamics models incorporating temperature control methods is of great importance for molecular dynamics practitioners. In this paper, we study the way in which biomolecular systems achieve thermal equilibrium. In unthermostatted (constant energy) and Nosé-Hoover dynamics simulations, correct partition of energy is not observed on a typical MD simulation timescale. We discuss the practical use of numerical schemes based on Nosé-Hoover chains, Nosé-Poincaré and recursive multiple thermostats (RMT) [8], with particular reference to parameter selection, and show that RMT appears to show the most promise as a method for correct thermostatting. All of the MD simulations were carried out using a variation of the CHARMM package in which the Nosé-Poincaré, Nosé-Hoover Chains and RMT methods have been implemented.

1 Introduction

Molecular dynamics (MD) is an increasingly popular tool in chemistry, physics, engineering and biology. In many molecular simulations, the dynamics trajectory is used as a method of sampling a desired ensemble, for example to compute the average of some function of the phase space variables. In such cases it is important that the trajectory produce a representative collection of phase points for all variables of the model. A common ensemble used in biomolecular simulation is the NVT ensemble, which weights points of phase space according to the Gibbs density

$$\rho \propto e^{-\beta H}, \quad \beta = (k_B T)^{-1} ,$$

where H is the system Hamiltonian, k_B is Boltzmann's constant, and T is temperature. In normal practice, MD samples from the isoenergetic (microcanonical) ensemble, so some device must be employed to generate points from the NVT ensemble. The methods discussed in this article are based on construction of extended Hamiltonians whose microcanonical dynamics generate canonical sampling sequences (Nosé dynamics). Nosé [5] proposed a Hamiltonian of the form:

$$H^{Nose} = H\left(q, \frac{p}{s}\right) + \frac{p_s^2}{2Q} + N_f k_B T \ln s , \tag{1}$$

where Q is the Nosé mass, s is the thermostatting variable, p_s is its conjugate momentum, k_B is the Boltzmann constant and N_f gives the number of degrees of freedom in the system. Simulations are often conducted using a time-reversible but non-Hamiltonian formulation (Nosé-Hoover, [7]) that incorporates a correction of timescale (this time-transformation has some important implications for the stability of numerical methods). In a 1998 paper [4], a Hamiltonian time-regularized formulation was introduced along with reversible *and* symplectic integrators (see also [15, 16]). These methods show enhanced long term stability compared to Nosé-Hoover schemes. The Nosé-Poincaré schemes, as they are termed because of the use of a Poincaré time transformation, have been extended to NPT and other ensembles in several recent works [9–11].

In classical models of biomolecules, when thermostatting with schemes derived from Nosé's method, trapping of energy in subsystems can result in long equilibration times. The presence of many strongly coupled harmonic components of not too different frequency means that the systems should eventually equilibrate, but the equilibration time in all-atom models (including bond vibrations) nonetheless greatly exceeds the time interval on which simulation is performed (a few nanoseconds, in typical practice). The only way to be sure that an initial sample is properly equilibrated is to check that in subsequent runs, the individual momentum distributions associated to each degree of freedom are Maxwellian. This is typically not done in practice. To bring a given molecular system rapidly to equilibrium and maintain the system in that state to ensure good sampling of all degrees of freedom, it is necessary to employ a suitable thermostatting mechanism.

To illustrate the primary challenge that we will attempt to address in this paper, we have performed a molecular dynamics simulation of an unsolvated alanine dipeptide molecule using a representative molecular dynamics software package (CHARMM [1]). We used the Verlet method to perform a microcanonical simulation on the system and examined the convergence to thermal equilibrium in the "light" (H) and "heavy" (C,N,O) atoms. Note that because of the presence of conserved quantities (total linear momenta) the usual equipartition of energy does not hold; the modified formulas are given in Sect. 2. The details regarding the setup of this simulation can be found in Sect. 3.4. It is clear from these experiments that the adiabatic localization of energy is a significant cause for concern as seen in Figs. 1 and 2.

It might be thought that the energy trapping is a result of performing these simulations in vacuo, but this is not the case: similar problems have been verified by the authors for solvated models.[3] It might also be thought that the Nosé dynamics technique, in introducing a "global demon" which couples all degrees of freedom, would successfully resolve this issue. In fact, as seen in Fig. 2, this is not the case: although such methods successfully control the overall temperature of the system, the thermal distributions observed in light and heavy degrees of freedom using Nosé-Hoover

---

[3]The use of solvated models raises some additional issues regarding bond thermalization and the selection of parameters for some of our methods. These results will be reported elsewhere.

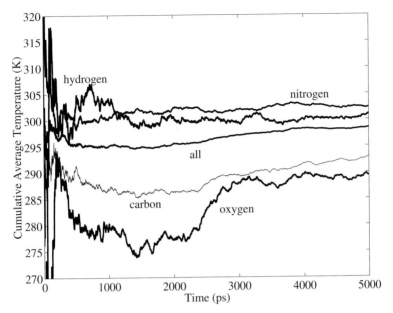

Figure 1. A 5ns trajectory for alanine dipeptide using the Verlet integrator clearly shows that equilibrium is not achieved on the indicated timescale. The plot shows the cumulative time-averaged temperatures for the entire system (all), and for each type of atom separately

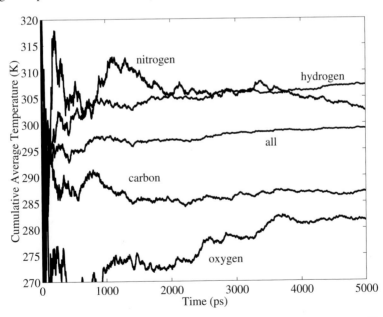

Figure 2. Cumulative average temperature, and temperature of subsystems computed by a 5ns trajectory for alanine dipeptide using the Nosé-Hoover option in CHARMM with $Q=0.3$ — as we report later, this is the optimal value for Q. Correct thermalization is clearly not achieved on this time scale

(and Nosé-Poincaré) are incorrect, as can be seen in Fig. 2. The system evidently does not have sufficient ergodicity to provide the correct energetic distribution on the timescale of interest.

Several techniques have been proposed to improve ergodicity in molecular simulations. In [12] a Nosé-Hoover chain method was developed which coupled additional thermostatting variables to the system degrees of freedom, retaining the property that integration over the auxiliary variables reduced sampling of the extended microcanonical phase space to canonical sampling of H. As this extension is based on Nosé-Hoover, it also sacrifices the Hamiltonian structure: the additional variables are introduced in such a way that the extended system is only time-reversible, so that methods based on this scheme cannot be reversible-symplectic. In [8, 13, 14] several new Hamiltonian-based multiple thermostat schemes have been developed. Nosé-Poincaré chains, described in [13] are the natural analogue of Nosé-Hoover chains. The more recent recursive multiple thermostat (RMT) schemes of [8] are a new departure, obtaining thermalization from a more complicated interaction of thermostat variables with the physical variables. A careful analysis of Nosé dynamics and RMT schemes for harmonic models was performed in [8]; arguments presented there and numerical evidence suggest that the formulation is potentially superior to other dynamical alternatives, including Nosé-Hoover chains, in obtaining well-equilibrated sampling sequences for the canonical ensemble. However, the results of [8] have so far only been verified for harmonic oscillators and coupled harmonic models.

The method of Gaussian moment thermostatting [6] also attempts to address incorrect thermalization of Nosé-Hoover methods, but was not considered here. In this paper we study the convergence to ensemble for chains and recursive methods applied to biomolecular models. We first discuss problem formulation and computation of temperature in all-atom biomolecular models.

2 Molecular Dynamics Formulation

In this article, we treat a classical all-atom N-body model. The Hamiltonian is of the form

$$H(q_1, q_2, \dots q_N, p_1, p_2, \dots p_N) = \sum_{i=1}^{N} \frac{p_i^2}{2m_i} + V(q_1, q_2, \dots q_N).$$

Here m_i represents the mass of the ith atom, $q_i \in \mathbf{R}^3$ and $p_i \in \mathbf{R}^3$ are Cartesian position and momentum vectors of the atomic point masses, and V represents potential energy. The potential energy function can be decomposed into a sum of terms, including pairwise (distance dependent) short-ranged Lennard-Jones potentials V_{LJ}, Coulombic potentials due to charges on the atoms V_C, and potential energies that describe the covalent bonding structure of the molecule, including $V_B^{(2)}, V_B^{(3)}, V_B^{(4)}$, representing 2-atom (length bond), 3-atom (angle bond), and 4-atom (dihedral angle) terms, respectively. In vacuum, with internal potentials only, the system described above is invariant under translations and rotations, thus it would admit six conserved quantities (linear and angular momentum). In practice, most biomolecular simulations also incorporate a collection of water molecules and are performed using periodic boundary conditions, meaning that the system is allowed to interact with copies of itself extended along the cubic lattice vectors. In this setting, the angular momentum conservation is broken, although the translational symmetry (and with it linear momentum conservation) is still present.

Although it is a slight abuse of language, we define the instantaneous temperature as the average kinetic energy per degree of freedom:

$$T_{\text{inst}}(p_1, p_2, \ldots p_N) = \left(\frac{1}{(3N-d)k_B} \right) \sum_{i=1}^{N} \frac{p_i^2}{2m_i} ,$$

where N is the number of atoms, k_B is the Boltzmann constant and $N_f \equiv 3N - d$ is the number of degrees of freedom (d is the number of conserved quantities in the dynamics). For example, in a simulation without periodic boundary conditions, common microcanonical integrators such as Verlet conserve both linear and angular momentum, so $d = 6$. In the presence of periodic boundary conditions we have $d = 3$. Standard Langevin dynamics, in which no quantities are conserved, has $d = 0$. The ergodic hypothesis of statistical mechanics states that time averages will (eventually) converge to ensemble averages:

$$\lim_{t \to \infty} \langle T_{\text{inst}} \rangle_t = \langle T \rangle_{\text{ensemble}} .$$

In this article we will need to discuss convergence to ensemble in different variables. When conserved quantities are present in a Hamiltonian system, the usual equipartition result must be adjusted. Indeed, we no longer observe equipartition, but there is still an appropriate partition of kinetic energies. Assuming that the trajectories sample from the canonical ensemble (in the microcanonical case, this is essentially equivalent to assuming ergodicity and a sufficiently large system) we have, assuming linear momentum conservation [18],

$$\left\langle \frac{p_{ix}^2}{2m_i} \right\rangle = \frac{M - m_i}{M} \frac{k_B T}{2} , \tag{2}$$

where $M = \sum_i m_i$, the sum of all particle masses in the system. A simple, direct proof of this result can be constructed by integrating the Nosé Hamiltonian partition function. In anticipation of work on solvated systems, we introduced a weak anisotropic interaction potential between a *single pair* of backbone atoms (the C-C bond at the N-terminus of alanine dipeptide) in our constant energy simulations of the form

$$\phi_{\text{anisotropy}}(q_i, q_j) = \frac{k_1}{2}(x_i - x_j)^2 + \frac{k_2}{2}(y_i - y_j)^2 + \frac{k_3}{2}(z_i - z_j)^2 .$$

(Modest values of the three constants were used: $k_1 = 0.1$, $k_2 = 0.15$, $k_3 = 0.2$.) This has the effect of breaking angular momentum conservation in vacuum simulations, while leaving linear momentum invariant, so that (2) can be used to compute subsystem temperatures.[4]

For trajectories generated according to the Nosé Hamiltonian, linear and angular momentum are conserved only weakly, in the sense that these quantities are conserved if initialized to zero, but will vary from any nonzero initial condition. In the simulations reported here, the linear momentum was set to be zero initially, and was conserved throughout the trajectories. The angular momentum was initially nonzero (but small, on the order of 10^{-3}), and was observed to vary over the course of the simulation. Hence the number of conserved quantities for all variants of Nosé-Hoover and Nosé-Poincaré was $d = 3$.

3 Thermostatting using Nosé-Hoover Chains, Nosé-Poincaré and RMT Methods

Because correct thermalization is not achieved by the use of the Verlet or Nosé-Hoover methods, as shown in Figs. 1 and 2, we studied the Hamiltonian Nosé-Poincaré method and some methods which are designed to achieve enhanced thermalization: Nosé-Hoover chains and RMT. It is well known [5, 8, 12] that the correct choice of parameters is essential if these methods are to thermostat the model correctly, and we here consider the different methods proposed for choosing them.

3.1 Nosé-Poincaré and RMT Methods

The Nosé-Poincaré method [4] involves direct symplectic discretization of the Hamiltonian

$$H^{NP} = s\left[H - H_0^{Nose}\right] .$$

Here H_0^{Nose} is the initial value of the Nosé Hamiltonian and leads to the Nosé-Poincaré Hamiltonian

$$H^{NP}(q,s,p,p_s) = s\left(H\left(q,\frac{p}{s}\right) + \frac{p_s^2}{2Q} + N_f k_B T \ln s - H_0\right) . \tag{3}$$

The numerical method proposed in [4] was the generalized leapfrog method. Although this method is typically implicit, in the case of the Nosé-Poincaré system the numerical challenge reduces to solving a scalar quadratic equation, and it is this variation which we implemented in CHARMM.

---

[4]The method for calculation of instantaneous kinetic temperature for a subsystem has not been widely reported in the literature, but is needed in practice. In particular, both old and new variants of Nose-Hoover dynamics implemented in CHARMM have the option of thermostatting subsystems separately, thus necessitating the calculation of the temperature of subsystems and correct handling of the degrees of freedom. However, this seems to be done incorrectly in CHARMM (version c31b1): the first specified subsystem s, with N^s atoms, is assigned $g^s = 3N^s - 6$ degrees of freedom, with each successive subsystem m being assigned the full $g^m = 3N^m$.

The RMT method [8] is produced by applying additional thermostats to the Nosé-Poincaré method recursively, with each additional thermostat acting on the previous one in addition to the original system. As discussed in [8], the stability of the numerical implementation of Nosé-Poincaré chains is not as good as the underlying Nosé-Poincaré method. The RMT method [8] corrects this deficiency while introducing a stronger coupling between bath and phase variables. With M thermostats, the recommended RMT formulation is:

$$
H^{RMT} = s_1 s_2 \cdots s_M \left[\sum_{i=1}^{N} \frac{p_i^2}{2m_i s_1^2 \cdots s_M^2} + V(q) \right.
$$
$$
+ \sum_{j=1}^{M-1} \frac{p_{s_j}^2}{2Q_j s_{j+1}^2 \cdots s_M^2} + \frac{p_{s_M}^2}{2Q_M} + N_f k_B T \ln s_1
$$
$$
\left. + \sum_{j=2}^{M} ((N_f + j - 1) k_B T \ln s_j + f_i(s_i)) - H_0 \right] ,
$$

where H_0 is chosen so that the initial value of H^{RMT} is zero. This formulation introduces a number of additional parameters, some of which are dependent on the choice of the $f_i(s_i)$. A recommended choice for these functions is

$$
f_i(s_i) = \frac{(a_j - s_j)^2}{2C_j} . \tag{4}
$$

The value a_i is chosen as the required average value of s_i, generally 1, as the additional term will operate as a negative feedback loop to minimize $(a_i - s_i)$, as can be seen from the equations of motion. In the context of Nosé-Poincaré chains the value of C_i, $i \geq 2$ can be estimated by considering the equations of motion for the ith thermostat. From this we see that s_i is driven by the changes in $p_{s_{i-1}}$. The purpose of the auxiliary function is to limit the excursions of s_i, which can be achieved if $ds_i/dp_{s_{i-1}}$ is a maximum at $s_i = a_i$. It was shown in [13] that C_i should satisfy,

$$
C_i \leq \frac{a_i^2}{8k_B T} . \tag{5}
$$

For RMT, the choice of C_i is less clear but numerical experiments indicate a similar upper bound on C_i. Experiments also suggest that the precise choice of C_i is not as critical to effectiveness of RMT as is selection of thermostat mass parameters.

We note that for small values of thermostat masses discussed below, the RMT discretization used in the numerical experiments reported here is limited by numerical stability to smaller timesteps than the Verlet method is able to use. Work on improving the stable timestep (and hence numerical efficiency) by enhancement of the RMT numerical integrator is ongoing.

3.2 Choice of the Nosé Mass: Ideas from the Literature

All methods based on the work of Nosé [5] have parameters. Although the proofs of canonical sampling (which assume ergodicity) are not strictly speaking dependent on their values, a careful choice is needed in order for good sampling to be observed in practice. In standard Nosé dynamics the only parameter is the Nosé mass Q in (1).

For the purpose of this discussion we will consider the time reparameterized variation of Nosé's scheme, the Nosé-Poincaré method [4, 16], which produces trajectories in real time and has the advantage of being Hamiltonian-based (3). Various attempts have been made to identify the optimal value for Q [5, 8, 9], and these are examined below.

Choice of Q Based on Total Kinetic Energy

In molecular dynamics the temperature of a system, at equilibrium, can be defined as,

$$T = \frac{1}{N_f k} \left\langle \sum_{i=1}^{N} \frac{\tilde{p}_i^2}{m_i} \right\rangle . \tag{6}$$

From this it is tempting to assume that if this criterion is met then the system must be at equilibrium. However thermostatting methods derived from Nosé's scheme are based on a "negative feedback loop" which controls the average kinetic energy such that (6) is satisfied. This can be seen by considering the equations of motion for the thermostat conjugate momentum,

$$\dot{p}_s = \sum_{i=1}^{N} \frac{p_i^2}{m_i s^2} - N_f k_B T . \tag{7}$$

Taking averages, assuming time averages of time derivatives disappear and substituting $\tilde{p}_i = p_i/s$ gives (6). From this we see that (6) is satisfied for all values of Q, whereas it is known that the correct sampling is not obtained unless Q is chosen correctly.

If we study (7) carefully we see that the method only guarantees that the total average kinetic energy is fixed, it does not indicate what happens to subsystems. From the equipartition theorem we have that,

$$\left\langle \frac{\tilde{p}_i^2}{m_i} \right\rangle = k_B T \quad \text{all } i . \tag{8}$$

If we look at individual subsystems we find that (8) is not satisfied for all Q, as seen in Fig. 3, and hence it is not possible to choose Q in this manner.

Choice of Q Based on Self-Oscillation Frequency

By utilizing linearization methods Nosé determined [5] that, for time re-parameterized thermostatting methods, the thermostat subsystem has a natural frequency

$$\omega_N = \sqrt{\frac{2N_f k_B T}{Q}} . \tag{9}$$

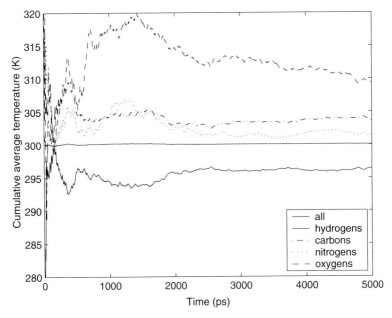

Figure 3. Average kinetic energy for subsystems of alanine dipeptide consisting of the hydrogen atoms, oxygen atoms, nitrogen atoms and carbon atoms for a 5ns simulation time with $Q = 2.0$ using the Nosé-Poincaré method

It has been proposed [5] that setting Q so that this frequency is close to some natural frequency in the original system would induce resonance between the system and thermostat, and hopefully produce ergodic behavior. Although the need to increase Q with the number of degrees of freedom N_f and the temperature T can be verified empirically, setting the value of Q according to (9) generally gives poor results. In fact it is observed that if the thermostat subsystem oscillates at its resonant frequency then it becomes decoupled from the system [8, 12], and poor sampling is obtained. For systems of harmonic oscillators it can be shown [8] that the optimum choice of Q does not coincide with this estimate. As an example, for a single harmonic oscillator of frequency 1 it has been found [8] that the optimum value of Q is around 0.3 where Nosé's estimate predicts $Q = 2$.

Choice of Q Based on Control of Average Thermostat Kinetic Energy

It can be shown [8], under the assumption of ergodicity, that the following holds,

$$\left\langle \frac{p_s^2}{Q} \right\rangle = k_B T , \tag{10}$$

where p_s is the thermostat conjugate momentum. For harmonic oscillators the value of this quantity is a good indicator of the optimum choice of Q [8].

The value of $\langle p_s^2/Q \rangle$ was examined for Q in the range 0.1 to 1000 for a model of alanine dipeptide, using the CHARMM molecular dynamics package. In this case the value of $\langle p_s^2/Q \rangle$ was found to be close to $k_B T$ for all values of Q tested, over a sufficiently long integration time, indicating that this method fails for complex systems of biomolecules.

Choice of Q Based on Kinetic Energy Distribution

In [9] it was proposed that the optimum value of Q could be obtained by comparing the resulting total energy distribution with the expected canonical distribution. For crystalline Aluminum it was found that a broad range of Q, 100 to 10000, gave good convergence to the correct distribution and it was concluded that, for this model, the value of Q was not critical.

Applying the same method to the alanine dipeptide produced similar results with Q in the range 0.1 to 1000 giving good total energy distributions. In this case we know that, for most of these values of Q, good sampling is not obtained when considering the average kinetic energy of subsystems.

3.3 A New Approach to Nosé Masses Based on Minimization of $\langle s \rangle$

We now propose a new scheme for choosing thermostat masses. This new method, which has produced the promising results reported below, will be examined in fuller detail in a related publication. It has been shown [8] that for systems of harmonic oscillators the average value of the thermostatting variable s, when sampling from the canonical ensemble, is

$$\langle s \rangle_c = \exp \left(\frac{H_0}{N_f k_B T} \right) \left(\frac{N_f}{N_f + 1} \right)^{\frac{2N_f+1}{2}}. \tag{11}$$

In the limit of large N_f, this is equivalent to

$$\langle s \rangle_c = \exp \left(\frac{H_0}{N_f k_B T} - 1 \right), \tag{12}$$

and this is found to be a good approximation for all N_f.

We can rearrange (3), given that H_0 is chosen such that $H_{NP} = 0$, as follows:

$$\ln s = \left(\frac{H_0 - H\left(q, \frac{p}{s}\right) - \frac{p_s^2}{2Q}}{N_f k_B T} \right). \tag{13}$$

As $Q \to \infty$ we expect that $\langle p_s^2/Q \rangle \to 0$ and $s \to \langle s \rangle$, and hence $\langle \ln s \rangle \to \ln \langle s \rangle$. We also note that, for harmonic oscillators, the following holds:

$$\langle KE \rangle = \langle PE \rangle, \tag{14}$$

where KE is the kinetic energy and PE is the potential energy. Taking averages of (13) and rearranging then gives,

$$\langle s \rangle_{Qlim} = \exp\left(\frac{H_0}{N_f k_B T} - 1\right). \tag{15}$$

Since (12) is the same as (15) the average value of s in the limit of large Q is the same as the average value of s when sampling from the canonical ensemble for systems of harmonic oscillators.

It is of interest to examine the average value of s as we decrease Q and the system is somewhere between these two regimes. We can highlight the difference between the two regimes by taking averages in (13). For the canonical ensemble we have,

$$\langle \ln s \rangle_c = \left(\frac{H_0 - \langle H\left(q, \frac{p}{s}\right)\rangle - \langle \frac{p_s^2}{2Q}\rangle}{N_f k_B T}\right),$$

$$= \left(\frac{H_0 - (N_f + \frac{1}{2})k_B T}{N_f k_B T}\right). \tag{16}$$

For the limit of large Q,

$$\langle \ln s \rangle_{Qlim} = \left(\frac{H_0 - N_f k_B T}{N_f k_B T}\right).$$

We note that that, although $\langle s \rangle_c = \langle s \rangle_{Qlim}$, we have $\langle \ln s \rangle_c < \langle \ln s \rangle_{Qlim}$. This is due to the probability distribution of s, which in the canonical ensemble is log-normal (in the limit of $N_f \to \infty$) giving rise to the difference in the $\ln s$ averages and funding the additional energy required by the $\langle p_s^2/2Q \rangle$ term.

If we let Q_c be the optimum value of Q for canonical sampling then our intermediate regime occurs when $\infty > Q > Q_c$, and here we expect that $\langle p_s^2/Q \rangle > 0$. Since the system is not sampling from the canonical ensemble s will not have the required log-normal distribution and $\langle \ln s \rangle$ will not provide the additional energy for $\langle p_s^2/2Q \rangle$. If we assume that, away from equilibrium, $\langle \ln s \rangle_* = \langle \ln s \rangle_{Qlim}$, then

$$\langle V(q) \rangle_* = \langle V(q) \rangle_{Qlim} - \langle p_s^2/2Q \rangle_*, \tag{17}$$

where the $*$ subscript represents the averages in the intermediate regime.

Reducing the average potential energy is harmonic oscillators is equivalent to reducing the average kinetic energy from (14) and hence the average value of s will have to increase to accommodate this (the average kinetic energy in the re-scaled momenta is fixed due to the feedback loop, but the reduction can be seen in the original momenta). From this we would expect that the average value of s would have minima where the system is sampling from the canonical ensemble and in the limit of large Q. Experiments with a harmonic oscillator show that this is indeed the case.

Although some of the assumptions above do not hold for typical molecular dynamics models, the critical requirement that s has a log-normal probability distribution is nonetheless verified in practice. From this we would expect that the average of

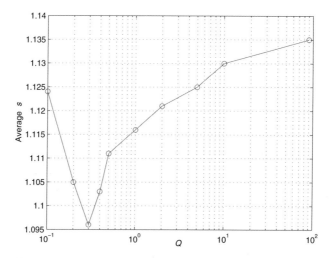

Figure 4. Average s against Q showing minimum at $Q_c = 0.3$

s will be minimized when sampling from the canonical ensemble. To verify this $\langle s \rangle$ was measured for various Q for a simulation of an alanine dipeptide molecule in the popular CHARMM package with the results in Fig. 4. The figure shows a well defined minimum, and suggests that the optimal value is around $Q = 0.3$. Simulations of the alanine dipeptide model were conducted at 300K with $Q = 0.3$. Average energies for hydrogen and heavy atoms are shown in Fig. 5. Although exact equipartition does not occur in this experiment, the results are the best obtained from numerous Nosé-Poincaré simulations.

3.4 Preliminary Experiments with RMT and Nosé-Hoover Chains

The results we describe here are obtained in all atom CHARMM simulations [1, 2] of the alanine dipeptide in vacuum, with nonbonded terms computed without cutoffs. The starting structures were equilibrated with a 50 ps heating cycle, followed by 200ps of equilibration at 300K. The Verlet, Nosé-Hoover and Nosé-Hoover Chain trajectories were generated with a 1 fs timestep, the accepted standard for biomolecules. We used a somewhat conservative timestep of 0.02 fs for RMT. To ensure that the comparison between RMT and Nosé-Hoover methods is valid, we have verified that the the smaller timesteps used in RMT did not improve the thermostatting properties of the Nosé-Hoover methods.

In Sect. 3.3, Fig. 5, we saw that, for the correct choice of parameters, the Nosé-Poincaré method produces much better results than the Verlet or Nosé-Hoover method. With chains and RMT methods additional parameters need to be selected. Our methodology is to select the value of the first Nosé mass by the "minimum average thermostat variable" scheme. Our experience leads us to choose subsequent masses to be close to, but greater than, the previous mass according to the formula

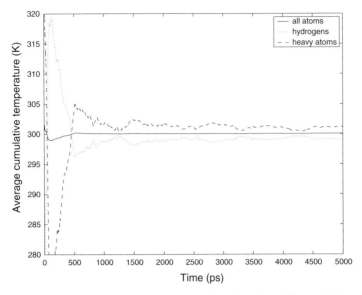

Figure 5. Average kinetic energy for subsystems of alanine dipeptide consisting of the hydrogen atoms and the heavy atoms for a 5ns simulation time with $Q = 0.3$ using the Nosé-Poincaré method

$$Q_{i+1} \approx \frac{3}{2}Q_i .\qquad(18)$$

For the RMT method the C_i were selected as $C_i = 1/16k_BT$ in accordance with (5). The results for the RMT method are seen in Fig. 6, where good results are obtained for the suggested choice of parameters.

We also implemented Nosé-Hoover chains in CHARMM, using an explicit integrator described by Jang and Voth [3]. The results for the Nosé-Hoover chains are seen in Fig. 7 where, even with optimum values for the Nosé masses, the thermalization is extremely slow. Results for many other values of the parameters were either similar or worse.

4 Conclusions

It is clear from the preceding experiments that the choice of the parameters applicable to Nosé methods is critical in obtaining good sampling. We observe that, even for methods which are expected to enhance the ergodicity, the range of parameter values which yield good results is small. Experiments with Nosé-Hoover chains indicate that equipartition does not occur far from the optimal choice of Nosé mass, and similar results are seen for RMT.

The timestep limitation which we have encountered illustrates the need for more careful numerical treatment of the thermostat variables in RMT. It is generally believed that solvated models do not display the equipartition problems reported here,

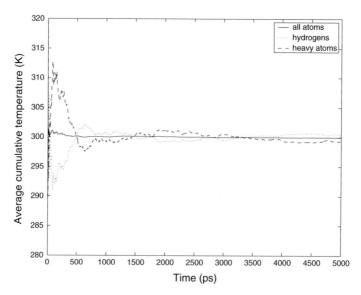

Figure 6. Cumulative average temperature, and temperature of subsystems computed by a 5ns trajectory for alanine dipeptide using the RMT method implemented in CHARMM, with optimum $Q = 0.3$

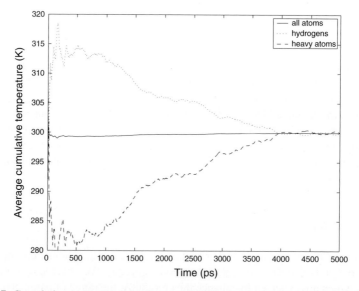

Figure 7. Cumulative average temperature, and temperature of subsystems computed by a 5ns trajectory for alanine dipeptide using Nosé-Hoover chains in CHARMM with optimum $Q = 0.3$. The system requires a long time (4ns) to reach equipartition

but initial experiments by the authors indicate otherwise. These issues will be addressed in a subsequent article.

Many methods for determining the optimum value of the thermostat mass in Nosé dynamics have been proposed. Our examination of these methods has shown that, at best, most provide a poor indicator of the optimum value in the setting of biomolecules. Further study of the method presented here based on the "minimum average thermostat value" will be undertaken to provide a better understanding of this technique in the molecular dynamics setting.

Acknowledgments

This work was undertaken during visits by EB to the Mathematical Modelling Centre at the University of Leicester, and by all authors to the Institute for Mathematics and its Applications at the University of Minnesota. EB acknowledges the NIH under grant 1 R15 GM065126-01A1 as well as the donors to the Petroleum Research Fund, administered by the ACS, for partial support of this project. BL acknowledges the support of the Leverhulme Foundation and EPSRC Grant GR-R24104. The authors gratefully acknowledge helpful discussions with Frederic Legoll (ENPC).

References

[1] B.R. Brooks, R.E. Bruccoleri, B.D. Olafson, D.J. States, S. Swaminathan and M. Karplus, CHARMM: A program for macromolecular energy, minimization, and dynamics calculations, J. Comp. Chem, 4 (1983), pp. 187–217

[2] A.D. MacKerell Jr., D. Bashford, M. Bellott, R.L. Dunbrack Jr., J. Evanseck, M.J. Field, S. Fischer, J. Gao, H. Guo, S. Ha, D. Joseph, L. Kuchnir, K. Kuczera, F.T.K. Lau, C. Mattos, S. Michnick, T. Ngo, D.T. Nguyen, B. Prodhom, W.E. Reiher III, B. Roux, M. Schlenkrich, J. Smith, R. Stote, J. Straub, M. Watanabe, J. Wiorkiewicz-Kuczera, D. Yin and M. Karplus, An all-atom empirical potential for molecular modeling and dynamics of proteins, J. Phys. Chem., 102 (1998), pp. 3586–3616

[3] S. Jang and G. Voth, Simple reversible molecular dynamics algorithms for Nosé-Hoover chain dynamics, J. Chem. Phys. 110: 3263 (1999)

[4] S.D. Bond and B.B. Laird and B.J. Leimkuhler, The Nosé-Poincaré method for constant temperature molecular dynamics, J. Comp. Phys., 151, 114, 1999

[5] S. Nosé, A molecular dynamics method for simulation in the canonical ensemble, Mol. Phys., 52, 255, 1984

[6] Yi Liu and Mark E. Tuckerman, Generalized Gaussian moment thermostatting: A new continuous dynamical approach to the canonical ensemble, J. Chem. Phys, 112, 1685-1700, 2000

[7] W. G. Hoover, Canonical dynamics: equilibrium phase-space distributions, Phys. Rev. A 31, 1695 (1985)

[8] B. J. Leimkuhler and C. R. Sweet, Hamiltonian formulation for recursive multiple thermostats in a common timescale, SIADS, 4, No. 1, pp. 187–216, 2005

[9] B.B. Laird and J.B. Sturgeon, Symplectic algorithm for constant-pressure molecular dynamics using a Nosé-Poincaré thermostat, J. Chem. Phys., 112, 3474, 2000

[10] Hernandez, E. Metric-tensor flexible-cell algorithm for isothermal-isobaric molecular dynamics simulations J. Chem. Phys., 115, 10282-10291, 2001

[11] Dahlberg, Laaksonen and Leimkuhler, in preparation

[12] G.J. Martyna and M.L. Klein and M. Tuckerman, Nosé-Hoover chains: The canonical ensemble via continuous dynamics, J. Chem. Phys., 97, 2635, 1992

[13] B.J. Leimkuhler and C.R. Sweet, The canonical ensemble via symplectic integrators using Nosé and Nosé-Poincaré chains, J. Chem. Phys., 121, 1, 2004

[14] B.B. Laird and B.J. Leimkuhler, A generalized dynamical thermostatting technique, Phys. Rev. E, 68, 16704, 2003

[15] S. Nosé, An improved symplectic integrator for Nosé-Poincaré thermostat, J. Phys. Soc. Jpn, 70, 75, 2001

[16] C.P. Dettmann, Hamiltonian for a restricted isoenergetic thermostat, Phys. Rev. E, 60, 7576, 1999

[17] C.R. Sweet, Hamiltonian thermostatting techniques for molecular dynamics simulation, Ph.D. Dissertation, University of Leicester, 2004

[18] B. Laird, private communication.

The Targeted Shadowing Hybrid Monte Carlo (TSHMC) Method

Elena Akhmatskaya[1] and Sebastian Reich[2]

[1] Fujitsu Laboratories of Europe Ltd (FLE), Hayes Park Central, Hayes End Road, Hayes UB4 8FE, United Kingdom,
`Elena.Akhmatskaya@uk.fujitsu.com`

[2] Institut für Mathematik, Universität Potsdam, Postfach 60 15 53, D-14415 Potsdam, Germany
`sreich@math.uni-potsdam.de`

Abstract. Following IZAGUIRRE & HAMPTON [14], HOROWITZ [13], and ATTARD [1] as well as work of one of the authors on dissipative particle dynamics [4] and modified equations [23], we suggest a modified Metropolis criterion and a more flexible momentum update to improve the acceptance rate and the flexibility of the thermal coupling in standard hybrid Monte Carlo simulations.

1 Introduction

The starting point of any (classical) molecular simulation is a system of N particles, which interact through both long and short range forces through Newton's second law:

$$\dot{\mathbf{r}}_i = \mathbf{p}_i/m_i, \tag{1}$$

$$\dot{\mathbf{p}}_i = \mathbf{F}_i, \qquad i = 1, \ldots, N \tag{2}$$

where m_i is the mass of particle i with position vector $\mathbf{r}_i = (x_i, y_i, z_i)^T \in \mathbb{R}^3$ and momentum $\mathbf{p}_i = m_i \dot{\mathbf{r}}_i \in \mathbb{R}^3$. It is also assumed that the force acting on the ith particle is conservative, i.e., there is a potential energy function $V(\mathbf{r}_1, \ldots, \mathbf{r}_N)$ such that

$$\mathbf{F}_i = -\nabla_{\mathbf{r}_i} V.$$

Molecular dynamics can be performed under various ensembles. The most popular ensembles are (i) constant number of particles, constant energy, and constant volume (NVE) ensemble, (ii) constant number of particles, constant pressure, and constant temperature (NPT) ensemble, and (iii) constant number of particles, constant volume, and constant temperature (NVT) ensemble. For the purpose of this paper we will restrict ourselves to NVT ensemble simulations using the hybrid Monte

Carlo (HMC) method of DUANE, KENNEDY, PENDLETON & ROWETH [6]. However, NVE simulation techniques play an important role in the design of the HMC
method.

The main contribution of this paper is the development of a new HMC method
that allows for (i) a modified momentum update and (ii) leads to a much improved
acceptance rate. That part of the report relies on theoretical results for symplectic
time-stepping methods (see, for example, BENETTIN & GIORGILLI [2]), the recently
introduced shadow hybrid Monte Carlo (SHMC) method of IZAGUIRRE & HAMP-
TON [14], the modified Monte-Carlo method suggested by HOROWITZ [13] (see also
ATTARD [1]), and work of one of the authors on dissipative particle dynamics (DPD)
(see, for example, [4]) and modified equations [23].

2 Description of the Basic HMC Method

The HMC method offers an elegant and efficient way to turn an NVE simulation into
a sampling method for the NVT ensemble. Let us assume that we have a numerical
method that time-steps the equations (1)-(2). We first randomly sample all momenta
$\mathbf{p}_i$ according to the Boltzmann distribution

$$\rho_{\text{Boltzmann}} \sim e^{-\beta \|\mathbf{p}_i\|^2 / 2m_i}.$$

Here $\beta = 1/k_B T$ denotes the inverse temperature.

We next apply a MD simulation for a fixed number of time-steps and denote the
resulting update from the current positions and momenta $(\mathbf{r}_i, \mathbf{p}_i)$, $i = 1, \ldots, N$, to a
new configuration $(\mathbf{r}'_i, \mathbf{p}'_i)$, $i = 1, \ldots, N$, by

$$\Psi : (\mathbf{r}, \mathbf{p}) \to (\mathbf{r}', \mathbf{p}'),$$

where $\mathbf{r}$ is the collection of the N particle position vectors $\mathbf{r}_i$ and $\mathbf{p}$ is the associated
momentum vector.

The new configuration is accepted with a probability of

$$\min \left(1, \exp \left[-\beta \left\{(\mathbf{p}')^T \mathbf{M}^{-1} \mathbf{p}'/2 + V(\mathbf{r}') - \mathbf{p}^T \mathbf{M}^{-1} \mathbf{p}/2 - V(\mathbf{r})\right\}\right]\right). \qquad (3)$$

The sequence of steps:

1. randomly re-sample momenta from the Boltzmann distribution,
2. generate a new configuration $(\mathbf{r}', \mathbf{p}')$,
3. accept the new configuration according to the Metropolis criterion (3),

is now repeated as often as necessary to sample properly from an NVT ensemble. As
shown by MEHLIG, HEERMANN & FORREST [21], it is essential for the method to
work properly that the map Ψ is time-reversible and volume preserving.

A number of modifications to Step 1 and Step 3 have been suggested. In particular, IZAGUIRRE & HAMPTON [14] implemented a modified Metropolis criterion (3) to further increase the acceptance rate of HMC. Their approach relies on

recent advances for symplectic integration methods and their theoretical understanding. Furthermore, HOROWITZ [13] (see also [1]) pointed out that it is not necessary to completely re-sample the momenta under Step 1 of a HMC scheme. However, the scheme is not widely used due to its frequent reversal of momenta in case of non-negligible rejection rates.

Combining these two specific modifications with work of one of the authors (see, e.g., [4]) on dissipative particle dynamics (DPD) [7, 11], we will derive yet another class of HMC methods, which we will describe in §4 and which we wish to promote for further use in NVT simulations.

The results of the paper can be easily extended to Hamiltonian systems with holonomic constraints by using the SHAKE extension [26] of the standard Störmer-Verlet time-stepping method [9, 18].

3 Störmer-Verlet Time-Stepping Method and Modified Hamiltonian

The most widely used numerical method for MD of type (1)-(2) is the Störmer-Verlet/leapfrog method, which is written here in the velocity/momentum formulation:

$$\mathbf{p}^{n+1/2} = \mathbf{p}^n - \frac{\Delta t}{2}\nabla_{\mathbf{r}}V(\mathbf{r}^n), \tag{4}$$

$$\mathbf{r}^{n+1} = \mathbf{r}^n + \Delta t\mathbf{M}^{-1}\mathbf{p}^{n+1/2}, \tag{5}$$

$$\mathbf{p}^{n+1} = \mathbf{p}^{n+1/2} - \frac{\Delta t}{2}\nabla_{\mathbf{r}}V(\mathbf{r}^{n+1}), \tag{6}$$

where Δt is the step-size.

The popularity of the Störmer-Verlet method is due to its simplicity and its remarkable conservation properties. We next outline a particular implication of its conservation of symplecticity [9, 18]. Namely, one can find a time-dependent Hamiltonian function $\tilde{\mathcal{H}}(\mathbf{r}, \mathbf{p}, 2\pi t/\Delta t)$, which is 2π-periodic in its third argument such that the solution of

$$\dot{\mathbf{r}} = +\nabla_{\mathbf{p}}\tilde{\mathcal{H}}(\mathbf{r}, \mathbf{p}, 2\pi t/\Delta t),$$
$$\dot{\mathbf{p}} = -\nabla_{\mathbf{r}}\tilde{\mathcal{H}}(\mathbf{r}, \mathbf{p}, 2\pi t/\Delta t),$$

with initial conditions $\mathbf{r}(0) = \mathbf{r}^n$ and $\mathbf{p}(0) = \mathbf{p}^n$ is exactly equivalent to $(\mathbf{r}^{n+1}, \mathbf{p}^{n+1})$ at $t = \Delta t$. (See the papers by KUKSIN & PÖSCHEL [17] and MOAN [22] for the mathematical details.)

This statement is not entirely satisfactory as it is well known that energy is *not* conserved for time-dependent Hamiltonian problems. However, as first pointed out by NEISHTADT [24], the time-dependence in $\tilde{\mathcal{H}}$ averages itself out up to negligible terms of size $\mathcal{O}(e^{-c/\Delta t})$ for sufficiently small step-sizes Δt (here $c > 0$ is a constant which depends on the particular problem). Hence the Störmer-Verlet method is the nearly exact solution [2] of a Hamiltonian problem with *time-independent* Hamiltonian $\hat{\mathcal{H}}_{\Delta t}(\mathbf{r}, \mathbf{p})$. This time-independent Hamiltonian possesses an asymptotic expansion in the step-size Δt of the form

$$\hat{\mathcal{H}}_{\Delta t} = \mathcal{H} + \Delta t^2 \delta \mathcal{H}_2 + \Delta t^4 \delta \mathcal{H}_4 + \Delta t^6 \delta \mathcal{H}_6 + \ldots, \tag{7}$$

with

$$\mathcal{H}_2 = \frac{1}{12} \mathbf{p}^T \mathbf{M}^{-1} [D_{\mathbf{rr}} V(\mathbf{r})] \mathbf{M}^{-1} \mathbf{p} + \frac{1}{24} [\nabla_{\mathbf{r}} V(\mathbf{r})]^T \mathbf{M}^{-1} \nabla_{\mathbf{r}} V(\mathbf{r}),$$

where $D_{\mathbf{rr}} V(\mathbf{r})$ denotes the Hessian matrix of the potential energy V. Expressions for the higher-order correction terms $\delta \mathcal{H}_i$, $i = 4, 6, \ldots$, can be found using the Baker-Campbell-Hausdorff (BCH) formula (see, e.g., [9, 18]).

A practical algorithm for assessing energy conservation with respect to a modified Hamiltonian has been proposed by SKEEL & HARDY [27]. We will follow the modified equation approach of MOORE & REICH [23], which is particularly suited to the Störmer-Verlet method.

The fact that the modified energy $\hat{\mathcal{H}}_{\Delta t}$ is essentially preserved exactly under the Störmer-Verlet method has implications for HMC simulations. Namely, the quasi-exact conservation of $\hat{\mathcal{H}}_{\Delta t}$ under the Störmer-Verlet method allows one to accept almost all candidate moves with regard to a modified canonical ensemble (see IZAGUIRRE & HAMPTON [14]). We outline the details in the following section.

4 A New Method: Targeted Shadowing Hybrid Monte Carlo (TSHMC)

A high acceptance rate is a desirable property of any Monte Carlo scheme. In fact, one of the reasons for the introduction of the HMC method was its vastly superior acceptance rate over standard Monte Carlo methods. However the acceptance rate of HMC degrades with the size of the simulated molecular system. Furthermore, in light of the modified Hamiltonian, it would appear that essentially no rejections are necessary at all for a symplectic integration method such as Störmer-Verlet. In fact, that is indeed the case up to a small rejection rate caused by the truncation of (7) after a finite number of terms. A practical HMC algorithm based on modified Hamiltonians was first proposed by IZAGUIRRE & HAMPTON [14]. We will describe below a variant of their SHMC method with two important modifications:

(i) a simplified evaluation of the modified energy (Hamiltonian),
(ii) a modified and more flexible momentum update.

Note that time averages need to include the factor

$$w^m = e^{\beta(\hat{E}_{\Delta t}^m - E^m)}, \tag{8}$$

where E^m is the value of the given energy after completion of the mth SHMC/TSHMC step and $\hat{E}^m$ is the modified energy, respectively, i.e., averages of an observable A are computed according to the formula:

$$\langle A \rangle = \frac{\frac{1}{M} \sum_{m=1}^{M} A(\mathbf{r}^m, \mathbf{p}^m) w^m}{\frac{1}{M} \sum_{m=1}^{M} w^m}.$$

This is a standard re-weighting procedure for simulations in modified ensembles.

It is essential for the SHMC/TSHMC method that the modified energy can be evaluated inexpensively. This rules out a direct evaluation according to the asymptotic expansion (7), since it would require the evaluation of higher-order derivatives of the potential energy function V. However, it turns out that one can approximate the modified energy to any order *without* any evaluation of higher-order derivatives of V. We will outline the basic idea in the following subsection.

4.1 Evaluation of the Modified Energy

We now describe the approach of MOORE & REICH [23] for approximating modified Hamiltonians. The Störmer-Verlet method is first expressed in its positions only leapfrog formulation

$$\mathbf{M} \frac{\mathbf{r}^{n+1} - 2\mathbf{r}^n + \mathbf{r}^{n-1}}{\Delta t^2} = -\nabla_{\mathbf{r}} V(\mathbf{r}^n).$$

We next assume that there is a smooth function $\mathbf{r}(t)$ such that $\mathbf{r}(t_n) = \mathbf{r}^n$ for all time-steps t_n of interest. Taylor expansion of $\mathbf{r}(t)$ about t_n readily yields the following well-known expression for the local truncation error formula for the second-order central difference approximation:

$$\frac{\mathbf{r}^{n+1} - 2\mathbf{r}^n + \mathbf{r}^{n-1}}{\Delta t^2} = \ddot{\mathbf{r}}(t_n) + \frac{\Delta t^2}{12} \mathbf{r}^{(4)}(t_n) + \mathcal{O}(\Delta t^4).$$

Since, by assumption, $\mathbf{r}(t_n) = \mathbf{r}^n$, we find that the smooth function $\mathbf{r}(t)$ has to satisfy the (in second-order) modified equation

$$\mathbf{M}\ddot{\mathbf{r}} + \frac{\Delta t^2}{12} \mathbf{M}\mathbf{r}^{(4)} = -\nabla_{\mathbf{r}} V(\mathbf{r}) \tag{9}$$

up to terms of order Δt^4. Higher-order modified equations can be easily found. We restrict the discussion to the second-order modification for simplicity.

We now multiply the whole equation (9) by $\dot{\mathbf{r}}^T$ from the left to obtain, after a rearrangement of term, the scalar equation

$$\frac{d}{dt} \left[\frac{1}{2} \dot{\mathbf{r}}^T \mathbf{M}\dot{\mathbf{r}} + V(\mathbf{r}) \right] = -\frac{\Delta t^2}{12} \dot{\mathbf{r}}^T \mathbf{M}\mathbf{r}^{(4)}.$$

A remarkable observation is that the term on the right hand side of the last equation can also be written as a total time derivative, i.e.

$$\frac{\Delta t^2}{12} \dot{\mathbf{r}}^T \mathbf{M}\mathbf{r}^{(4)} = \frac{\Delta t^2}{12} \frac{d}{dt} \left[\dot{\mathbf{r}}^T \mathbf{M}\mathbf{r}^{(3)} - \frac{1}{2} \ddot{\mathbf{r}}^T \mathbf{M}\ddot{\mathbf{r}} \right].$$

We may conclude that the modified energy

$$\hat{E}_{\Delta t} = \frac{1}{2}\dot{\mathbf{r}}^T\mathbf{M}\dot{\mathbf{r}} + V(\mathbf{r}) + \frac{\Delta t^2}{12}\left[\dot{\mathbf{r}}^T\mathbf{M}\mathbf{r}^{(3)} - \frac{1}{2}\ddot{\mathbf{r}}^T\mathbf{M}\ddot{\mathbf{r}}\right] \tag{10}$$

is preserved to fourth-order in the step-size Δt along numerical solutions computed by the Störmer-Verlet/leapfrog method.

For a numerical verification of the modified energy $\hat{E}_{\Delta t}$, we need to approximate the time derivatives by finite difference approximations of sufficiently high order. For the given modified energy (10), this requires that the first-order time derivative in $\dot{\mathbf{r}}^T\mathbf{M}\dot{\mathbf{r}}/2$ needs to be discretized to fourth-order in Δt while the remaining time derivatives need only be approximated to second-order (due to the prefactor of $\Delta t^2/12$). We use the fourth-order finite difference formula

$$\dot{\mathbf{r}}(t_n) \approx \frac{\mathbf{r}^{n+1} - \mathbf{r}^{n-1}}{2\Delta t} - \frac{\mathbf{r}^{n+2} - 2\mathbf{r}^{n+1} + \mathbf{r}^{n-1} - \mathbf{r}^{n-2}}{12\Delta t}$$

as well as the second-order formulas

$$\ddot{\mathbf{r}}(t_n) \approx \frac{\mathbf{r}^{n+1} - 2\mathbf{r}^n + \mathbf{r}^{n-1}}{\Delta t^2}$$

and

$$\mathbf{r}^{(3)}(t_n) \approx \frac{\mathbf{r}^{n+2} - 2\mathbf{r}^{n+1} + \mathbf{r}^{n-1} - \mathbf{r}^{n-2}}{12\Delta t^3}.$$

It follows that we require the five coordinate approximations $\{\mathbf{r}^{n+i}\}_{i=-2,-1,0,1,2}$ to evaluate the modified energy (10) at time t_n to the desired fourth-order accuracy. Given $\mathbf{r}^0$ and $\dot{\mathbf{r}}^0$ at the start of a MD simulation, this implies the additional computation of "past" positions $\mathbf{r}^{-1}$ and $\mathbf{r}^{-2}$. Making use of the time-reversibility of the Newtonian equations of motion, those positions can be computed by integrating the equations forward in time over two time-steps with initial values $\mathbf{r}^0$ and $-\dot{\mathbf{r}}^0$. In summary, given any pair of initial conditions $(\mathbf{r}^0, \dot{\mathbf{r}}^0)$, the Störmer-Verlet/leapfrog method assigns a modified energy $\hat{E}_{\Delta t}^0$ by the procedure just described. For all later times t_n, the modified energy $\hat{E}_{\Delta t}^n$ is computed "on the fly". (At the end of the simulation interval we have to take two additional time-steps.)

The standard Metropolis criterion (3) is now replaced by

$$\min\left(1, \exp\left[-\beta\{\hat{E}_{\Delta t}(\mathbf{r}', \mathbf{v}') - \hat{E}_{\Delta t}(\mathbf{r}, \mathbf{v})\}\right]\right), \tag{11}$$

where $(\mathbf{r}, \mathbf{v}) = (\mathbf{r}^0, \mathbf{v}^0)$ and $(\mathbf{r}', \mathbf{v}')$ is the numerical solution obtained at a given time t_N (i.e., after N Störmer-Verlet integration steps with step-size Δt). Here we used the notation $\mathbf{v} = \dot{\mathbf{r}} = \mathbf{M}^{-1}\mathbf{p}$, $\mathbf{v}' = \dot{\mathbf{r}}' = \mathbf{M}^{-1}\mathbf{p}'$. If the proposed move is rejected, then the simulation is continued with $\mathbf{r}$ and negated momenta $-\mathbf{p}$ [13].

Following the HMC analysis of MEHLIG, HEERMANN & FORREST and HOROWITZ [13, 21], it follows that the Störmer-Verlet method combined with the Metropolis criterion (11) satisfies detailed balance and preserves the canonical ensemble for the modified energy $\hat{E}_{\Delta t}$.

The negation of the momenta in case of rejection is needed to satisfy detailed balance. Contrary to the guided Monte Carlo method [13] such a reversal of momenta is very infrequent under the TSHMC method and should not impact on the sampling efficiency of the method. At the same time, the same argument leads to the expectation that ignoring momentum negation after a rejected step will not alter the simulation results significantly. In fact, the simulations carried out in §5 used no such momentum negation.

4.2 Alternative Momentum Updates

HOROWITZ pointed out in [13] that it is not necessary to completely re-sample the momenta under Step 1 of a standard HMC scheme. Instead one takes the set of given momenta $\mathbf{p}$ and modifies it by a vector $\boldsymbol{\xi} \in \mathbb{R}^{3N}$ to obtain a new set given by

$$\mathbf{p}' = \mathbf{p} + \sigma \boldsymbol{\xi}. \tag{12}$$

Here $\sigma > 0$ is a free parameter and $\boldsymbol{\xi}$ is sampled from the Boltzmann distribution $\rho_{\text{Boltzmann}}$, i.e., $\boldsymbol{\xi}$ is a vector of independent Gaussian random variables with mean zero and variance $k_B T$. Smaller values of σ lead to smaller perturbations in the momenta. The new set of momentum vectors $\mathbf{p}'$ is accepted with a probability of

$$\min \left(1, \exp \left[-\beta \left\{ (\mathbf{p}')^T \mathbf{M}^{-1} \mathbf{p}'/2 - \mathbf{p}^T \mathbf{M}^{-1} \mathbf{p}/2 \right\} \right] \right).$$

This momentum update replaces the above Step 1 in a standard HMC method. Step 2 is then started from the given position vector $\mathbf{q}$ and the accepted momentum vector, which we denote again by $\mathbf{p}$.

Following COTTER & REICH [4], we suggest a further generalization of the update (12):

$$\mathbf{p}' = \mathbf{p} + \sigma \sum_{k=1}^{K} \nabla_{\mathbf{r}} h_k(\mathbf{r}) \xi_k, \tag{13}$$

where σ and $\boldsymbol{\xi} = (\xi_1, \dots, \xi_K)^T \in \mathbb{R}^K$ are defined as before, and the functions $h_k(\mathbf{r})$, $k = 1, \dots, K$, can be chosen quite arbitrarily. The particular choice

$$h_k(\mathbf{r}) = \phi(r_{ij}), \qquad r_{ij} = \|\mathbf{r}_i - \mathbf{r}_j\|,$$

$k = 1, \dots, (N-1)N/2$, ϕ a given function of inter-particle distances r_{ij}, transforms (13) into an update very similar to what is used in dissipative particle dynamics (DPD) [7, 11] (see also MA & IZAGUIRRE [20] for another application to MD). One attractive feature of such an update is its conservation of linear and angular momenta:

$$\sum_{i=1}^{N} \mathbf{p}_i = \sum_{i=1}^{N} \mathbf{p}'_i, \qquad \sum_{i=1}^{N} \mathbf{r}_i \times \mathbf{p}_i = \sum_{i=1}^{N} \mathbf{r}'_i \times \mathbf{p}'_i.$$

One can also set $K = 3N$ and

$$h_i(\mathbf{r}) = x_i, \qquad h_{i+N}(\mathbf{r}) = y_i, \qquad h_{i+2N}(\mathbf{r}) = z_i,$$

$i = 1, \ldots, N$, in (13), which leads back to (12). Many more choices are feasible and of potential use. For example, one could apply the update only to particles near the boundary of the simulation domain or to certain subunits of the molecular system. Again, the strength of the coupling is controlled by the parameter σ.

Given a new set of momenta $\mathbf{p}'$, we need to evaluate the corresponding modified energy $\hat{E}_{\Delta t}(\mathbf{r}, \mathbf{v}')$, $\mathbf{v}' = \mathbf{M}^{-1}\mathbf{p}'$. This step requires time-stepping the equations of motion two steps forward and backward in time and, hence, two additional force field evaluations are needed. We then apply (11) in its slightly modified form:

$$\min\left(1, \exp\left[-\beta\{\hat{E}_{\Delta t}(\mathbf{r}, \mathbf{v}') - \hat{E}_{\Delta t}(\mathbf{r}, \mathbf{v})\}\right]\right). \qquad (14)$$

It is again easily verified that the momentum update (13) combined with the Metropolis criterion (14) satisfies detailed balance and preserves the canonical density corresponding to $\hat{E}_{\Delta t}$. Hence we may conclude that the TSHMC method (without reweighting) constitutes a Markov chain Monte Carlo method which samples from the canonical density $\rho_{\text{canonical}} \sim \exp(-\beta\hat{E}_{\Delta t})$.

The targeted shadowing hybrid Monte Carlo (TSHMC) method may now be summarized as follows. One alternates between constant energy MD steps, which are accepted according to the Metropolis criterion (11) with modified energy (10), and a partial DPD-type velocity resampling according to (13) and Metropolis criterion (14). The concatenation of two Markov processes with identical invariant probability density functions produces another Markov process with the same invariant probability density function. It should be noted that the computation of averages in the original ensemble requires the weight factors (8).

5 Computer Experiment

The suggested approach was implemented and tested on a Linux cluster for an alanin side chain analog. This system contains 900 water molecules with a total of 2705 atoms. We apply the OPLS-AA [16] forcefield parameters for alanin and the original TIP3P model [15] for water molecules. Electrostatic interactions were treated using a particle-mesh Ewald summation (PME) method [5, 8] and periodic boundary conditions for a truncated octahedron box were applied.

We performed the simulation using three different techniques: TSHMC, standard HMC and traditional MD. All three approaches used GROMACS 3.2.1 [19] to perform the necessary molecular dynamics simulation steps. The system was initially equilibrated for 200 ps and then run for 1 ns at a temperature of 298 K. In the traditional MD approach the temperature was coupled to a heat bath of 298 K with a coupling time constant of 0.1 ps using the Nosé-Hoover procedure [12, 25]. Integration of the equations of motion was performed using the Störmer-Verlet algorithm and all bonds were constrained using the LINCS [10] algorithm with a distance constraint of 10e-6 Å. We used a step-size of $\Delta t = 2$ fs in the traditional MD method.

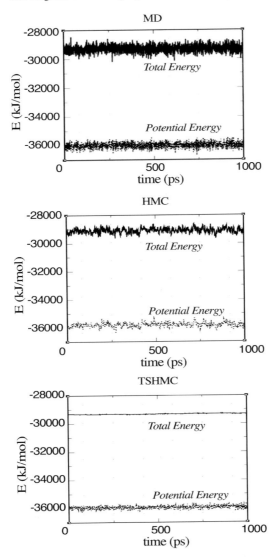

Figure 1. Total and potential energies for MD, HMC, and TSHMC

To investigate the effect of step-size and length of MD simulations per HMC step on sampling efficiency of HMC and TSHMC, we ran both algorithms using three different step-sizes ($\Delta t = 1$ fs, 2 fs, 2.5 fs) and two different MD simulation lengths (150 and 1000 time steps, respectively). The fourth-order modified Hamiltonian was used in the TSHMC approach. The parameter σ in (12) was set to 0.1, which led to a rejection rate of about 40% in the momentum update within the TSHMC sim- ulation. Total and potential energies for the three simulation approaches are com- pared in Fig. 1. The average energies are in good agreement within the simulation

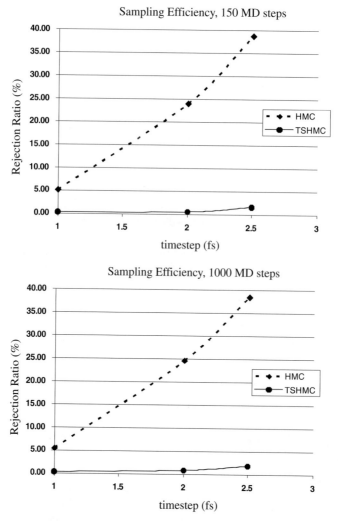

Figure 2. Sampling efficiency of HMC and TSHMC as a function of step-size and number of MD steps (per HMC step)

accuracy. As expected, TSHMC demonstrates much smoother energy profiles than the other two simulation methods due to higher-order energy conservation in the modified Hamiltonian. The magnitudes of energy fluctuations in both HMC approaches are significantly smaller than those observed for the standard constant temperature MD simulations. The sampling efficiency of HMC and TSHMC as a function of step-size and number of MD steps (per HMC step) is presented in Fig. 2. Observed rejection rates are almost identical for both choices of the MD simulation length, which suggests that the MD simulation length per HMC step does not affect much the acceptance/rejection rate at least for this particular model problem. On the contrary, the step size Δt has a pronounced impact on the number of accepted config-urations in HMC and TSHMC. The number of rejected configurations increases with increasing time steps in both HMC methods. However, the increase is visibly faster in standard HMC. In TSHMC the rejection rate is still rather low, around 2% for the largest tested step-size Δt. This is 20 times less than the rejection rate observed in the corresponding standard HMC simulation.

6 Conclusion

We have suggested and implemented a modified HMC algorithm, which, in its most general form, can be viewed as a thermodynamically consistent implementation of dissipative particle dynamics (DPD). (See, for example, [3] for inconsistency prob-lems with known time-stepping method for DPD.) In addtion, we have also inves-tigated the effect of modified energies to increase the acceptance rate of standard HMC. The results for our particular formulation of the modified energies are in line with the observations of IZAGUIRRE & HAMPTON [14] for the SHMC method. We re-emphasize that the TSHMC method is different from SHMC in two regards:

(i) a simplified evaluation of modified energies for the Störmer-Verlet method,
(ii) a thermodynamically consistent DPD-type (partial) momentum update.

Acknowledgment

We would like to thank Radford Neal for pointing us to the reference [13].

References

[1] P. Attard.   Stochastic molecular dynamics: A combined Monte-Carlo and molecular dynamics technique for isothermal simulations. *J. Chem. Phys.*, 116:9616–9619, 2002.

[2] G. Benettin and A. Giorgilli. On the Hamiltonian interpolation of near to the identity symplectic mappings with application to symplectic integration algo-rithms. *J. Stat. Phys.*, 74:1117–1143, 1994.

[3] G. Besold, I. Vattulainen, M. Karttunen, and J.M. Polson. Towards better in-tegrators for dissipative particle dynamics. *Phys. Rev. E*, 62:R7611–R7614, 2000.

[4] C.J. Cotter and S. Reich. An extended dissipative particle dynamics model. *Europhys. Lett.*, 64:723–729, 2003.

[5] T. Darden, D. York, and L. Pedersen. Particle mesh Ewald: an N - log(N) method for Ewald sums in large systems. *J. Comput. Phys.*, 98:10089–10092, 1993.

[6] S. Duane, A.D. Kennedy, B.J. Pendleton, and D. Roweth. Hybrid Monte-Carlo. *Phys. Lett. B*, 195:216–222, 1987.

[7] P. Español and P.B. Warren. Statistical mechanics of dissipative particle dynamics. *Europhys. Lett.*, 30:191–196, 1995.

[8] U. Essmann, L. Perera, M. L. Berkowitz, T. Darden, H. Lee, and L. G. Pedersen. A smooth particle mesh Ewald potential. *J. Comput. Phys.*, 103:8577–8592, 1995.

[9] E. Hairer, Ch. Lubich, and G. Wanner. *Geometric Numerical Integration.* Springer-Verlag, Berlin Heidelberg, 2002.

[10] B. Hess, H. Bekker, H.J.C. Berendsen, and J.G.E.M. Fraaije. LINCS: A linear constraint solver for molecular dynamics simulations. *J. Comput. Chem.*, 18:1463–1472, 1997.

[11] P.J. Hoogerbrugge and J.M.V.A. Koelman. Simulating microscopic hydrodynamic phenomena with dissipative particle dynamics. *Europhys. Lett*, 19:155–160, 1992.

[12] W.G. Hoover. Canonical dynamics: Equilibrium phase-space distributions. *Phys. Rev. A*, 31:1695–1697, 1985.

[13] A.M. Horowitz. A generalized guided Monte-Carlo algorithms. *Phys. Lett. B*, 268:247–252, 1991.

[14] J.A. Izaguirre and S.S. Hampton. Shadow Hybrid Monte Carlo: An efficient propagator in phase space of macromolecules. *J. Comput. Phys.*, 200:581–604, 2004.

[15] W. L. Jorgensen, J. Chandrasekhar, J. D. Madura, R. W. Impey, and M. L. Klein. Comparison of simple potential functions for simulating liquid water. *J. Comput. Phys.*, 79:926–935, 1983.

[16] W. L. Jorgensen, D. S. Maxwell, and J. Tirado-Rives. Development and testing of the OPLS all-atom force field on conformational energetics and properties of organic liquids. *J. Am. Chem. Soc.*, 118:11225–11236, 1996.

[17] S. Kuksin and J. Pöschel. On the inclusion of analytic symplectic maps in analytic Hamiltonian flows and its applications. In S. Kuksin, V. Lazutkin, and J. Pöschel, editors, *Seminar on Dynamical Systems (St. Petersburg, 1991)*, volume 12 of *Progr. Nonlinear Differential Equations Appl.*, pages 96–116, Basel, 1994. Birkhäuser Verlag.

[18] B. Leimkuhler and S. Reich. *Simulating Hamiltonian Dynamics.* Cambridge University Press, Cambridge, 2005.

[19] E. Lindahl, B. Hess, and D Spoel. GROMACS 3.0: A package for molecular simulation and trajectory analysis. *J. Mol. Modeling*, 7:305–317, 2001.

[20] Q. Ma and J.A. Izaguirre. Targeted modified impulse - a multiscale stochastic integrator for long molecular dynamics simulations. *Multiscale Modeling and Simulation*, 2:1–21, 2003.

[21] B. Mehlig, D.W. Heermann, and B.M. Forrest. Hybrid Monte Carlo method for condensed-matter systems. *Phys. Rev. B*, 45:679–685, 1992.

[22] P.C. Moan. On the KAM and Nekhoroshev theorems for symplectic integrators and implications for error growth. *Nonlinearity*, 17:67–83, 2004.

[23] B.E. Moore and S. Reich. Backward error analysis for multi-symplectic integration methods. *Numer. Math.*, 95:625–652, 2003.

[24] A.I. Neishtadt. The separation of motions in systems with rapidly rotating phase. *J. Appl. Math. Mech.*, 48:133–139, 1984.

[25] S. Nosé. A molecular dynamics method for simulations in the canonical ensemble. *Mol. Phys.*, 52:255–268, 1984.

[26] J.P. Ryckaert, G. Ciccotti, and H.J.C. Berendsen. Numerical integration of the cartesian equations of motion of a system with constraints: molecular dynamics of n-alkanes. *J. Comput. Phys.*, 23:327–341, 1977.

[27] R.D. Skeel and D.J. Hardy. Practical construction of modified Hamiltonians. *SIAM J. Sci. Comput.*, 23:1172–1188, 2001.

The Langevin Equation for Generalized Coordinates

Reinier L. C. Akkermans

Accelrys Ltd, 334 Cambridge Science Park, Cambridge, CB4 0WN, United Kingdom,
ReinierA@Accelrys.com

Abstract. Using the projection operator formalism we derive the generalized Langevin equation for a subset of generalized coordinates obtained from a full set of Cartesian coordinates by a canonical transformation. The resulting equations of motion treat positions and momenta on equal footing, with dissipative terms that involve time correlations amongst both the random forces and the "random velocities". We consider point transformations in more detail, and, in the special case of a bath position-independent metric tensor, obtain the ordinary Langevin involving just random force correlations.

1 Introduction

Statistical mechanical methods to reduce a description of matter by integrating out degrees of freedom have become standard since the early work of Kubo, Zwanzig and Mori.[1, 2] Tracing the electronic degrees of freedom in a quantum mechanical description, for example, leads formally to an effective Hamiltonian as a function of the nuclei positions and momenta, parametrized by the thermodynamic state at which the contraction was made. Not only the thermodynamics can be written down formally, but also the dynamics can be resolved in terms of the well-known memory kernel in a Langevin framework, which represents the delayed effect the integrated degrees of freedom have on the remaining variables. The projection operator formalism provides a clear and unambiguous way to derive such a reduced description.[3]

In recent years, the idea of reduced descriptions has drawn renewed interest in the field of soft condensed matter. For example, the description of polymer melts, or colloidal suspensions, is naturally carried out at a mesoscopic level, in which particles represent entities on a scale much larger than the atomic, or microscopic, scale. On the mesoscopic scale, the degrees of freedom that are averaged out enter the Hamiltonian via a state-dependent interaction (which, in turn, is often contracted further, to a state-dependent pair interaction). Although the state-dependency is, of course, well-known since the early days, its consequences are currently subject to active research.[4–7] In the above context it is natural to refer to the variables of interest as *solute* variables, and those that are averaged out as *bath* variables. We shall maintain this terminology in the current work, keeping in mind that generalized coordinates may bare little resemblance to a true solute/bath system.

To evaluate the solute Hamiltonian and the dissipative interactions, the method of constrained dynamics is particularly well suited.[8] In this method the solute positions are constrained in a microscopic system. Under this condition, the average constraint force is related to (and in some cases equal to) the gradient of the solute Hamiltonian with respect to the solute positions. Moreover, the constraint force fluctuations serve as a first estimate to the fluctuating force acting on the solute. From the time correlation of the fluctuating force, the dissipative terms can then be obtained using the fluctuation-dissipation theorem.

The above strategy is straightforward if the solute positions are simple (e.g., linear) functions of the microscopic degrees of freedom, such that the metric tensor is independent of the bath positions, and the average constraint force equals the derivative of the solute Hamiltonian. An example we reported previously, is the coarse-graining of a polymer chain where the solute positions are the centres-of-mass of short segments of the chain.[8] For general solute coordinates, however, the metric tensor will be position-dependent. It is now well-known that this leads to a bias in the sampling of the average force by constrained dynamics.[9, 10] The potential of mean constraint force lacks the entropy associated with the zero velocities of the constrained positions.[11] Since this entropy is the average of a fluctuating quantity, the fluctuations, in principle, will affect the dynamics of the solute variables, and should, therefore, appear in the equation of motion. To the best of our knowledge, however, a systematic derivation of the equations of motion for solute coordinates in this general case has not been published. In the following we shall therefore revisit the projection operator formalism[3, 12] in terms of generalized solute coordinates.

Adelman[13] does derive the Langevin equation for the generalized *momenta*, using his molecular-time-scale-Langevin framework. In essence, in this method the reduced equations of motion are written as an equivalent harmonic chain, and linear response theory is used to describe the dissipation. Physical arguments then lead to an approximated equation of motion of the solute momenta. The conjugated equations of motion for the generalized *positions*, however, were not mentioned explicitly. More recently, Gelin[14] addressed the Fokker-Planck equation in the space of generalized coordinates. His interest is in inertial effects in constrained Brownian dynamics of polymers. Both papers consider point transformations only. Finally, two papers[15, 16] appeared recently, addressing the canonical formulation of the Langevin equation. Here, the starting point is the Kac-Zwanzig model, i.e., a system coupled to a bath of harmonic oscillators. By generalizing the coupling Hamiltonian to include momentum-couplings, the authors obtained a symmetric form of the equation of motion of the solute variables.

The present work differs from the above studies in that the equations of motion for *both* conjugated coordinates are *derived* from a more complete description, *without* approximations or physical reasoning, and for *arbitrary* canonical transformations. Point transformations are considered as a special case. The resulting equation is an exact reformulation of the Hamilton equations of motion, and the canonical Langevin equation is obtained after taking the Markov approximation.

The outline of this report is as follows: In Sect. 2 we set the notation for a canonical transformation to generalized coordinates. In Sect. 3 we introduce a projection operator, which performs a thermodynamical average over a subset of coordinates. Substituting this operator in the Liouville equation, we obtain the generalized Langevin equation for the complimentary set of variables. In Sect. 4 we work out the case for point transformations in more detail. Conclusions in Sect. 5 finalize the report.

A word on notation: where appropriate, we shall use Latin capitals for variables referring to the solute coordinates, whereas lower case characters refer to the remaining, bath variables. Greek lower case characters are reserved to denote the microscopic degrees of freedom. For example, R, r, ρ, stand for (vectors containing) the positions of the solute, bath, and microscopic position, respectively. Secondly, to avoid the burden of indices, we shall employ a matrix notation, at the expense of introducing superscripts T for transpose. We use the convention that all vectors are column vectors, and use ∂_R to denote the column-vector operator with i-th element $\frac{\partial}{\partial R_i}$. For example, $(\partial_\pi^T \eta)\partial_\rho$ is the operator $\sum_i \frac{\partial \eta}{\partial \pi_i} \frac{\partial}{\partial \rho_i}$, whereas $\dot{P}\dot{R}^T$ is a matrix with (i,j) element equal to $\dot{P}_i \dot{R}_j$. Finally, infinitesimals are abbreviated as $dR \equiv \prod_i dR_i$.

2 Generalized Coordinates

As a starting point, consider a system of particles with Cartesian positions ρ, conjugated momenta π, and associated masses as entries in the diagonal matrix μ, moving in a potential energy field $\upsilon(\rho)$. The Hamiltonian $\eta(\rho, \pi)$ of the system is

$$\eta = \upsilon + \tfrac{1}{2}\pi^T \mu^{-1}\pi , \tag{1}$$

and the equations of motion are given by

$$\dot{\rho} = \partial_\pi \eta = \mu^{-1}\pi, \quad \dot{\pi} = -\partial_\rho \eta = -\partial_\rho \upsilon . \tag{2}$$

We shall transform the coordinates (ρ, π) to two sets of conjugated variables: (R, P) respectively (r, p), referred to as solute and bath variables. For the moment, all variables are regarded as functions of both positions ρ and momenta π. In Sect. 4 we shall consider point transformations in more detail. For a general, time-independent, transformation to be canonical, the functions $R(\rho, \pi), P(\rho, \pi), r(\rho, \pi)$, and $p(\rho, \pi)$ must satisfy the (Maxwell) conditions[17]

$$\begin{aligned}
\partial_\rho R^T &= (\partial_P \pi^T)^T , & \partial_\rho P^T &= -(\partial_R \pi^T)^T , \\
\partial_\pi R^T &= -(\partial_P \rho^T)^T , & \partial_\pi P^T &= (\partial_R \rho^T)^T , \\
\partial_\rho r^T &= (\partial_p \pi^T)^T , & \partial_\rho p^T &= -(\partial_r \pi^T)^T, \\
\partial_\pi r^T &= -(\partial_p \rho^T)^T, & \partial_\pi p^T &= (\partial_r \rho^T)^T .
\end{aligned} \tag{3}$$

It is worth noting that the Jacobian matrix of the transformation satisfying (3) has determinant ± 1, hence the relations $R(\rho, \pi)$ etc. are invertible, and the phase space volume is invariant under this transformation, that is $d\rho d\pi = dRdPdrdp$.

Using (3) and (2) it is easily shown that the Hamiltonian η also governs the equations of motion of the solute and bath variables. For example,

$$\dot{R} = (\partial_\pi R^T)^T \dot{\pi} + (\partial_\rho R^T)^T \dot{\rho} = (\partial_P \rho^T)\partial_\rho \eta + (\partial_P \pi^T)\partial_\pi \eta = \partial_P \eta \ . \quad (4)$$

Likewise

$$\dot{P} = -\partial_R \eta \ , \quad \dot{r} = \partial_p \eta \ , \quad \dot{p} = -\partial_r \eta \ .$$

In the following we shall often replace the (implicit) time derivative by the operator

$$\mathcal{L} \equiv \partial_t = \dot{\rho}^T \partial_\rho + \dot{\pi}^T \partial_\pi \ ,$$

which we refer to as the Liouville operator, although, strictly speaking, the latter would be $-i\mathcal{L}$. We shall decompose the Liouville operator as $\mathcal{L} = \mathcal{L}_{RP} + \mathcal{L}_{rp}$ where

$$\mathcal{L}_{RP} \equiv \dot{R}^T \partial_R + \dot{P}^T \partial_P, \quad \mathcal{L}_{rp} \equiv \dot{r}^T \partial_r + \dot{p}^T \partial_p \ . \quad (5)$$

3 Generalized Langevin Equation

Let $f(\rho, \pi) = f(R, P, r, p)$ be an arbitrary function of the coordinates. The equation of motion $\dot{f} = \mathcal{L}f$ leads with initial condition $f(0) = f_0$ to $f(t) = \exp(\mathcal{L}t)f_0$. We now ask for the evolution on average over a canonically distributed bath. Let us therefore introduce the projection operator[1]

$$\mathcal{P}f \equiv \langle f \rangle = \int d\tilde{r}d\tilde{p} \, e^{\beta(H-\eta)} f \ , \quad (6)$$

where $d\tilde{r}d\tilde{p} = drdp/h^n$; Planck's constant h raised to the number of bath variables n, renders this infinitesimal dimensionless. Through this projection the (inverse) temperature of the bath, β, enters the description. The bath partition function is incorporated through $H(R, P)$, given by

$$H = -\tfrac{1}{\beta} \ln(\int d\tilde{r}d\tilde{p} \, e^{-\beta\eta}) \ , \quad (7)$$

which acts as a Hamiltonian to the solute, in the sense that from (6) and (4),

$$\langle \dot{R} \rangle = \partial_P H, \quad \langle \dot{P} \rangle = -\partial_R H \ . \quad (8)$$

We shall use $\mathcal{P}$ in the identity for the evolution operator

$$e^{\mathcal{L}t} = e^{\mathcal{L}t}\mathcal{P} + e^{(\mathcal{I}-\mathcal{P})\mathcal{L}t}(\mathcal{I} - \mathcal{P}) + \int_0^t d\tau \, e^{\mathcal{L}(t-\tau)}\mathcal{P}\mathcal{L}e^{(\mathcal{I}-\mathcal{P})\mathcal{L}\tau}(\mathcal{I} - \mathcal{P}) \ ,$$

which is readily verified by differentiation ($\mathcal{I}$ is the unit-operator: $\mathcal{I}f = f$). The value of f at time t thus decomposes as $f(t) = f_c(t) + f_f(t) + f_d(t)$, comprising a "drift" (c), "fluctuation" (f) and "dissipation" (d) part, where

$$f_c(t) \equiv e^{\mathcal{L}t}\langle f_0 \rangle \ ,$$
$$f_f(t) \equiv e^{(\mathcal{I}-\mathcal{P})\mathcal{L}t}(f_0 - \langle f_0 \rangle) \ ,$$
$$f_d(t) \equiv \int_0^t d\tau \, e^{\mathcal{L}(t-\tau)}\langle \dot{f}_f(\tau) \rangle \ .$$

Notice that f_c evolves by the ordinary operator $\exp(\mathcal{L}t)$, but f_f evolves in a non-Hamiltonian way, ensuring $\langle f_f(t) \rangle = 0$ at any $t > 0$.

The kernel in the dissipative term should be read as $\langle \dot{f}_f(\tau) \rangle = \langle \mathcal{L} f_f(\tau) \rangle$, which, by (6), (5), and (4), noticing that $\partial_R^T \dot{R} = -\partial_P^T \dot{P}$, simplifies to

$$\langle \dot{f}_f(\tau) \rangle = \langle \mathcal{L}_{RP} f_f(\tau) \rangle = \langle \partial_R^T [\dot{R} f_f(\tau)] \rangle + \langle \partial_P^T [\dot{P} f_f(\tau)] \rangle. \tag{9}$$

Partial differentiating the first term leads to

$$\langle \partial_R^T [\dot{R} f_f(\tau)] \rangle = \int d\tilde{r} d\tilde{p} \, e^{\beta(H-\eta)} \partial_R^T [\dot{R} f_f(\tau)]$$
$$= \partial_R^T [\int d\tilde{r} d\tilde{p} \, e^{\beta(H-\eta)} \dot{R} f_f(\tau)] - \int d\tilde{r} d\tilde{p} \, [\partial_R^T e^{\beta(H-\eta)}] \dot{R} f_f(\tau)$$
$$= (\partial_R^T - \beta \partial_R^T H) \langle \dot{R} f_f(\tau) \rangle + \beta \langle (\partial_R^T \eta) \dot{R} f_f(\tau) \rangle.$$

Similarly for the last term in (9),

$$\langle \partial_P^T [\dot{P} f_f(\tau)] \rangle = (\partial_P^T - \beta \partial_P^T H) \langle \dot{P} f_f(\tau) \rangle + \beta \langle (\partial_P^T \eta) \dot{P} f_f(\tau) \rangle.$$

Using (4) and (8), the evolution of f can thus be written as

$$f(t) = f_c(t) + f_f(t) + \int_0^t d\tau \, e^{\mathcal{L}(t-\tau)}$$
$$\times [(\partial_R^T + \beta \dot{P}_c^T) \langle \dot{R}_f f_f(\tau) \rangle + (\partial_P^T - \beta \dot{R}_c^T) \langle \dot{P}_f f_f(\tau) \rangle], \tag{10}$$

noticing that in the correlation of the random variable f_f with a function g, only the random part of g survives, i.e., $\langle g f_f \rangle = \langle g_f f_f \rangle$.

Whereas this holds for arbitrary functions f, we shall be interested in the equations of motion of the solute variables. Substituting $f = R$, respectively P, we obtain

$$\begin{cases} \dot{R}(t) = \dot{R}_c(t) + \dot{R}_f(t) + \int_0^t d\tau \, e^{\mathcal{L}(t-\tau)} \\ \quad \times [(\partial_R^T + \beta \dot{P}_c^T) \langle \dot{R}_f \dot{R}_f^T(\tau) \rangle + (\partial_P^T - \beta \dot{R}_c^T) \langle \dot{P}_f \dot{R}_f^T(\tau) \rangle]^T, \\ \dot{P}(t) = \dot{P}_c(t) + \dot{P}_f(t) + \int_0^t d\tau \, e^{\mathcal{L}(t-\tau)} \\ \quad \times [(\partial_R^T + \beta \dot{P}_c^T) \langle \dot{R}_f \dot{P}_f^T(\tau) \rangle + (\partial_P^T - \beta \dot{R}_c^T) \langle \dot{P}_f \dot{P}_f^T(\tau) \rangle]^T, \end{cases}$$

We have thus obtained the exact evolution of the solute variables in terms of the drift variables $(\dot{R}_c, \dot{P}_c)$ and the fluctuation variables $(\dot{R}_f, \dot{P}_f)$. The dissipative terms are expressed in the correlation among the fluctuations, which may be looked upon as a generalized fluctuation-dissipation theorem. Notice that the above equations of motion are invariant under the canonical transformation $(R, P) \to (P, -R)$.

Equation (10) is also a convenient starting point for the Fokker-Planck equation describing the evolution of the probability density in the space spanned by the variables R and P. Although the derivation of this equation is straightforward, is it too lengthy to reproduce in this short communication. The main findings are velocity derivatives in the drift part of the Fokker-Planck equation, which contain the correlation functions that appear in (10).

To arrive at a more tractable set of equations we shall make a number of approximations. First, we assume that the drift components can be obtained as if the projection and evolution operators would commute, e.g., $\dot{R}_c(t) = \partial_P H(t)$. Secondly, we assume that the partial derivatives of the correlation functions can be neglected with respect to the remaining terms, e.g, $\partial_R^T \langle \dot{R}_f \dot{P}_f^T(\tau) \rangle \ll \beta \dot{R}_c^T \langle \dot{R}_f \dot{P}_f^T(\tau) \rangle$. Thirdly, we assume that the correlation functions are stationary (invariant under time translation and inversion), e.g., $\langle \dot{R}_f(t - \tau) \dot{P}_f^T(t) \rangle = \langle \dot{R}_f \dot{P}_f^T(\tau) \rangle$, omitting, as above, the time dependency if evaluated at $t = 0$. Under the above approximations, the equations of motion have the structure of a generalized Langevin equation,

$$
\begin{cases}
\dot{R} = \partial_P H + \dot{R}_f \\
\quad - \beta \int_0^t d\tau \left[\langle \dot{R}_f(\tau) \dot{R}_f^T \rangle \partial_R H(t - \tau) + \langle \dot{R}_f(\tau) \dot{P}_f^T \rangle \partial_P H(t - \tau) \right], \\
\dot{P} = -\partial_R H + \dot{P}_f \\
\quad + \beta \int_0^t d\tau \left[\langle \dot{P}_f(\tau) \dot{R}_f^T \rangle \partial_R H(t - \tau) + \langle \dot{P}_f(\tau) \dot{P}_f^T \rangle \partial_P H(t - \tau) \right],
\end{cases}
\tag{11}
$$

omitting the explicit time dependence, except for the last term.

If furthermore the correlation functions decay on a time scale on which $\partial_R H$ and $\partial_P H$ do not change appreciably, we arrive at the Langevin equation,

$$
\begin{cases}
\dot{R} = \partial_P H + \dot{R}_f \\
\quad - \beta [\int_0^\infty d\tau \, \langle \dot{R}_f(\tau) \dot{R}_f^T \rangle] \partial_R H - \beta [\int_0^\infty d\tau \, \langle \dot{R}_f(\tau) \dot{P}_f^T \rangle] \partial_P H, \\
\dot{P} = -\partial_R H + \dot{P}_f \\
\quad - \beta [\int_0^\infty d\tau \, \langle \dot{P}_f(\tau) \dot{R}_f^T \rangle] \partial_R H - \beta [\int_0^\infty d\tau \, \langle \dot{P}_f(\tau) \dot{P}_f^T \rangle] \partial_P H.
\end{cases}
\tag{12}
$$

Because of the separation of time scales, we can extend the integration to infinity. A strict separation of time scales, of course, can only be achieved by assuming that the correlation functions are δ-correlated. Notice that this equation still has the canonical structure, in the sense that it is invariant under the transformation $(R, P) \rightarrow (P, -R)$.

In summary, under the approximations specified above, we have arrived at a Langevin equation for the solvent variables, either including memory effects (11) or without (12). Both conjugated variables are subject to dissipative terms. To solve the set of equations in time, we only need to determine the free energy H, and the fluctuation parts. In the next section we explore a special case which gives a more physical interpretation to the terms involved, and allows us to further simplify the equations.

4 Point Transformations

Whereas the previous holds for arbitrary, time-independent, canonical transformations, we shall now consider transformations of the position space only, .i.e., contact or point transformations. In this special case it is possible to further elaborate on the equations and obtain conditions under which the ordinary Langevin equation is obtained. For point transformations we have, in addition to (3),

$$\partial_\pi R^T = 0, \quad \partial_\pi r^T = 0 . \tag{13}$$

(13) and (3) imply that the microscopic momenta $\pi(R, r, P, p)$ are linear combinations of the solute and bath momenta,

$$\pi = (\partial_\rho R^T)P + (\partial_\rho r^T)p .$$

The coefficients are the entries of the Jacobian matrix associated with the transformation $\rho \to (R, r)$. For any but the simplest (i.e., linear) transformations, these coefficients will depend on the position coordinates.

Substituting this into (1) we obtain the Hamiltonian

$$\eta = \upsilon + \tfrac{1}{2}P^T(\partial_\rho R^T)^T \mu^{-1}(\partial_\rho R^T)P + P^T(\partial_\rho R^T)^T \mu^{-1}(\partial_\rho r^T)p$$
$$+ \tfrac{1}{2}p^T(\partial_\rho r^T)^T \mu^{-1}(\partial_\rho r^T)p$$
$$\equiv \upsilon + \tfrac{1}{2}P^T X P + P^T Y p + \tfrac{1}{2}p^T Z p , \tag{14}$$

where the last line defines the metric matrices $X(R, r)$, $Y(R, r)$, and $Z(R, r)$. The equations of motion (4) are

$$\begin{cases} \dot{R} = XP + Yp , \\ \dot{P} = -\partial_R \upsilon - \tfrac{1}{2}(\partial_R X) : PP^T - (\partial_R Y) : pP^T - \tfrac{1}{2}(\partial_R Z) : pp^T , \\ \dot{r} = Zp + Y^T P , \\ \dot{p} = -\partial_r \upsilon - \tfrac{1}{2}(\partial_r X) : PP^T - (\partial_r Y) : pP^T - \tfrac{1}{2}(\partial_r Z) : pp^T , \end{cases}$$

where the semi-colon indicates a double contraction, e.g. $[(\partial_R Y) : pP^T]_i = \sum_j \sum_k \frac{\partial Y_{jk}}{\partial R_i} p_k P_j$. Note that, apart from the conservative forces, the momenta also change in time via centrifugal forces.

The solute Hamiltonian (7) can be simplified by integrating over the momenta p to give

$$H = -\tfrac{1}{\beta} \ln\left[\tfrac{1}{h^n} \int dr\, e^{-\beta(\upsilon + \frac{1}{2}P^T X P)} \int dp\, e^{-\beta(P^T Y p + \frac{1}{2}p^T Z p)}\right]$$
$$= -\tfrac{1}{\beta} \ln\left[\int d\tilde{r}\, \frac{1}{\sqrt{|mZ|}} e^{-\beta(\upsilon + \frac{1}{2}P^T W P)}\right] ,$$

where $W = X + YZ^{-1}Y^T$. We have used that βZ is positive definite. The infinitesimal $d\tilde{r} = dr/\Lambda^n$ is reduced by the length unit $\Lambda = h/\sqrt{2\pi m/\beta}$, with m an arbitrary constant with unit of mass, introduced to reduce the inverse mass units of Z.

Using either the above expression in (8), or the full Hamiltonian (14) in (4), it is readily shown that the average change in positions and momenta obey

$$\begin{cases} \langle \dot{R} \rangle = \langle X \rangle P + \langle Y p \rangle = \langle W \rangle P \,, \\ \langle \dot{P} \rangle = -\langle \partial_R v \rangle - \frac{1}{2} \langle \partial_R X \rangle : P P^T - \langle (\partial_R Y) : p P^T \rangle \\ \qquad - \frac{1}{2} \langle (\partial_R Z) : p p^T \rangle \\ \qquad = -\langle \partial_R v \rangle - \frac{1}{2} \langle \partial_R W \rangle : P P^T - \frac{1}{2} \frac{1}{\beta} \langle \partial_R \ln |Z| \rangle \,. \end{cases}$$

Comparing the two equations of motion, we observe that, since X, Y, and Z are, in general, functions of the bath variables, the random part of the velocity does not necessarily vanish, even at $t = 0$: $\dot{R}_f = \dot{R} - \langle \dot{R} \rangle = (W - \langle W \rangle)P$. As a consequence, its autocorrelation, and the cross-correlation with the random force have to be retained in the Langevin equation.

Regarding the metric matrix Y, it has been shown by Darve and Pohorille[10] that the variables r can be chosen such that this matrix vanishes for each point R. As a consequence, we arrive at the conclusion that, under the above-specified choice of bath coordinates, the random velocities vanish if the mass metric matrix associated with the solute variables, X, is independent of the bath coordinates. Such is obviously true, in case of a linear transformation of the microscopic positions. If X is independent of r, or its dependence can be otherwise neglected, we obtain the ordinary Langevin equation,

$$\begin{cases} \dot{R} = X P, \\ \dot{P} = -\langle \partial_R v \rangle + \dot{P}_f + \beta \int_0^t d\tau \, \langle \dot{P}_f(\tau) \dot{P}_f^T \rangle X P(t - \tau) \\ \quad \rightarrow -\langle \partial_R v \rangle + \dot{P}_f + [\beta \int_0^\infty d\tau \, \langle \dot{P}_f(\tau) \dot{P}_f^T \rangle X] P \,. \end{cases}$$

The last line holds in the Markov limit, in which we recognize the friction matrix in square brackets, in particular if $X = M^{-1} I$, with M the mass associated with the solute variables. Entirely the same reasoning explains the absence of velocity derivatives in the diffusion terms of the associated Fokker-Planck equation.

We should notice that in the case X is independent of r, and the bath positions are chosen such that $Y = 0$, although the random velocities vanish, the metric matrix may still affect the random forces, through the matrix Z. The latter dependence vanishes if Z is also independent of the bath positions, or if Z is independent of the solute positions. Clearly, the metric matrix effects are absent if both X and Z are independent of both the bath and solute positions, which is the true for a linear transformation of the position space.

We conclude at this point that the ordinary, asymmetric, Langevin equation corresponds to the equations of motion of a subset of variables obtained via a linear point transformation. For more complicated (i.e. non-linear) point transformation, or general canonical transformations such as described in Sect. 2, the equations of motion of the subset of variables obey a more general, symmetric, Langevin equation. The last class of variables include important cases such as dihedral angles, Euler coordinates, action angle variables, spherical coordinates, and the like.

5 Conclusions

The Langevin equation originated as an equation of motion to describe the phenomenon of random motion of a pollen particle. In its original form, the effect of the bath only persists through the friction coefficient and the random force, subject to the fluctuation-dissipation theorem. The projection operator formalism has subsequently shown that an equation of motion for the solute with the same structure can be derived, by starting from a full Hamiltonian description, and subsequently averaging out the bath variables.

In this context, we hope to have achieved two things. First, to provide an explanation as to why the ordinary Langevin equation has a typical asymmetric structure, with positions and momenta treated on different footing, i.e., the change in momenta being subject to random noise, whereas the change in position is not. By looking upon the solute positions as obtained via a point transformation of the microscopic coordinates, we can state that if the mass metric matrix associated with the solute is independent of the bath coordinates, such an asymmetric equations of motion naturally arises, whereas for a general canonical transformation there is actually no reason why the Langevin equation should treat positions and momenta differently. The latter is, of course, expected, since the Hamilton-Jacobi framework only deals with pairs of conjugated variables, leaving the distinction between them as one of nomenclature.

Secondly, we have stated the considerations before applying the Langevin equation to coordinates other than Cartesian coordinates. Being an equation of phenomenological nature, it is tempting, and not uncommon, to write down the ordinary Langevin equation for, say, consecutive dihedral angles in a protein backbone, whereas the current work shows that this is not necessarily the equation of motion of those coordinates. This situation is not unlike the calculation of the free energy from molecular dynamics in which the solute positions (in this context often referred to as reaction coordinates) are constrained. In that case, it is well known that the potential of mean constrain force differs from the free energy along the reaction coordinate by an entropy term, since the system does not explore the solute-velocity space. This entropy term contains the mass metric matrix, similar to the effect on the drift force shown in this work.

From free energy calculation by constrained molecular dynamics it is known that metric matrix effects can be rather small, compared to the potential of mean constrain force. Not much is known about the fluctuations. To analyse the dynamics of generalized coordinates, one option is to approximate the random velocities and forces by their fluctuations from the average in a constrained dynamics simulation. Time correlating the fluctuations, integrating over time, then leads to the friction matrix required to solve the Langevin equation.

Acknowledgements

The author would like to thank Patrick Warren, Andrea Ferrante, and Wim Briels for discussions and the European Commission for financial support through a Marie Curie Industry Host Fellowship under contract number HPMI-CT-1999-00028.

References

[1] Evans, D. J., Morriss, G. P.: Statistical Mechanics of Nonequilibrium Liquids. Academic Press, London (1990)

[2] Gardiner, C. W.: Handbook of Stochastic Methods for Physics, Chemistry and the Natural Sciences. Springer Series in Synergetics, **13**. Springer-Verlag, Heidelberg (2004).

[3] Grabert, H.: Projection operator techniques in nonequilibrium statistical mechanics. Springer-Verlag, Berlin (1982)

[4] Karttunen, M., Vattulainen, I., Lukkarinen, A. (Eds.): Novel Methods in Soft Matter Simulations. Lecture Notes in Physics, **640**. Springer-Verlag, Heidelberg (2004)

[5] Louis, A. A.: Beware of density dependent pair potentials. J. Phys.: Condens. Matter, **14**, 9187–9206 (2002)

[6] Akkermans, R. L. C., Briels, W. J.: A structure-based coarse-grained model for polymer melts. J. Chem. Phys., **114**, 1020–1032 (2001)

[7] Akkermans, R. L. C., Briels, W. J.: Coarse-grained interactions in polymer melts: a variational approach. J. Chem. Phys., **115**, 6210–6219 (2001)

[8] Akkermans, R. L. C., Briels, W. J.: Coarse-grained dynamics of one chain in a polymer melt. J. Chem. Phys., **113**, 6409–6422 (2000)

[9] den Otter, W. K., Briels, W. J.: The calculation of free-energy differences by constrained molecular-dynamics simulations. J. Chem. Phys., **109**, 4139–4146 (1998)

[10] Darve, E., Pohorille, A.: Calculating free energies using average force. J. Chem. Phys., **115**, 9169–9183 (2001)

[11] Schlitter, J., Klähn, M.: The free energy of a reaction coordinate at multiple constraints: a concise formulation. Mol. Phys., **101**, 3439–3443 (2003)

[12] Deutch, J. M., Oppenheim, I.: Molecular theory of Brownian motion for several particles. J. Chem. Phys., **54**, 3547–3555 (1971)

[13] Adelman, S. A.: Generalized Langevin theory for many-body problems in chemical dynamics: The method of partial clamping and formulation of the solute equations of motion in generalized coordinates. J. Chem. Phys., **81**, 2776–2788 (1984)

[14] Gelin, M. F.: Inertial effects in the Brownian dynamics with rigid constraints. Macromol. Theory Simul., **8**, 529–543 (1999)

[15] Kerr, W. C., Graham, A. J.: Generalized phase space version of Langevin equations and associated Fokker-Planck equations. Eur. Phys. J. B, **15**, 305–311 (2000)

[16] Cépas, O., Kurchan, J.: Canonically invariant formulation of Langevin and Fokker-Planck equations. Eur. Phys. J. B, **2**, 221–223 (1998)
[17] Goldstein, H.: Classical Mechanics. Narosa Publishing House, New Delhi (1996)

Metastability and Dominant Eigenvalues
of Transfer Operators

Wilhelm Huisinga[1] and Bernd Schmidt[2]

[1] Free University of Berlin, Department of Mathematics and Computer Science, Arnimallee 2-6, D-14195 Berlin, Germany
[2] Max Planck Institute for Mathematics in the Sciences, Inselstr. 22, D-04103 Leipzig, Germany

Abstract. Metastability is an important characteristic of molecular systems, e.g., when studying conformation dynamics, computing transition paths or speeding up Markov chain Monte Carlo sampling methods. In the context of Markovian (molecular) systems, metastability is closely linked to spectral properties of transfer operators associated with the dynamics. In this article, we prove upper and lower bounds for the metastability of a state-space decomposition for reversible Markov processes in terms of dominant eigenvalues and eigenvectors of the corresponding transfer operator. The bounds are explicitly computable, sharp, and do not rely on any asymptotic expansions in terms of some smallness parameter, but rather hold for arbitrary transfer operators satisfying a reasonable spectral condition.

Key words: Molecular dynamics, Markov processes, spectral properties, Rayleigh-trace, lower and upper bounds on metastability

1 Introduction

There are many problems in physics, chemistry, or biology where the length and time scales corresponding to the microscopic descriptions (given in terms of some stochastic or deterministic dynamical system) and the resulting macroscopic effects differ by many orders of magnitude. Rather than resolving all microscopic details one is often interested in characteristic features on a macroscopic level (e.g., phase transitions, conformational changes of molecules, climate changes, etc.). A typical mathematical example is the long-time limit behavior, where invariant measures or limit cycles are established characteristic objects (e.g., [27, 29]). Metastability is another important characteristic which is related to the long time behavior of the dynamical system. It refers to the property that the dynamics is likely to remain within a certain part of the state space for a long period of time until it eventually exits and transits to some other part of the state space. There are well-established links of metastability to, e.g., exit times [3, 16, 17], eigenvalues of transfer operators or generators [3, 7, 9, 10, 35], phase transitions [2, 6], reduced Markovian approximations [23, 24, 35], averaging [40], and many other areas.

There exist several characterizations of metastability in the literature (see, e.g., [2, 7, 35, 37]). There are at least two different conceptual approaches to metastability. (1) A subset C is called metastable if the fraction of systems in C (measured w.r.t. some pre-specified probability measure) whose trajectory exits C during some pre-defined microscopic time span is significantly small. (2) A subset C is called metastable, if with high probability a typical trajectory stays within C longer than some macroscopic time span. Thus, in broad terms, you may either observe an ensemble of systems for a short time or a typical system for a long time to characterize metastability. We will restrict our attention to the ensemble approach, the use of which was motivated by a molecular application (conformation dynamics, see [13, 34, 35]) where the probability measure is given by the canonical ensemble or Boltzmann distribution while the observation time span is linked to the experimental setting.

We will assume that the dynamical system is given in terms of some reversible Markov chain with invariant measure μ. Equivalently, we may specify the dynamics in terms of the associated transfer operator P acting on $L^2(\mu)$ and being self-adjoint due to reversibility. There is a classical connection between invariant (stable) subsets and degeneracy of the maximal eigenvalue 1 of P. The degeneracy of 1 is just the number of invariant subsets of the state space (see e.g. [12, 25]). Analogously, to each eigenvalue close to 1 there corresponds an almost invariant or metastable subset of the state space, see e.g. [8–10].

Pursuing this analogy, there is a large amount of literature relating metastability to eigenvalues of transfer operators or generators corresponding to the underlying Markov process. However, the theoretical investigations are either restricted to the finite dimensional state space case (and thus related to stochastic matrices or Laplace matrices), e.g., [20–22, 28, 38] or stated asymptotically in terms of some smallness parameter, e.g., [16]. General state space non-asymptotic results are much more rare and may be found in the setting of exit times [4, 5] or in the setting of symmetric Markov semigroups [7, 8, 37]. To our knowledge, even for the finite dimensional state space case, there are no lower bounds on the metastability (in the ensemble characterization) of a finite number of subsets in terms of eigenvalues known. It is our aim to derive an upper and in particular a lower bound on the metastability of an arbitrary decomposition from spectral properties of the transfer operator P. Such bounds are not only of theoretical interest but also of algorithmic relevance, e.g., in the context of dynamical clustering [9, 14].

The paper is organized as follows: In Sect. 2, we review stochastic models for molecular dynamics to motivate our study of metastability. In Sect. 3, we introduce the set-up including the definition of metastability and its transfer operator formulation. In Sect. 4, a variational formula for the Rayleigh-trace of self-adjoint operators is reviewed, which is crucial in the proofs of our results. We prove upper and lower bounds for the metastability of arbitrary decompositions of the phase space under some quite general spectral assumption on P. Finally, in Sect. 5, we state some examples illustrating the sharpness and usefulness of the bounds.

2 Markovian Molecular Dynamics

This section introduces three popular Markovian models for molecular dynamics. For simplicity, we assume that all masses are equal to one.

Hamiltonian System with Randomized Momenta. The Hamiltonian system with randomized momenta is a reduced dynamics defined on the position space and derived from the deterministic Hamiltonian system by "randomizing the momenta" and integrating for some fixed observation time span τ [33]. Let Φ^t denote the flow corresponding to the deterministic Hamiltonian system

$$\dot{q} = p, \qquad \dot{p} = -\nabla_q V(q) ,$$

defined on the phase space Γ, and let $\Pi_q : \Gamma \to \Omega$ denote the projection onto the position space Ω. Then, the *Hamiltonian system with randomized momenta* is the discrete time Markov process $Q_n = \{Q_n\}_{n \in \mathbf{N}}$ satisfying

$$Q_{n+1} = \Pi_q \Phi^\tau(Q_n, P_n); \qquad n \in \mathbf{N} ,$$

where τ is some fixed observation time span, and $\{P_n\}$ is an i.i.d. sequence distributed according to the canonical distribution of momenta $\mathcal{P} \propto \exp(-\beta p^2/2)$ with inverse temperature $\beta = 1/(k_B T)$. As shown in [33], the Markov process is reversible w.r.t. the positional canonical ensemble $\mu \propto \exp(-\beta V(q))$. For further details, in particular comments on τ, see [33, 35, 36]. We finally remark that the Hamiltonian system with randomized momenta is closely related to the hybrid Monte Carlo method.

Langevin Equation. The most popular model for an open system stochastically interacting with its environment is the *Langevin equation* (e.g., [32])

$$\dot{q} = p, \qquad \dot{p} = -\nabla_q V(q) - \gamma p + \sigma \dot{W}$$

corresponding to some friction constant $\gamma > 0$ and external force $F_{\text{ext}} = \sigma \dot{W}$ defined in terms of a standard $3N$-dimensional Brownian motion W. The Langevin equation defines a continuous time Markov process on the phase space Γ whose invariant distribution is given by the canonical ensemble $\mu \propto \exp(-\beta H(q, p))$ with Hamiltonian H describing the internal energy of the system. The inverse temperature β is linked to the friction and the stochastic excitation via $\beta = 2\gamma/\sigma^2$.

Smoluchowski Equation. As a reduced model of the Langevin equation, we introduce the Smoluchowski equation

$$\gamma \dot{q} = -\nabla_q V(q) + \sigma \dot{W} .$$

It is derived from the Langevin equation by considering the high friction limit $\gamma \to \infty$ [30, Theorem 10.1]. In contrast to the Langevin equation, the Smoluchowski dynamics defines a reversible Markov process; its stationary distribution is given by $\mu \propto \exp(-\beta V(q))$ with $\beta = 2\gamma/\sigma^2$.

The above models describe the dynamics of a single system. On the contrary, the evolution of densities v_t w.r.t. μ is described by the action of the so-called semigroup of transfer operators $\{P_t\}$, i.e., $v_t = P_t v_0$ where v_0 denotes the initial density. For the Hamiltonian system with randomized momenta we obtain

$$Pv(q) = \int_{\mathbf{R}^d} v(\Pi_q \Phi^{-\tau}(q,p)) \mathcal{P}(p) dp \, ,$$

while for the Langevin and Smoluchowski equation the semigroup is given by $P_t v = \exp(t\mathcal{L})v$ where $\mathcal{L}$ denotes the infinitesimal generator defined by the corresponding Fokker-Planck equation (e.g., [25, 32, 35]). In general, the spectrum $\sigma(P)$ of P is contained in the unit circle (the modulus of every eigenvalue is less or equal 1) and symmetric w.r.t. the real axis; moreover, $1 \in \sigma(P)$. Exploiting spectral properties of the transfer operators (eigenvalues close to 1 and their eigenvectors) we will be able to identify a decomposition of the state space into metastable subsets.

3 Markov Chains, Transfer Operators, and Metastability

We now state the general setting in terms of which the above specified models of molecular dynamics are specific cases. Throughout let $X = (X_n)_{n \in \mathbf{N}}$ denote a homogeneous Markov chain on the state space $\mathcal{X}$ with transition kernel

$$p(x, A) = \mathbf{P}[X_1 \in A | X_0 = x] \tag{1}$$

for all $x \in \mathcal{X}$ and all subsets $A \subset \mathcal{X}$ contained in the σ–algebra $\mathcal{A}$. Consider a probability measure ν on $\mathcal{X}$, and assume that the Markov chain is initially distributed according to ν, i.e., $X_0 \sim \nu$ meaning

$$\mathbf{P}[X_0 \in A] = \nu(A) \tag{2}$$

for all $A \in \mathcal{A}$. Then, the Markov chain at time $k > 0$ is distributed according to

$$\mathbf{P}[X_k \in A | X_0 \sim \nu] = \mathbf{P}_\nu[X_k \in A] =: \nu_k(A).$$

The time-evolution of probability measures $\{\nu_k\}$ can be described by the transfer operator P acting on the space of bounded measures on $(\mathcal{X}, \mathcal{A})$ via

$$P\nu(A) = \mathbf{P}_\nu[X_1 \in A] = \int_{\mathcal{X}} p(x, A)\nu(dx) \, . \tag{3}$$

Assume that the Markov chain exhibits a unique invariant probability measure μ, i.e., $P\mu = \mu$ and define the weighted Hilbert space of measurable functions

$$L^2(\mu) = \{f : \mathcal{X} \to \mathbf{R} : \|f\|^2 = \int_{\mathcal{X}} |f(x)|^2 \mu(dx) < \infty\}$$

with inner product given by

$$\langle f, g \rangle = \int_X f(x)g(x)\mu(dx) .$$

If μ is the invariant probability measure of P, then $\nu_0 \ll \mu$ implies $\nu_k \ll \mu$ [31, Chapter 4]. Hence we may consider P as an operator on $L^2(\mu)$ acting on probability measures that are absolutely continuous w.r.t. μ according to

$$\int_A Pv(x)\mu(dx) = \int_X p(x, A)v(x)\mu(dx) .$$

In the sequel, we assume that the Markov chain X is reversible, hence the transition kernel satisfies

$$\mu(dx)p(x, dy) = \mu(dy)p(y, dx) .$$

As a consequence, P is self-adjoint on $L^2(\mu)$. We now introduce the notion of the transition probabilities between subsets (see [25, 35]) in terms of which metastability will be defined:

Definition 1. *Let $A, B \subset X$ denote measurable subsets of the state space.*

1. The transition probability from A to B is defined to be the conditional probability

$$p(A, B) = \mathbf{P}_\mu[X_1 \in B | X_0 \in A] = \frac{1}{\mu(A)} \int_A p(x, B)\mu(dx) ,$$

if $\mu(A) > 0$ and $p(A, B) = 0$ otherwise. Hence, the transition probability quantifies the dynamical fluctuations within the invariant distribution μ.
2. A subset $A \in \mathcal{A}$ is called invariant if $p(A, A) = 1$.
3. A subset $A \in \mathcal{A}$ is called metastable if $p(A, A) \approx 1$.

Hence, metastability is almost invariance. Requiring the transition probability to be "close to 1" is obviously a vague statement, however, in most applications, we are interested in a decomposition into the most metastable subsets, which eliminates the problem of interpreting "close to 1". Instead, we have to determine the number of subsets we are looking for. This is done by examining the spectrum of the transfer operator P. Alternatively, we could determine a cascade of decompositions with an increasing number of metastable subsets.

It is easy to see that the transition probability between subsets can be rewritten in terms of the inner product $\langle \cdot, \cdot \rangle$ according to

$$p(A, B) = \frac{\langle P\mathbf{1}_A, \mathbf{1}_B \rangle}{\langle \mathbf{1}_A, \mathbf{1}_A \rangle} , \tag{4}$$

where $\mathbf{1}_A$ denotes the characteristic function of the subset A.

Consider a decomposition of the state space $\mathcal{X}$ into mutually disjoint subsets $\mathcal{D} = \{A_1, \ldots, A_n\}$. Then,

$$m(\mathcal{D}) = p(A_1, A_1) + \ldots + p(A_n, A_n)$$

can be thought of as a measure of metastability of the decomposition $\mathcal{D}$. In general, $0 \leq m(\mathcal{D}) \leq n$ with $m(\mathcal{D}) = n$ if all subsets are invariant. It is our aim to get upper and lower bounds on the metastability of the decomposition in terms of eigenvalues and corresponding eigenvectors of the transfer operator.

We are interested in situations where the spectrum of the transfer operator satisfies the following

Assumption S: The transfer operator $P : L^2(\mu) \rightarrow L^2(\mu)$ is self–adjoint and exhibits n eigenvalues

$$\lambda_n \leq \ldots \leq \lambda_2 < \lambda_1 = 1$$

counted according to their multiplicity. The corresponding set of μ–orthonormal eigenvectors will be denoted by $\{v_n, \ldots, v_1\}$. Furthermore, the spectrum $\sigma(P)$ of P satisfies

$$\sigma(P) \subset [a, b] \cup \{\lambda_n, \ldots, \lambda_2, 1\}$$

for some constants a, b satisfying $-1 < a \leq b < \lambda_n$. In this sense, the eigenvalues $\lambda_1, \ldots, \lambda_n$ are called dominant.

In particular, Assumption S is satisfied if the underlying Markov chain is reversible and geometrically or V-uniformly ergodic (see, e.g., [25, Thm. 4.31]) which is always the case if the state space is finite dimensional (and the Markov chain reversible). Moreover, fixing some time span $\tau > 0$ the Assumption S is satisfied for (i) the Hamiltonian system with randomized momenta with periodic boundary conditions and some smooth potential, (ii) the Smoluchowski equation with periodic boundary conditions and some smooth potential, or for so-called bounded systems with smooth potentials satisfying a suitable growth condition at infinity. For further details, also regarding the Langevin equation, see [25, 33]. Note that reversibility and "simple" ergodicity (space and time average coincide) is not sufficient to guarantee Assumption S.

4 Upper and Lower Bounds

This section proves upper and lower bounds on the metastability of an arbitrary decomposition of the state space.

Recall that by Rayleigh's Principle the kth largest eigenvalue λ_k for $1 \leq k \leq n$, is given by the variational formula

$$\lambda_k = \max\{\langle Pw, w\rangle : w \in L^2(\mu), \|w\|_2 = 1, w \perp v_1 \ldots, v_{k-1}\},$$

where $\perp$ denotes orthogonality w.r.t. the inner product $\langle \cdot, \cdot \rangle$, and v_i is the eigenvector corresponding to λ_i. The above variational formula can be generalized (for our purpose) in the following way: Consider a finite dimensional subspace U of $L^2(\mu)$ with orthonormal basis $(\varphi_1, \ldots, \varphi_n)$. Then, for a self-adjoint operator P on $L^2(\mu)$ the Rayleigh-trace w.r.t. U is defined as

$$\mathrm{Tr}_U P = \sum_{i=1}^{n} \langle P\varphi_i, \varphi_i\rangle.$$

Note that this definition is independent of the particular choice of the orthonormal basis (see, e.g., [1]).

Theorem 1. *Assume that* $P : L^2(\mu) \to L^2(\mu)$ *is a self adjoint transfer operator satisfying Assumption S. Then*

$$\lambda_n + \ldots + \lambda_1$$
$$= \max\{Tr_U P : U \text{ is } n\text{-dimensional subspace}\}$$
$$= \max\left\{\sum_{i=1}^{n} \langle P\varphi_i, \varphi_i\rangle : (\varphi_1, \ldots, \varphi_n) \text{ is orthonormal system}\right\}.$$

The above theorem is actually known to hold for every self-adjoint bounded operator P on a Hilbert space H [1]. The generalized Rayleigh Principle can be exploited to prove upper bounds on the metastability of some (arbitrary) partition $A_1, \ldots, A_n$ of the state space $\mathcal{X}$ that satisfies $\mu(A_k) > 0$ for $k = 1, \ldots n$. Recall that the orthogonal projection $Q : L^2(\mu) \to L^2(\mu)$ onto $\mathrm{span}\{1_{A_1}, \ldots, 1_{A_n}\}$ is defined as

$$Qv = \sum_{k=1}^{n} \frac{\langle v, 1_{A_k}\rangle}{\langle 1_{A_k}, 1_{A_k}\rangle} 1_{A_k} = \sum_{k=1}^{n} \langle v, \chi_{A_k}\rangle \chi_{A_k}$$

with

$$\chi_{A_k} = \frac{1_{A_k}}{\sqrt{\langle 1_{A_k}, 1_{A_k}\rangle}}$$

for $k = 1, \ldots, n$ and for every $v \in L^2(\mu)$. Our central result is

Theorem 2. *Consider some transfer operator* $P : L^2(\mu) \to L^2(\mu)$ *satisfying Assumption S. Then the metastability of an arbitrary decomposition* $\mathcal{D} = \{A_1, \ldots, A_n\}$ *of the state space can be bounded from above by*

$$p(A_1, A_1) + \ldots + p(A_n, A_n) \leq 1 + \lambda_2 + \ldots + \lambda_n,$$

while it is bounded from below by

$$1 + \rho_2 \lambda_2 + \ldots + \rho_n \lambda_n + c \leq p(A_1, A_1) + \ldots + p(A_n, A_n)$$

where $\rho_j = \|Qv_j\|^2 = \langle Qv_j, Qv_j \rangle \in [0, 1]$ and

$$c = a\left(1 - \rho_2 + \ldots + 1 - \rho_n\right).$$

In particular, we have $c \geq 0$ if $\sigma(P) \subset [0, 1]$.

The coefficients ρ_j in Theorem 2 measure how close the eigenfunctions v_j restricted to each of the subsets of the decomposition are to the constant function. The more constant the eigenfunctions are on the subsets, the larger the ρ_j are and thus the larger the lower bound is. Hence, one can try to maximize the lower bound in seeking for a decomposition such that the eigenfunctions, restricted to each subset, are as constant as possible. This strategy is used algorithmically to identify a decomposition into metastable subsets ([9, 14, 36]).

Proof. Upper bound: Since $p(A_k, A_k) = \langle P\chi_{A_k}, \chi_{A_k} \rangle$ using (4) and the definition of χ_{A_k} we have

$$\sum_{k=1}^{n} p(A_k, A_k) = \sum_{k=1}^{n} \langle P\chi_{A_j}, \chi_{A_j} \rangle . \tag{5}$$

By Theorem 1, the right hand side of (5) is less than or equal to $\lambda_1 + \ldots + \lambda_n$ since $\{\chi_{A_1}, \ldots, \chi_{A_n}\}$ is an orthonormal basis of $\mathrm{span}\{\mathbf{1}_{A_1}, \ldots, \mathbf{1}_{A_n}\}$.

Lower bound: Denote by $\Pi : L^2(\mu) \rightarrow \mathrm{span}\{v_1, \ldots, v_n\}$ the orthogonal projection onto the subspace spanned by the maximal eigenvectors, and set $\Pi^\perp = \mathrm{Id} - \Pi$. Then

$$\sum_{j=1}^{n} p(A_j, A_j)$$

$$= \sum_{j=1}^{n} \langle (P - a\mathrm{Id})\chi_{A_j}, \chi_{A_j} \rangle + \sum_{j=1}^{n} a \langle \chi_{A_j}, \chi_{A_j} \rangle$$

$$= \sum_{j=1}^{n} \langle ((P - a\mathrm{Id})\Pi + (P - a\mathrm{Id})\Pi^\perp)\chi_{A_j}, (\Pi + \Pi^\perp)\chi_{A_j} \rangle$$

$$+ an$$

$$= \sum_{j=1}^{n} \langle (P - a\mathrm{Id})\Pi\chi_{A_j}, \Pi\chi_{A_j} \rangle$$

$$+ \sum_{j=1}^{n} \langle (P - a\mathrm{Id})\Pi^\perp\chi_{A_j}, \Pi^\perp\chi_{A_j} \rangle + an .$$

The first two terms of the right hand side can be further analyzed:

$$\sum_{j=1}^{n} \left\langle (P - a\mathrm{Id})\Pi \chi_{A_j} , \Pi \chi_{A_j} \right\rangle$$

$$= \sum_{j=1}^{n} \left\langle \sum_{k=1}^{n} (\lambda_k - a)\langle \chi_{A_j}, v_k \rangle v_k , \sum_{l=1}^{n} \langle \chi_{A_j}, v_l \rangle v_l \right\rangle$$

$$= \sum_{j=1}^{n} \sum_{k=1}^{n} (\lambda_k - a)\langle \chi_{A_j}, v_k \rangle^2$$

$$= \sum_{k=1}^{n} (\lambda_k - a)\langle Qv_k, Qv_k \rangle .$$

Now $\left\langle (P - a\mathrm{Id})\Pi^{\perp} \chi_{A_j} , \Pi^{\perp} \chi_{A_j} \right\rangle$ is non–negative since $P - a\mathrm{Id}$ is non-negative definite according to the assumptions made. Hence,

$$\sum_{j=1}^{n} p(A_j, A_j) \geq \sum_{k=1}^{n} \lambda_k \rho_k + a \sum_{k=1}^{n} (1 - \rho_k),$$

which completes the proof since invariance of μ implies $v_1 = \mathbf{1}_\mathcal{X}$ and thus $\rho_1 = \|Qv_1\|^2 = 1$.

Our main result, Theorem 2, does hold for an arbitrary transfer operator satisfying Assumption S. Note that for the finite state space case Assumption S is trivially satisfied (and to the best of our knowledge even for finite stochastic matrices the lower bound is new). We did not need to introduce any asymptotic smallness parameter κ, say, in order to prove asymptotic results for $\kappa \to 0$. This is a remarkable difference to other approaches. Moreover, the lower bound is computable explicitly given some decomposition of the state space. Hence, comparing the lower and upper bound one is able to "judge" the quality of the decomposition. Moreover, we remark that if $\{\varphi_1, \ldots, \varphi_n\}$ is an arbitrary orthonormal basis of $\mathrm{span}\{\mathbf{1}_{A_1}, \ldots, \mathbf{1}_{A_n}\}$, then

$$\kappa_1 \lambda_1 + \ldots + \kappa_n \lambda_n + c \leq p(A_1, A_1) + \ldots + p(A_n, A_n) , \tag{6}$$

where $\kappa_j = |\langle v_j, \varphi_j \rangle|^2 \in [0, 1]$ which follows from

$$\rho_j = \langle Qv_j, Qv_j \rangle = \sum_{k=1}^{n} |\langle v_j, \varphi_k \rangle|^2 \geq |\langle v_j, \varphi_j \rangle|^2 = \kappa_j .$$

In some situations, we additionally know that P is positive, for instance, if we consider the case of $P = P_\tau$, where $(P_t)_{t \geq 0}$ is a semigroup of transfer operators, and τ is some fixed time. (This is the case, e.g., for the Langevin and Smoluchowski dynamics discussed in Sect. 2 but not for the Hamiltonian system with randomized momenta.) Then, we can state:

Corollary 3 *Consider a reversible homogeneous continuous-time Markov process* $X = (X_t)_{t \in [0,\infty)}$ *and its corresponding semigroup of transfer operators* $P_t : L^2(\mu) \to L^2(\mu)$. *If* $P = P_\tau$ *satisfies Assumption S for some fixed* $\tau > 0$, *then*

$$1 + \rho_2 \lambda_2 + \ldots + \rho_n \lambda_n \le p(A_1, A_1) + \ldots + p(A_n, A_n) \le \lambda_1 + \ldots + \lambda_n, \quad (7)$$

where λ_k denote eigenvalues of the operator P_τ.

Proof. Simply note that $P = P_\tau$ is positive since $P = P_{\tau/2} P_{\tau/2}$, and apply Theorem 2 (with $a = 0$).

5 Illustrative Examples

The first example proves that both the lower and the upper bound of Theorem 2 are sharp.

Example 1. Let $\mathcal{X} = \{0, 1, 2\}$ and the transition probability P be given by

$$P = \begin{pmatrix} 0.90 & 0.05 & 0.05 \\ 0.05 & 0.05 & 0.90 \\ 0.05 & 0.90 & 0.05 \end{pmatrix}.$$

Clearly P is ergodic, and since it is symmetric, the measure μ given by $\mu(\{0\}) = \mu(\{1\}) = \mu(\{2\}) = 1/3$ is invariant. The eigenvalues λ_j and corresponding eigenvectors v_j are calculated to be

$$\lambda_1 = 1, \quad \lambda_2 = 0.85, \quad \lambda_3 = -0.85$$

and (we do not need v_3)

$$v_1 = \begin{pmatrix} 1 \\ 1 \\ 1 \end{pmatrix}, \quad v_2 = \frac{1}{\sqrt{2}} \begin{pmatrix} 2 \\ -1 \\ -1 \end{pmatrix}.$$

Consider the partition $(A_1, A_2) = (\{0, 1\}, \{2\})$. The resulting metastability is given by $p(A_1, A_1) + p(A_2, A_2) = 0.525 + 0.05 = 0.575$, which is bounded from above by $1 + \lambda_2 = 1.85$. Calculating the lower bound from Theorem 2, we obtain (here the correction term is $c = -0.6375$)

$$0.575 = 1 + \rho_2 \lambda_2 + c \le p(A_1, A_1) + p(A_2, A_2) = 0.575$$

which proves that the lower bound is sharp.

Now consider the partition $(A_1, A_2) = (\{0\}, \{1, 2\})$. The resulting metastability is given by

$$p(A_1, A_1) + p(A_2, A_2) = 0.90 + 0.95 = 1.85$$

which in this case is equal to both the upper and lower bound

$$1 + \lambda_2 = 1.85 \text{ and } 1 + \rho_2 \lambda_2 + c = 1.85$$

since $\rho_1 = \rho_2 = 1$ and $c = 0$. This additionally proves that the upper bound is sharp, too. Note that although $\lambda_3 = -0.85$ is large negative, the correction term c does not necessarily result in some lower bound that underestimates the metastability of the decomposition. The above example particularly demonstrates the need for a correction of the sum $1 + \rho_2\lambda_2 + \ldots + \rho_n\lambda_n$ by c in order to get a correct lower bound stated in Theorem 2.

We next illustrate for a more advanced system that the lower bound mimics the behavior of the metastability for different decompositions of the state space.

Example 2. Consider the Smoluchowski dynamics

$$\gamma\dot{q} = -\mathrm{grad}V(q)t + \sigma\dot{W}_t$$

within a "perturbed" three well potential (see Fig. 1)

$$V(q) = 0.01\left(q^6 - 30q^4 + 234q^2 + 14q + 100 + \right. \tag{8}$$
$$\left. 30\sin(17q) + 26\cos(11q)\right)$$

for given parameters $\gamma = 2$ and $\sigma^2 = 2\gamma/\beta$ with inverse temperature β. For a fixed observation time span $\tau = 1$ we discretize the transfer operator P_τ (for details of the discretization see [25, 35]). In view of the potential function V shown in Fig. 1, we would expect to exist three metastable subsets, each corresponding to one of the wells, if the inverse temperature is not too small. Hence, we choose $L < R$ and decompose the state space into the subsets

$$A_1 = (-\infty, L], A_2 = (L, R], A_3 = (R, \infty)$$

The purpose of this example is to illustrate the dependence of the metastability $m(\mathcal{D})$ of the decomposition $\mathcal{D} = \{A_1, A_2, A_3\}$ on the location of the boundaries L and R and, moreover, to show that the lower bound (in this case) behaves similar to the true metastability of the decomposition for different choices of L, R.

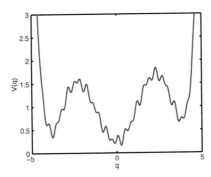

Figure 1. Graph of the perturbed three well potential V defined in (8)

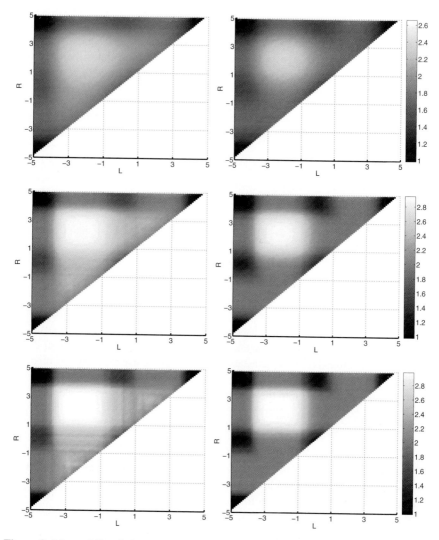

Figure 2. Metastability (left column) and lower bound (right column) corresponding to the perturbed three–well potential. From top to bottom increasing metastability due to increasing inverse temperature $\beta = 1$ (top), $\beta = 3$ (middle) and $\beta = 5$ (bottom)

Figure 2 shows the calculated metastability $m(\mathcal{D})$ and the lower bound according to Theorem 2 for three different values of inverse temperature. The case $\beta = 1$ mimics the situation of moderate metastability, while the case $\beta = 7$ corresponds to high metastability. As can be seen from Fig. 2, the lower bound is a good indicator for the actual metastability of the decomposition. Moreover, the shape of the "metastability landscape" gets more plateau-like for larger β, in particular, the identification

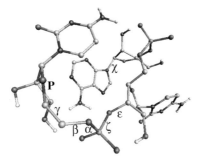

Figure 3. Configuration of the trinucleotide $r(ACC)$ in a ball–and–stick representation. The Greek symbols indicate some of the important dihedral angles of the molecule

of the optimal decomposition becomes more and more difficult (and numerically ill-conditioned) for larger β.

The third example demonstrates the application to a small biomolecule.

Example 3. Consider the triribonucleotide r(ACC) model system (see Fig. 3) in vacuum. Its physical representation ($N = 70$ atoms) is based on the GROMOS96 extended atom force field [41]. In the first step, the canonical ensemble has been sampled by means of the Hamiltonian system with randomized momenta [18, 36]. Then, based on some coarse-grained clustering of the resulting sampling point, the transfer operator has be discretized. Solving the eigenvalue problem for the discretized transfer operator, a cluster of eight eigenvalues with a significant gap to the remaining part of the spectrum showed up:

k	1	2	3	4	5	6	7	8	9	...
λ_k	1.000	0.999	0.989	0.974	0.963	0.946	0.933	0.904	0.805	...

Then, the almost constant level structure of eigenvectors corresponding to the cluster of eigenvalues near 1 was exploited and eight conformations characterized by their statistical weight within the canonical ensemble (the probability of the conformation) and metastability were identified (see [18, 36] for details):

conformation	1	2	3	4	5	6	7	8
statistical weight	0.107	0.011	0.116	0.028	0.320	0.038	0.285	0.095
metastability	0.986	0.938	0.961	0.888	0.991	0.949	0.981	0.962

The above table clearly illustrates that metastability and the statistical weight within the stationary ensemble are two independent properties. Of course, in most biochemical applications one likely is interested to identify metastable conformations with a relevant statistical weight.

6 Outlook

The detection and identification of metastability in dynamical systems is an active field of research. The existence of metastability is exploited to avoid so-called trapping problems in Markov chain Monte Carlo simulations (e.g., Uncoupling-Coupling techniques [15]), to identify molecular conformations (e.g., [35]) or to extend standard averaging schemes [24]. Recent studies indicate a relation between metastable subsets and hidden states in so-called Hidden Markov models [19]. The concept of metastability is also exploited to identify multiple transition paths between different metastable conformations. Besides molecular applications, the concept of metastability is applied, e.g., to economics [39], climate modelling [26], or astronomy [11].

Acknowledgements

W.H. acknowledges financial support by the DFG Research Center MATHEON, Berlin.

References

[1] C. Bandle. *Isoperimetric Inequalities and Applications*. Pitman, Boston, London, Melbourne, 1980.

[2] Anton Bovier, Michael Eckhoff, Véronique Gayrard, and Markus Klein. Metastability in stochastic dynamics of disordered mean–field models. *Probab. Theor. Rel. Fields*, 119:99–161, 2001.

[3] Anton Bovier, Michael Eckhoff, Véronique Gayrard, and Markus Klein. Metastability and low lying spectra in reversible Markov chains. *Com. Math. Phys.*, 228:219–255, 2002.

[4] Anton Bovier, Michael Eckhoff, Véronique Gayrard, and Markus Klein. Metastability in reversible diffusion processes I: Sharp estimates for capacities and exit times. *J. Eur. Math. Soc.*, 6:399–424, 2004.

[5] Anton Bovier, Véronique Gayrard, and Markus Klein. Metastability in reversible diffusion processes II: Precise estimates for small eigenvalues. *J. Eur. Math. Soc.*, 7:69–99, 2005.

[6] Edward Brian Davies. Metastability and the Ising model. *J. Statist. Phys*, 27:657–675, 1982.

[7] Edward Brian Davies. Metastable states of symmetric Markov semigroups I. *Proc. London Math. Soc.*, 45(3):133–150, 1982.

[8] Edward Brian Davies. Metastable states of symmetric Markov semigroups II. *J. London Math. Soc.*, 26(2):541–556, 1982.

[9] Peter Deuflhard, Wilhelm Huisinga, Alexander Fischer, and Christof Schütte. Identification of almost invariant aggregates in reversible nearly uncoupled Markov chains. *Lin. Alg. Appl.*, 315:39–59, 2000.

[10] Michael Dellnitz and Oliver Junge. On the approximation of complicated dynamical behavior. *SIAM J. Num. Anal.*, 36(2):491–515, 1999.

[11] M. Dellnitz, O. Junge, W.S. Koon, F. Lekien, M.W. Lo, J.E. Marsden, K. Padberg R. Preis, S.D. Ross, and B. Thiere. Transport in dynamical astronomy and multibody problems. *International Journal of Bifurcation and Chaos*, 2005. to appear.

[12] J. L. Doob. *Stochastic Processes*. John Wiley & Sons, Inc., New York, 1953.

[13] P. Deuflhard and Ch. Schütte. Molecular conformation dynamics and computational drug design. In J.M. Hill and R. Moore, editors, *Applied Mathematics Entering the 21st Century, Proc. ICIAM 2003*, pages 91–119, Sydney, Australia, 7-11 July 2004.

[14] P. Deuflhard and M. Weber. Robust Perron cluster analysis in conformation dynamics. *Lin. Alg. Appl. — Special issue on Matrices and Mathematical Biology. In Dellnitz, S. Kirkland, M. Neumann and C. Schütte (eds.)*, 398C:161–184, 2004.

[15] Alexander Fischer. *An Uncoupling–Coupling Method for Markov chain Monte Carlo simulations with an application to biomolecules*. PhD thesis, Free University Berlin, 2003.

[16] M.I. Freidlin and A.D. Wentzell. *Random Perturbations of Dynamical Systems*. Springer, New York, 1984. Series in Comprehensive Studies in Mathematics.

[17] C. W. Gardiner. *Handbook of Stochastic Methods*. Springer, Berlin, 2nd enlarged edition edition, 1985.

[18] Wilhelm Huisinga, Christoph Best, Rainer Roitzsch, Christof Schütte, and Frank Cordes. From simulation data to conformational ensembles: Structure and dynamic based methods. *J. Comp. Chem.*, 20(16):1760–1774, 1999.

[19] I. Horenko, E. Dittmer, A. Fischer, and Ch. Schütte. Automated model reduction for complex systems exhibiting metastability. *Multiscale Modeling and Simulation*, 2005. submitted.

[20] R. Hassin and M. Haviv. Mean passage times and nearly uncoupled Markov chains. *SIAM J. Disc. Math.*, 5(3):386–397, 1992.

[21] Bruce Hendrickson and Robert Leland. An improved spectral graph partitioning algorithm for mapping parallel computations. *SIAM J. Sci. Comput.*, 16(2):452–469, 1995.

[22] D. J. Hartfiel and C. D. Meyer. On the structure of stochastic matrices with a subdominant eigenvalue near 1. *Lin. Alg. Appl.*, 272:193–203, 1998.

[23] Wilhelm Huisinga, Sean Meyn, and Christof Schütte. Phase transitions & metastability in Markovian and molecular systems. *Ann. Appl. Probab.*, 14:419–458, 2004.

[24] Wilhelm Huisinga, Christof Schütte, and Andrew M. Stuart. Extracting macroscopic stochastic dynamics: Model problems. *Comm. Pure Appl. Math.*, 56:234–269, 2003.

[25] Wilhelm Huisinga. *Metastability of Markovian systems: A transfer operator approach in application to molecular dynamics*. PhD thesis, Freie Universität Berlin, 2001.

[26] P. Imkeller and A. Monahan. Conceptual stochastic climate models. *Stochastics and Dynamics*, 2(3):311–326, 2002.

[27] Andrzej Lasota and Michael C. Mackey. *Chaos, Fractals and Noise*, volume 97 of *Applied Mathematical Sciences*. Springer, New York, 2nd edition, 1994.

[28] C. D. Meyer. Stochastic complementation, uncoupling Markov chains, and the theory of nearly reducible systems. *SIAM Rev.*, 31:240–272, 1989.

[29] S.P. Meyn and R.L. Tweedie. *Markov Chains and Stochastic Stability*. Springer, Berlin, 1993.

[30] Edward Nelson. *Dynamical Theories of Brownian Motion*. Mathematical Notes. Princeton Uni. Press, 1967.

[31] D. Revuz. *Markov Chains*. North–Holland, Amsterdam, Oxford, 1975.

[32] Hannes Risken. *The Fokker-Planck Equation*. Springer, New York, 2nd edition, 1996.

[33] Ch. Schütte. *Conformational Dynamics: Modelling, Theory, Algorithm, and Application to Biomolecules*. Habilitation Thesis, Fachbereich Mathematik und Informatik, Freie Universität Berlin, 1998.

[34] Ch. Schütte, A. Fischer, W. Huisinga, and P. Deuflhard. A direct approach to conformational dynamics based on hybrid Monte Carlo. *J. Comput. Phys., Special Issue on Computational Biophysics*, 151:146–168, 1999.

[35] Christof Schütte and Wilhelm Huisinga. Biomolecular conformations can be identified as metastable sets of molecular dynamics. In P. G. Ciarlet, editor, *Handbook of Numerical Analysis*, volume Special Volume Computational Chemistry, pages 699–744. North–Holland, 2003.

[36] Christof Schütte, Wilhelm Huisinga, and Peter Deuflhard. Transfer operator approach to conformational dynamics in biomolecular systems. In Bernold Fiedler, editor, *Ergodic Theory, Analysis, and Efficient Simulation of Dynamical Systems*, pages 191–223. Springer, 2001.

[37] Gregory Singleton. Asymptotically exact estimates for metastable Markov semigroups. *Quart. J. Math. Oxford*, 35(2):321–329, 1984.

[38] Alistair Sinclair. *Algorithms for Random Generation and Counting – A Markov Chain Approach*. Progress in Theoretical Computer Science. Birkhäuser, 1993.

[39] John Stachurski. Stochastic growth with increasing returns: Stability and path dependence. *Studies in Nonlinear Dynamics & Econometrics*, 7(2):1104–1104, 2003.

[40] Christof Schütte, Jessika Walter, Carsten Hartmann, and Wilhelm Huisinga. An averaging principle for fast degrees of freedom exhibiting long-term correlations. *SIAM Multiscale Modeling and Simulation*, 2:501–526, 2004.

[41] W. F. van Gunsteren, S. R. Billeter, A. A. Eising, P. H. Hünenberger, P. Krüger, A. E. Mark, W. R. P. Scott, and I. G. Tironi. *Biomolecular Simulation: The GROMOS96 Manual and User Guide*. vdf Hochschulverlag AG, ETH Zürich, 1996.

Part IV

Computation of the Free Energy

Free Energy Calculations in Biological Systems. How Useful Are They in Practice?

Christophe Chipot

Equipe de dynamique des assemblages membranaires, UMR CNRS/UHP 7565, Institut nancéien de chimie moléculaire, Université Henri Poincaré, BP 239, 54506 Vandœuvre–lès–Nancy cedex, France

Abstract. Applications of molecular simulations targeted at the estimation of free energies are reviewed, with a glimpse into their promising future. The methodological milestones paving the road of free energy calculations are summarized, in particular free energy perturbation and thermodynamic integration, in the framework of constrained or unconstrained molecular dynamics. The continuing difficulties encountered when attempting to obtain accurate estimates are discussed with an emphasis on the usefulness of large–scale numerical simulations in non–academic environments, like the world of the pharmaceutical industry. Applications of the free energy arsenal of methods is illustrated through a variety of biologically relevant problems, among which the prediction of protein–ligand binding constants, the determination of membrane–water partition coefficients of small, pharmacologically active compounds — in connection with the blood–brain barrier, the folding of a short hydrophobic peptide, and the association of transmembrane α–helical domains, in line with the "two–stage" model of membrane protein folding. Current strategies for improving the reliability of free energy calculations, while making them somewhat more affordable, and, therefore, more compatible with the constraints of an industrial environment, are outlined.

Key words: Free energy calculations, molecular dynamics simulations, drug design, protein folding, protein recognition and association

1 Introduction

To understand fully the vast majority of chemical and biochemical processes, a close examination of the underlying free energy behavior is often necessary [1]. Such is the case, for instance, of protein–ligand binding constants and membrane–water partition coefficients, that are of paramount important in the emerging field of *de novo*, rational drug design, and cannot be predicted reliably and accurately without the knowledge of the associated free energy changes. The ability to determine *a priori* these physical constants with a reasonable level of accuracy, by means of statistical simulations, is now within reach. Developments on both the software and the hardware fronts have contributed to bring free energy calculations at the level of similarly robust and well–characterized modeling tools, while widening their field of applications. Yet, in spite of the tremendous progress accomplished since the first published calculations some twenty years ago [2–5], the accurate estimation of free energy changes in large, physically and biologically realistic molecular assemblies still constitutes a challenge for modern theoretical chemistry. Taking advantage of massively parallel architectures, cost–effective, "state–of–the–art" free energy calculations can provide a convincing answer to help rationalizing experimental observations, and, in some instances, play a predictive role in the development of new leads for a specific target.

In the first section of this chapter, the methodological background of free energy calculations is developed, focusing on the methods that are currently utilized to determine free energy differences. Next, four biologically relevant applications are presented, corresponding to distinct facets of free energy simulations. The first one delves into the use of these calculations in *de novo* drug design, through the estimation of protein–ligand binding affinities and water–membrane partition coefficients. The somewhat more challenging application of molecular dynamics (MD) simulations and free energy methods to validate the three–dimensional structure of a G protein–coupled receptor (GPCR) is shown next. Understanding of the intricate physical phenomena that drive protein folding by means of free energy calculations is also reported here, followed by an investigation of the reversible association of transmembrane (TM) α–helices in a membrane mimetic. Conclusions on the role played by free energy calculations in the molecular modeling community are drawn with a prospective look into their future.

2 Methodological Background

In the canonical, (N, V, T), ensemble, the Helmholtz free energy is defined by [6]:

$$A = -\frac{1}{\beta} \ln Q_{\text{NVT}} \tag{1}$$

$\beta = 1/k_B T$, where k_B is the Boltzmann constant and T is the temperature of the N–particle system. Q_{NVT} is its $6N$–dimensional partition function:

$$Q_{\text{NVT}} = \frac{1}{h^{3N} N!} \int \int \exp\left[-\beta \mathcal{H}(\mathbf{x}, \mathbf{p}_x)\right] \, d\mathbf{x} \, d\mathbf{p}_x \tag{2}$$

where $\mathcal{H}(\mathbf{x}, \mathbf{p}_x)$ is the classical Hamiltonian describing the system. In (2), integration is carried out over all atomic coordinates, $\{\mathbf{x}\}$, and momenta, $\{\mathbf{p}_x\}$. The normalization factor reflects the measure of the volume of the phase space through the Planck constant, h, and the indistinguishable nature of the particles, embodied in the factorial term, $N!$. In essence, the canonical partition function constitutes the corner stone of the statistical mechanical description of the ensemble of particles. From a phenomenological perspective, it may be viewed as a measure of the thermodynamic states accessible to the system in terms of spatial coordinates and momenta. It can be further restated in terms of energies:

$$Q_{\text{NVT}} = \frac{1}{h^{3N} N!} \int \varrho[\mathcal{H}(\mathbf{x}, \mathbf{p}_x)] \exp\left[-\beta \mathcal{H}(\mathbf{x}, \mathbf{p}_x)\right] \mathrm{d}\mathcal{H}(\mathbf{x}, \mathbf{p}_x) \qquad (3)$$

where $\varrho[\mathcal{H}(\mathbf{x}, \mathbf{p}_x)]$ is the so–called density of states accessible to the system of interest.

The definition of the partition function may be utilized to introduce the concept of probability distribution to find the system in the unique microscopic state characterized by positions $\{\mathbf{x}\}$ and momenta $\{\mathbf{p}_x\}$:

$$\mathcal{P}(\mathbf{x}, \mathbf{p}_x) = \frac{1}{h^{3N} N!} \frac{1}{Q_{\text{NVT}}} \exp\left[-\beta \mathcal{H}(\mathbf{x}, \mathbf{p}_x)\right] \qquad (4)$$

A logical consequence of this expression is that low–energy regions of the phase space will be sampled predominantly, according to their respective Boltzmann weight [7].

2.1 Free Energy Perturbation

Returning to the original definition (1) of the free energy, and using the identity:

$$\int \int \exp\left[+\beta \mathcal{H}(\mathbf{x}, \mathbf{p}_x)\right] \exp\left[-\beta \mathcal{H}(\mathbf{x}, \mathbf{p}_x)\right] \mathrm{d}\mathbf{x}\, \mathrm{d}\mathbf{p}_x = h^{3N} N! \qquad (5)$$

it follows that:

$$A = -\frac{1}{\beta} \ln \frac{1}{h^{3N} N!} \int \int \exp\left[-\beta \mathcal{H}(\mathbf{x}, \mathbf{p}_x)\right] \mathrm{d}\mathbf{x}\, \mathrm{d}\mathbf{p}_x$$

$$= +\frac{1}{\beta} \ln \int \int \exp\left[+\beta \mathcal{H}(\mathbf{x}, \mathbf{p}_x)\right] \mathcal{P}(\mathbf{x}, \mathbf{p}_x)\, \mathrm{d}\mathbf{x}\, \mathrm{d}\mathbf{p}_x$$

$$= +\frac{1}{\beta} \ln \langle \exp\left[+\beta \mathcal{H}(\mathbf{x}, \mathbf{p}_x)\right] \rangle \qquad (6)$$

This expression illuminates the fast growth of $\exp\left[+\beta \mathcal{H}(\mathbf{x}, \mathbf{p}_x)\right]$ with the total energy, $\mathcal{H}(\mathbf{x}, \mathbf{p}_x)$, of the system. It should, therefore, be expected that the weight of the high–energy regions of phase space be significant when evaluating the integral. Yet, as hinted by (4), in simulations of finite length, sampling of these regions is likely to be insufficient to guarantee a correct estimate of A. In most instances, evaluation of accurate absolute free energies from statistical simulations is not possible. The latter may, however, give access to free energy differences between two well–delineated thermodynamic states, provided that a reaction coordinate can be defined to characterize the pathway that connects these two states. In this context, the Hamiltonian, $\mathcal{H}(\mathbf{x}, \mathbf{p}_x)$, describing the transformation is made a function of the reaction coordinate, or "coupling parameter", λ [8]. Conventionally, λ varies between 0 and 1 when the system goes from the initial state, a, to the final state, b, characterized, respectively, by the Hamiltonians $\mathcal{H}(\mathbf{x}, \mathbf{p}_x; \lambda_a) = \mathcal{H}(\mathbf{x}, \mathbf{p}_x; \lambda = 0)$ and $\mathcal{H}(\mathbf{x}, \mathbf{p}_x; \lambda_b) = \mathcal{H}(\mathbf{x}, \mathbf{p}_x; \lambda = 1)$. In practice, λ can correspond to a variety of reaction coordinates, ranging from a simple distance to determine a potential of mean force (PMF) to non–bonded parameters in the so–called "alchemical transformations" or *in silico* point mutations [4, 5].

Within this framework, the canonical partition function defined in (2) now depends explicitly on the coupling parameter, and so does the free energy:

$$\Delta A_{a\to b} = A(\lambda_b) - A(\lambda_a) = -\frac{1}{\beta} \ln \frac{Q_{\mathrm{NVT}}(\lambda_b)}{Q_{\mathrm{NVT}}(\lambda_a)} \qquad (7)$$

Combining the above with the definition of the partition function, and introducing the identity (5), it follows that:

$$\Delta A_{a\to b} = -\frac{1}{\beta} \ln \langle \exp\{-\beta\, [\mathcal{H}(\mathbf{x}, \mathbf{p}_x; \lambda_b) - \mathcal{H}(\mathbf{x}, \mathbf{p}_x; \lambda_a)]\}\rangle_{\lambda_a} \qquad (8)$$

Here, $\langle \cdots \rangle_{\lambda_a}$ denotes an ensemble average over configurations representative of the initial state, a. Validity of perturbation formula (8) only holds for small changes between the initial state, a, and the final state, b, of the transformation [9]. At this stage, the condition of *small changes* should be clarified, as it has often been misconstrued in the past. It does not imply that the free energies characteristic of a and b be sufficiently close, but rather that the corresponding configurational ensembles overlap appropriately to guarantee the desired accuracy [10, 11]. In other words, it is expected that the density of states, $\varrho[\mathcal{H}(\mathbf{x}, \mathbf{p}_x)]$, describing the transformation from a to b be narrow enough — *viz.* typically on the order of $1/\beta$, to ascertain that, when multiplied by the exponential term of (3), the resulting distribution be located in a region where ample statistical data have been collected. In most circumstances, however, single–step transformations between rather orthogonal states are unlikely to fulfill this requirement. To circumvent this difficulty, the reaction pathway is broken down into a number of physically meaningless intermediate states connecting a to b, so that between any two contiguous states, the condition of overlapping ensembles is satisfied [12]. The interval separating these intermediate states, which corresponds to selected fixed values of the coupling parameter, λ, is often referred to as "window". It should be reminded that the vocabulary *window* adopted in perturbation theory is distinct from that utilized in "umbrella sampling" (US) simulations [13], where it denotes a range of values taken by the reaction coordinate. For a series of a N intermediate states, the total free energy change for the transformation from a to b is expressed as a sum of $N - 1$ free energy differences [12]:

$$\Delta A_{a\to b} = -\frac{1}{\beta} \sum_{k=1}^{N-1} \ln \langle \exp\{-\beta\, [\mathcal{H}(\mathbf{x}, \mathbf{p}_x; \lambda_{k+1}) - \mathcal{H}(\mathbf{x}, \mathbf{p}_x; \lambda_k)]\}\rangle_{\lambda_k} \qquad (9)$$

Assessing the ideal number of intermediate states, N, between a and b evidently depends upon the nature of the system that undergoes the transformation. The condition of overlapping ensembles should be kept in mind when setting N, remembering that the choice of $\delta\lambda = \lambda_{k+1} - \lambda_k$ ought to correspond to a perturbation of the system. A natural choice consists in using a number of windows that guarantees a reasonably similar free energy change between contiguous intermediate states. The consequence of this choice is that the width of the consecutive windows connecting a to b may be different.

Performing "alchemical transformations" calls for the definition of topologies that describe the initial and the final states of the mutation. To this end, a single– or a dual–topology paradigm [17] can be employed, as shown in Fig. 1. Each approach has its advantages and inherent drawbacks. To circumvent numerical instabilities in the trajectory caused by the concurrent scaling of the charges, the van der Waals and selected internal parameters, transformation of electrostatic and non–electrostatic terms are usually decoupled in the single–topology approach, thus, giving access to the corresponding free energy contributions. This scheme requires two distinct simulations, when the dual–topology paradigm only involves one. The latter, how-ever, does not provide much information about the contributions that drive the global free energy change — albeit contributions should be interpreted with great care, be-cause, in contrast with the net free energy, which is a state function, the former are path–dependent. The dual–topology approach has been recognized to be generally more sensitive to so–called "end–point catastrophes", when $\lambda \to 0$ or 1, than the

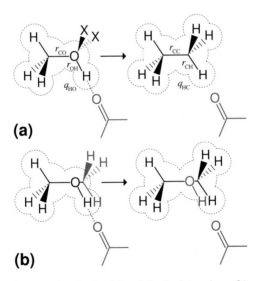

(a)

(b)

Figure 1. Difference between the single– (**a**) and the dual–topology (**b**) paradigms illustrated in the case of the methanol to ethane "alchemical transformation". In the single–topology approach, a common topology for the initial and the final states of the transformation is designed. Non–bonded parameters are scaled as λ varies from 0 to 1. Modification of the oxygen atom into a carbon one imposes that the chemical bond be shrunk, using, for instance, a PMF–type calculation [14–16]. In the dual–topology paradigm, the initial and the final states coexist , yet without "seeing" each other. The interaction energy of these topologies with their environment is scaled as λ goes from 0 to 1

single–topology paradigm. A number of schemes have been devised to circumvent this problem, among which the use of windows of decreasing width as λ tends towards 0 or 1. Introduction of a soft–core potential [18] to eliminate the singularities at 0 or 1 perhaps constitutes the most elegant method proposed hitherto.

2.2 Thermodynamic Integration

Closely related to the free energy perturbation (FEP) expression, thermodynamic integration (TI) restates the free energy difference between state a and state b as a finite difference [8, 19]:

$$\Delta A_{a \to b} = A(\lambda_b) - A(\lambda_a) = \int_{\lambda_a}^{\lambda_b} \frac{\mathrm{d} A(\lambda)}{\mathrm{d} \lambda} \, \mathrm{d}\lambda \tag{10}$$

Combining with the definition of the canonical partition function (2), it follows that:

$$\frac{\mathrm{d} A(\lambda)}{\mathrm{d} \lambda} = \frac{\int \frac{\partial \mathcal{H}(\mathbf{x}, \mathbf{p}_x; \lambda)}{\partial \lambda} \exp -\beta \mathcal{H}(\mathbf{x}, \mathbf{p}_x; \lambda) \, \mathrm{d}\mathbf{x} \, \mathrm{d}\mathbf{p}_x}{\int \exp -\beta \mathcal{H}(\mathbf{x}, \mathbf{p}_x; \lambda) \, \mathrm{d}\mathbf{x} \, \mathrm{d}\mathbf{p}_x} \tag{11}$$

Consequently, the integrand can be written as an ensemble average:

$$\boxed{\Delta A_{a \to b} = \int_{\lambda_a}^{\lambda_b} \left\langle \frac{\partial \mathcal{H}(\mathbf{x}, \mathbf{p}_x; \lambda)}{\partial \lambda} \right\rangle_\lambda \mathrm{d}\lambda} \tag{12}$$

In sharp contrast with the FEP method, the criterion of convergence here is the appropriate smoothness of $A(\lambda)$. Interestingly enough, assuming that the variation of the kinetic energy between states a and b can be neglected, the derivative of $\Delta A_{a \to b}$ with respect to some reaction coordinate, ξ, is equal to, $- \langle F_\xi \rangle_\xi$, the average of the force exerted along ξ, hence, the concept of PMF [20].

2.3 Unconstrained Molecular Dynamics and Average Forces

Generalization of the classical definition of a PMF, $w(r)$, based on the pair correlation function, $g(r)$, is not straightforward. For this reason, the free energy as a function of reaction coordinate ξ will be expressed as:

$$A(\xi) = -\frac{1}{\beta} \ln \mathcal{P}(\xi) + A_0 \tag{13}$$

where $\mathcal{P}(\xi)$ is the probability distribution to find the system at a given value, ξ, along that reaction coordinate:

$$\mathcal{P}(\xi) = \int \delta[\xi - \xi(\mathbf{x})] \, \exp[-\beta \mathcal{H}(\mathbf{x}, \mathbf{p}_x)] \, \mathrm{d}\mathbf{x} \, \mathrm{d}\mathbf{p}_x \tag{14}$$

Equation (13) corresponds to the classic definition of the free energy in methods like US, in which external biasing potentials are included to ensure a uniform distribution $\mathcal{P}(\xi)$. To improve sampling efficiency, the complete reaction pathway is broken down into "windows", or ranges of ξ, wherein individual free energy profiles are determined. The latter are subsequently pasted together using, for instance, the self–consistent weighted histogram analysis method (WHAM) [21].

For a number of years, the first derivative of the free energy with respect to the reaction coordinate has been written as [22]:

$$\frac{dA(\xi)}{d\xi} = \left\langle \frac{\partial V(\mathbf{x})}{\partial \xi} \right\rangle_\xi \tag{15}$$

This description is erroneous because ξ and $\{\mathbf{x}\}$ are evidently not independent variables [23]. Furthermore, it assumes that kinetic contributions can be safely omitted. This may not always be necessarily the case. For these reasons, a transformation of the metric is required, so that:

$$\mathcal{P}(\xi) = \int |J| \, \exp[-\beta V(\mathbf{q}; \xi)] \, d\mathbf{q} \int \exp[-\beta \mathcal{T}(\mathbf{p}_x)] d\mathbf{p}_x \tag{16}$$

Introduction in the first derivative of probability $\mathcal{P}(\xi)$, it follows that the kinetic contribution vanishes in $dA(\xi)/d\xi$:

$$\frac{dA(\xi)}{d\xi} = -\frac{1}{\beta} \frac{1}{\mathcal{P}(\xi)} \int \exp[-\beta V(\mathbf{q}; \xi^*)] \, \delta(\xi^* - \xi) \times$$

$$\left\{ -\beta |J| \frac{\partial V(\mathbf{q}; \xi^*)}{\partial \xi} + \frac{\partial |J|}{\partial \xi} \right\} d\mathbf{q} \, d\xi^* \tag{17}$$

After back transformation into Cartesian coordinates, the derivative of the free energy with respect to ξ can be expressed as a sum of configurational averages at constant ξ [24]:

$$\boxed{\frac{dA(\xi)}{d\xi} = \left\langle \frac{\partial V(\mathbf{x})}{\partial \xi} \right\rangle_\xi - \frac{1}{\beta} \left\langle \frac{\partial \ln |J|}{\partial \xi} \right\rangle_\xi = -\langle F_\xi \rangle_\xi} \tag{18}$$

In this approach, only $\langle F_\xi \rangle_\xi$ is the physically meaningful quantity, unlike the instantaneous components, F_ξ, from which it is evaluated. Moreover, it should be clearly understood that neither F_ξ nor $-\partial \mathcal{H}(\mathbf{x}, \mathbf{p}_x)/\partial \xi$ are fully defined by the sole choice of the reaction coordinate.

In practice, F_ξ is accumulated in bins of finite size $\delta\xi$ and provides an estimate of $dA(\xi)/d\xi$. After a predefined number of observables are accrued, the adaptive biasing force (ABF) [25, 26] is applied along the reaction coordinate:

$$\mathbf{F}^{ABF} = \nabla \tilde{A} = -\langle F_\xi \rangle_\xi \, \nabla \xi \tag{19}$$

As sampling proceeds, $\nabla \widetilde{A}$ is progressively refined. Evolution of the system along ξ is governed mainly by its self–diffusion properties. It is apparent from the present description that this method is significantly more effective than US or its variants, because no *a priori* knowledge of the free energy hypersurface is required to define the necessary biasing potentials that will guarantee uniform sampling along ξ. The latter can easily become intricate in the case of qualitatively new problems, in which variation of the free energy behavior cannot be guessed with the appropriate accuracy. Often misconstrued, it should be emphasized that, whereas ABF undoubtedly improves sampling dramatically along the reaction coordinate, efficiency suffers, like any other free energy method, from orthogonal degrees of freedom in the slow manifolds.

2.4 When is Enough Sampling Really Enough?

When is the trajectory long enough to assume safely that the results are converged? is a recurrent question asked by modelers performing free energy calculations. Assessing the convergence properties and the error associated to a free energy calculation often turns out to be daunting task. Sources of errors likely to be at play are diverse, and, hence, will modulate the results differently. The choice of the force field parameters undoubtedly affects the results of the simulation, but this contribution can be largely concealed by the statistical error arising from insufficient sampling. Paradoxically, exceedingly short free energy calculations employing inadequate non–bonded parameters may, nonetheless, yield the correct answer [14]. Under the hypothetical assumption of an optimally designed potential energy function, quasi non–ergodicity scenarios constitute a common pitfall towards fully converged simulations.

Appreciation of the statistical error has been devised following different schemes. Historically, the free energy changes for the $\lambda \rightarrow \lambda + \delta\lambda$ and the $\lambda \rightarrow \lambda - \delta\lambda$ perturbations were computed simultaneously to provide the hysteresis between the forward and the reverse transformations. In practice, it can be shown that when $\delta\lambda$ is sufficiently small, the hysteresis of such "double–wide sampling" simulation [27] becomes negligible, irrespective of the amount of sampling generated in each window — as would be the case in a "slow–growth" calculation [28]. A somewhat less arguable point of view consists in performing the transformation in the forward, $a \rightarrow b$, and in the reverse, $b \rightarrow a$, directions. Micro–reversibility imposes that, in principle, $\Delta A_{b \rightarrow a} = -\Delta A_{a \rightarrow b}$ — see for instance Fig. 2. Unfortunately, forward and reverse transformations do not necessarily share the same convergence properties. Case in point, the insertion and deletion of a particle [29]: Whereas the former simulation converges rapidly towards the expected excess chemical potential, the latter never does. This shortcoming can be ascribed to the fact that configurations in which a cavity does not exist where a real atom is present are never sampled. In terms of density of states, this scenario would translate into ϱ_a embracing ϱ_b entirely, thereby ensuring a proper convergence of the forward simulation, whereas the same cannot be said for the reciprocal, reverse transformation. Estimation of errors based on forward and reverse simulations should, therefore, be considered with great care.

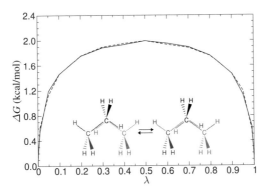

Figure 2. Free energy change characterizing the "zero–sum", for- ce–field–independent mutation of ethane into ethane in water (insert). Forward (*solid line*) and reverse (dashed line) profiles were obtained by dividing the reaction pathway into 40 windows, in which 16 ps of equilibration and 32 ps of data collection were sampled. The dual–topology paradigm was used to define a pseudo–propane molecule

In fact, appropriate combination of the two can be used profitably to improve the accuracy of free energy calculations [10].

In FEP, convergence may be probed by monitoring the time–evolution of the ensemble average (6). This is, however, a necessary, but not sufficient condition for convergence, because apparent plateaus of the ensemble average often conceal anomalous overlap of the density of states characterizing the initial, ϱ_a, and the final, ϱ_b, states [10, 11]. The latter should be the key–criterion to ascertain the local convergence of the simulation for those degrees of freedom that are effectively sampled. Statistical errors in FEP calculations may be estimated by means of a first–order expansion of the free energy:

$$\Delta A = -\frac{1}{\beta} \left\{ \ln \langle \exp\left[-\beta \Delta \mathcal{V}(\mathbf{x}; \lambda)\right]\rangle_\lambda \pm \frac{\delta \varepsilon}{\langle \exp\left[-\beta \Delta \mathcal{V}(\mathbf{x}; \lambda)\right]\rangle_\lambda} \right\} \qquad (20)$$

Here, $\delta\varepsilon$ is the statistical error on the ensemble average, $\langle \exp\left[-\beta \Delta \mathcal{V}(\mathbf{x}; \lambda)\right]\rangle_\lambda$, defined as:

$$\delta\varepsilon^2 = \frac{1 + 2\tau}{N} \left\{ \langle \exp\left[-2\beta \Delta \mathcal{V}(\mathbf{x}; \lambda)\right]\rangle_\lambda - \langle \exp\left[-\beta \Delta \mathcal{V}(\mathbf{x}; \lambda)\right]\rangle_\lambda^2 \right\} \qquad (21)$$

N is the number of samples accrued in the FEP calculation, and $(1 + 2\tau)$ is the sampling ratio of the latter [30].

Commonly, in US simulations, convergence is probed by verifying two criteria: (i) convergence of individual windows — the statistical error can be measured through a block–averaging over sub–runs — and (ii) appropriate overlap of the free energy profiles between adjacent windows — which gives rise to a systematic error. The latter can be evaluated by minimizing the difference in the curvature of two consecutive free energy profiles in the region where they overlap.

In the idealistic cases where a thermodynamic cycle can be defined — e.g. investigation of the conformational equilibrium of a short peptide through the $\alpha_R \rightarrow C_{7ax} \rightarrow \alpha_L \rightarrow \beta' \equiv (\beta, C_5, C_{7eq}) \rightarrow \alpha_R$ successive transformations — closure of the latter imposes that the sum of individual free energy contributions sum up to zero [31]. In principle, any deviation from this target should provide a valuable guidance to improve sampling efficiency. In practice, discrimination of the faulty transformation, or transformations, is cumbersome on account of possible mutual compensation or cancellation of errors.

As has been commented on previously, visual inspection of ϱ_a and ϱ_b indicates whether the free energy calculation has converged [10, 11]. Deficiencies in the overlap of the two distributions is also suggestive of possible errors, but it should be kept in mind that approximations like (20) only reflect the *statistical precision* of the computation, and evidently do not account for fluctuations in the system occurring over long time scales. In sharp contrast, the *statistical accuracy* is expected to yield a more faithful picture of the degrees of freedom that have been actually sampled. The safest route to estimate this quantity consists in performing the same free energy calculation, starting from different regions of the phase space — *viz.* the error is defined as the root mean square deviation over the different simulations [32]. Semantically speaking, the error measured from one individual run yields the statistical precision of the free energy calculation, whereas that derived from the ensemble of simulations provides its statistical accuracy.

3 Free Energy Calculations and Drug Design

One of the grand challenges of free energy calculations is their ability to play a predictive role in ranking ligands of potential pharmaceutical interest, agonist or antagonist, according to their relative affinity towards a given protein. The usefulness of such numerical simulations outside an academic environment can be assessed by answering the following question: Can free energy calculations provide a convincing answer faster than experiments are carried out in an industrial setting? Early encouraging results had triggered much excitement in the community, opening new vistas for *de novo*, rational drug design. They were, however, subsequently shattered when it was realized that accurate free energies would require considerably more computational effort than was appreciated hitherto. Beyond the fundamental need of appropriate sampling to yield converged ensemble averages — today, easily achievable in some favorable cases by means of inexpensive clusters of personal computers — the necessity of well–parameterized potential energy functions, suitable for non–peptide ligands, rapidly turns out to constitute a critical bottleneck for the routine use of free energy calculations in the pharmaceutical industry. Closely related to the parametrization of the force field, setting up free energy calculations — i.e. defining the alternate topology of the mutated moieties and the initial set of coordinates — is sufficiently time–consuming to be incompatible with the high–throughput requirements of industrial environments. In essence, this is where the paradox of free energy calculations lies: They have not yet come of age to be considered as black box routine jobs, but should evolve in that direction to become part of the arsenal of computational tools available to the pharmaceutical industry. Selection of potent ligand candidates in large data bases containing several millions of real or virtual molecules, employing a screening funnel that involves increasingly complex searching tools, from crude geometrical recognition to more sophisticated flexible molecular docking, offers new prospects for *de novo* drug design. In this pipeline of screening methods, free energy calculations should evidently be positioned at the very end, i.e. at the level of the optimization loop aimed at a limited number of ligands, which also checks adsorption–metabolism/toxicology (ADME/TOX) properties. Ideally, as selection in the funnel proceeds, the computational effort should remain constant — *viz.* the amount of CPU time necessary to perform free energy calculations on a few candidates is equivalent to that involved in the rough geometrical recognition over a number of molecules five to six orders of magnitude larger.

Perhaps the key to the generalized and routine use of free energy calculations for molecular systems of pharmacological relevance is the demonstration that this methodology can be applied fruitfully to problems that clearly go beyond the typical scope of an academic environment. Collaborative projects with the pharmaceutical industry provide such a framework. In the context of the search for therapeutic agents targeted at osteoporosis and other bone–related diseases, free energy calculations have been applied to complexes formed by the multi–domain protein ^{pp60}src kinase associated to non–peptide inhibitors. ^{pp60}src kinase is involved in signal transduction pathways and is implicated in osteoclast–mediated bone resorption [33]. Of particular interest, its SH2 domain, a common recognition motif of highly conserved protein sequence, binds preferentially phosphotyrosine (pY)–containing peptides. In most circumstances, the latter adopt an extended conformation mimicking a two–pronged plug that interacts at two distinct anchoring sites of the protein — i.e. the hydrophilic phosphotyrosine pocket and the hydrophobic pocket — separated by a flat surface [34, 35]. For instance, the prototypical tetrapeptide pYEEI — a sequence found on the PDGF receptor upon activation, appears to recognize the src SH2 domain with an appropriate specificity.

In silico point mutations have been performed on a series of non–peptide inhibitors — see Fig. 3, using the FEP methodology in conjunction with the dual–topology paradigm. All simulations have been carried in the isothermal–isobaric ensemble, using the program NAMD [36, 37]. The temperature and the pressure were fixed at 300 K and 1 atm, respectively, employing Langevin dynamics and the Langevin piston. To avoid possible end–point catastrophes, 33 windows of uneven width, $\delta\lambda$, were utilized to scale the interaction of the mutated moieties with their environment. The total length of each trajectory is equal to 1 ns, in the free and in the bound states. Forward and reverse simulations were run to estimate the statistical error, with the assumption that the two transformations have identical convergence properties. The structures of the protein–ligand complex were determined by x–ray crystallography [34, 35].

Compared with pYEEI, ligand **I1** adopts a very similar binding mode. The biphenyl moiety occupies the hydrophobic pocket entirely and the interaction of pY with the hydrophilic site is strong. The scaffold of the peptide forms steady van

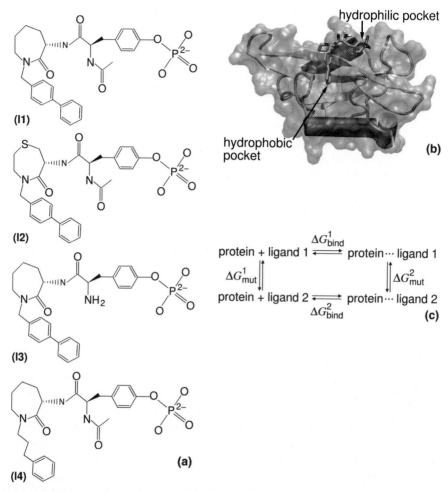

Figure 3. Non–peptide inhibitors of the SH2 domain of the ^{pp60}src kinase (**a**). Complex formed by the src SH2 domain and inhibitor I3 bound to the surface of the latter. Note the diphenyl moiety interacting with the hydrophobic pocket, while the pY motif is buried in the hydrophilic pocket (**b**). Thermodynamic cycle utilized to estimate relative protein–ligand binding free energies. Horizontal transformations are determined experimentally and generally are not amenable to statistical simulations. Vertical transformations correspond to "alchemical transformations" of the ligand in the free state (left) and in the bound state (right), so that $\Delta G^2_{bind} - \Delta G^1_{bind} = \Delta G^2_{mut} - \Delta G^1_{mut}$ (**c**)

der Waals contacts with the surface of the protein. A persistent water molecule is bridged between the carbonyl group of the lactam moiety and the amide –NH group of Lys^{62}. The relative binding free energies corresponding to the "alchemical transformations" of ligand I1 are gathered in Table 1. Overall, the agreement between the computational and the experimental estimates is good, within chemical accuracy.

Table 1. Point mutations of a series of non–peptide inhibitors of the SH2 domain of the ^{pp60}src kinase. The FEP methodology was employed in association with the dual–topology paradigm. Experimental binding free energies were determined using micro–calorimetry techniques

Transformation	Calculated $\Delta G^2_{\mathrm{mut}} - \Delta G^1_{\mathrm{mut}}$ (kcal/mol)	Experimental $\Delta G^2_{\mathrm{bind}} - \Delta G^1_{\mathrm{bind}}$ (kcal/mol)
I1 → I2	+0.6±0.4	+1.3
I1 → I3	+2.9±0.4	+2.2
I1 → I4	+1.9±0.4	+1.8

Replacement of one methylene group in the lactam scaffold by a sulfur atom — see Fig. 3, reduces the binding affinity by about an order of magnitude. No striking structural modification of the scaffold is observed, and the two–pronged plug motif is preserved. A closer look at the protein–ligand interface, however, reveals a loss of van der Waals contacts and unfavorable electrostatic interactions where the point mutation occurred, causing a somewhat increased flexibility in I2. Replacement of the amide group of I1 by an amino group decreases the binding affinity more significantly. Here again, the altered ligand, I3, remains perfectly anchored in the hydrophobic and hydrophilic pockets upon mutation. Yet, the interaction of the –NAc moiety with Arg^{14} vanishes as the former is modified into –NH$_2$, resulting in a weaker binding, *ca.* forty times less than for I1. Ligands I1 and I4 differ in their hydrophobic region, the first phenyl ring of the biphenyl moiety being replaced by two methylene groups. Close examination of the complex formed by the src SH2 domain and I4 reveals that the second phenyl ring of I1 and I4 superimpose nicely. Positioning of the peptide scaffold is, however, modified, so that –NAc no longer interacts with Arg^{14}, but rather with His^{60}. This structural change propagates to the pY ring, which adopts an alternative conformation, corresponding, overall, to a radically different binding mode of the ligand. It is important to underline that the "alchemical transformation" of I1 into I4 is able to capture the structural modifications in the protein–ligand association, and, hence, provide an accurate estimate of the relative free energy. This may be ascribed to the rapidly relaxing degrees of freedom of the short peptide, compatible with the time scale of the simulation.

Design of potent leads for a given target is only one element of the complex process that will eventually result in the release of a drug candidate suitable for clinical test phases. Downstream from the ranking of ligands according to their affinity for a protein, bioavailability and toxicity properties should be explored [38]. Of particular interest, the unassisted transport of pharmacologically active molecules across the membrane, in connection with the so–called blood–brain barrier (BBB), is an area where modeling techniques can provide a convincing answer, and, hence, complete the pipeline of *in silico* screening tools. Free energy methods may play a significant role in this effort by offering a detailed picture of the underlying energetics involved in the translocation of a drug from the aqueous medium to the hydrophobic core of the membrane. US calculations were used to simulate the passage of a series of pharmacologically relevant molecules across the interface formed by a lamella of

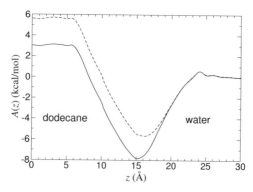

Figure 4. Transfer free energy of antipyrine across the water–dodecane interface. Uncorrected profile (dashed line) and profile incorporating the distortion energy (*solid line*), evaluated from $\Delta A_{\text{distort}} = \left\langle \psi | \hat{\mathcal{H}}_0 | \psi \right\rangle - \left\langle \psi_0 | \hat{\mathcal{H}}_0 | \psi_0 \right\rangle$, where $\hat{\mathcal{H}}_0$ and $|\psi_0\rangle$ are the unperturbed Hamiltonian and wave function. $|\psi\rangle$ is the SCRF wave function that includes the solvent contribution

dodecane in equilibrium with a lamella of water — a pertinent model for estimating partition coefficients [39]. Up to 20 ns were sampled to cover the whole reaction pathway, divided into sequentially overlapping windows to improve efficiency.

An example of such PMFs is depicted in Fig. 4, in the case of antipyrine, for which the net transfer free energy is equal to $+3.0\pm0.5$ kcal/mol, i.e. $\log_{10} P_{\text{wat}-\text{dod}}$ = -2.1, to be compared to the experimental value of -2.5 [40]. In most circumstances, modification of the environment and the ensuing change in the polarization of the solute is ignored when evaluating differential solvation properties. Traditionally, basis sets that inflate artificially the polarity of the molecule — e.g. 6–31G(d, p), have been utilized to parameterize the electrostatic term of the potential energy function, thereby compensating in an average sense for missing induction effects. Whereas the fitted sets of point charges are generally *ad hoc* for molecular simulations in an aqueous environment, the same cannot be said for non–polar media, in which the polarity of the solute is clearly exaggerated. A convenient framework to circumvent this difficulty is provided by the self–consistent reaction field (SCRF) method [41]. Atomic charges are derived from the electrostatic potential that accounts for the mutual polarization of the molecule by a dielectric continuum surrounding it. Upon translation from one medium to another, less polar one, the positive, reversible work corresponding to the overestimated polarity of the solute in the non–polar environment — i.e. the distortion energy [42], should be accounted for when determining the transfer free energy. In the example of antipyrine, this correction can be as large as *ca.* 2.5 kcal/mol (see Fig. 4), thus, demonstrating the necessity to incorporate this important contribution in solvation free energy calculations.

As indicated above, determination of accurate transfer free energies by means, for instance, of US simulations, is CPU demanding, and, hence, not compatible with the high–throughput requirements of industrial settings. These calculations should, therefore, be performed on the handful of drug candidates discriminated by the screening process. Faster approaches have been devised, however, based on quantum chemical calculations associated to an SCRF scheme for taking solvent effects into account. Aside from the electrostatic term, van der Waals contributions are usually evaluated from empirical formulation using the solvent accessible surface area (SASA) of the solute. This set of methods is substantially faster than statistical simulations, but, at the same time, only supply simple liquid–liquid partition coefficients, rather than the full free energy behavior characterizing the translocation of the molecule between the two media. This information may, however, turn out to be of paramount importance to rationalize biological phenomena. Such was the case, for instance, of general anesthesia by inhaled anesthetics, an interfacial process shown to result from the accumulation of anesthetics at the water–membrane interface of neuronal tissues [43].

4 Free Energy Calculations and Signal Transduction

The paucity of structural information available for membrane proteins has imparted a new momentum in the *in silico* investigation of these systems. The grand challenge of molecular modeling is to attain the microscopic detail that is often inaccessible to conventional experimental techniques. Of topical interest are seven transmembrane (TM) domain G–protein coupled receptors (GPCRs) [44], which correspond to the third largest family of genes in the human genome, and, therefore, represent privileged targets for *de novo* drug design. Full resolution by x–ray crystallography of the three–dimensional structure of bovine rhodopsin [45], the only GPCR structure to this date, has opened new vistas for the modeling of related membrane proteins. Unfortunately, crystallization of this receptor in its dark, inactive state precludes the use of the structure for homology modeling of GPCR–ligand activated complexes [46]. When neither theory nor experiment can provide atomic–level, three–dimensional structures of GPCRs, their synergistic combination offers an interesting perspective to reach this goal. Such a self–consistent strategy between experimentalists and modelers has been applied successfully to elucidate the structure of the human receptor of cholecystokinin (CCK1R) in the presence of an agonist ligand [47] — *viz.* a nonapeptide (CCK9) [48] of sequence Arg–Asp–S-Tyr–Thr–Gly–Trp–Met–Asp–Phe–NH$_2$, where S-Tyr stands for a sulfated tyrosyl amino acid. On the road towards a consistent *in vacuo* construction of the complex, site–directed mutagenesis experiments were designed to pinpoint key receptor–ligand interactions, thereby helping in the placement of TM α–helices and the docking of CCK9.

Whereas *in vacuo* models reflect the geometrical constraints enforced in the course of their construction, it is far from clear whether they will behave as expected when immersed in a realistic membrane environment. Accordingly, the model formed by CCK1R and CC9 was inserted in a fully hydrated palmitoyloleylphosphatidylcholine (POPC) bilayer, resulting in a system of 72,255 atoms, and the complete assembly was probed in a 10.5 ns MD simulation. Analysis of the trajectory reveals no apparent loss of secondary structure in the TM domain, and a distance root mean square deviation for the backbone atoms not exceeding 2 Å. More importantly, the crucial receptor–ligand interactions are preserved throughout the simulation — e.g. Arg^{336} with Asp^8 [49], and Met^{195} and Arg^{197} with $S-Tyr^3$ [50, 51]. Admittedly, such numerical experiments in essence only supply a qualitative picture of the structural properties of the molecular assembly and its integrity over the time scale explored. Free energy calculations go one step beyond by quantifying intermolecular interactions according to their importance, and, consequently, represent a tangible thermodynamic measure for assessing the accuracy of the model. Furthermore, this information is directly comparable to site–directed mutagenesis experiments utilized in the *in vacuo* construction of the receptor, thereby closing the loop of the modeling process.

To some extent, performing free energy calculations in such large molecular assemblies may be viewed as a bold and perhaps foolish leap of faith, considering the various possible sources of errors likely to affect the final result. Among the latter, attempting to reproduce free energy differences using a three–dimensional model in lieu of a well–resolved, experimentally determined structure casts the greatest doubts on the chances of success of this venture. Of equal concern, "alchemical transformations" involving charged amino acids are driven primarily by the solvation of the ionic moieties, resulting in large free energies, the difference of which, between the free and the bound states, is expected to be small. Assuming a validated model, which appears to be confirmed by the preliminary MD simulation, a key–question, already mentioned in this chapter, remains: When is "enough sampling" really enough? This conundrum should be, in fact, rephrased here as: Are the time scales characteristic of the slowest degrees of freedom in the system crucial for the free energy changes that are being estimated? For instance, is the mutation of the penultimate amino acid of CCK9 — *viz.* Asp^8 into alanine (see Fig. 5), likely to be affected by the slow collective motions of lipid molecules, or possible vertical and lateral motions of TM α– helices? Nanosecond MD simulations obviously cannot capture these events, which occur over significantly longer times. Yet, under the assumption that the replacement of an agonist ligand by an alternate one does not entail any noticeable rearrangement of the TM domain, current free energy calculations are likely to be appropriate for ranking ligands according to their affinity towards a given GPCR.

The FEP estimate of $+3.1\pm0.7$ kcal/mol for the D8A transformation agrees, indeed, very well with the site–directed mutagenesis experiments that yielded a free energy change equal to $+3.6$ kcal/mol. The *in silico* value was obtained from two runs of 3.4 ns each, in bulk water and in CCK1R, respectively, breaking the reaction path into 114 consecutive windows of uneven width, and using the dual–topology paradigm. Hénin, J.; Maigret, B.; Tarek, M.; Escrieut, C.; Fourmy, D.; Chipot, C. Probing a model of a GPCR/ligand complex in an explicit membrane environment. The human cholecystokinin-1 receptor. *Biophys. J.* 2005 (in press). The error was estimated from two distinct runs performed at 5.0 and 10.5 ns of the MD simulation. In contrast with an error derived from a first–order expansion of the free energy, which only reflects the statistical precision of the calculation — here, ±0.3 kcal/mol, repeating the simulation from distinct initial conditions accounts for fluctuations of the structure over longer time scales. Put together, while it is difficult to ascertain without ambiguity the correctness of the three–dimensional structure in the sole light of a limited

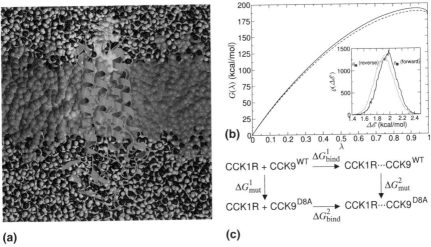

(a)   **(c)**

Figure 5. Human receptor of cholecystokinin (CCK1R) embedded in a fully hydrated POPC bilayer. The agonist ligand, nonapeptide CCK9, is highlighted (**a**). Free energy change for the mutation in CCK9 of Asp^8 into alanine: Transformation in the receptor (*solid line*) and in water (*dashed line*). Insert: Overlapping density of states characterizing adjacent states, at λ = 0.5 (**b**). Thermodynamic cycle utilized to estimate the relative receptor–ligand binding free energy for the D8A point mutation in CCK9 (**c**)

number of numerical experiments, it still remains that the host of observations accrued in theses simulations coincide nicely with the collection of experimental data. *De novo* development of new drug candidates for targets of unknown structure constitutes one of the greatest challenges faced today by the pharmaceutical industry. It is envisioned that the encouraging results presented herein for CCK1R will pave the way towards a more self–contained approach to drug design, emancipated from the requirement of well–resolved structures.

5 Free Energy Calculations and Peptide Folding

Capturing the underlying mechanisms that govern protein folding has been one of the holy grails of modern theoretical biophysics. Whereas the sequence of amino acids that forms the protein contains all the necessary information to determine the unique, compact structure of the chain under a given physiological condition, the understanding of the paths that lead to this native, generally biologically active structure remains fragmentary. Two classes of theoretical approaches for tackling the protein folding problem have been employed in recent years. In the first one, an all–atom representation of the protein and its environment and an empirically–based potential energy function are used to evaluate the intra– and intermolecular forces acting on the system. This description is often associated to an MD scheme for exploring the rugged conformational space of the solvated protein. Yet, ab initio folding by means of atomic–level MD simulations featuring both the protein and the solvent remain limited to short peptides and small proteins [52]. The reason for this limitation is rooted in the computational cost of such *in silico* experiments, which does not allow biologically relevant time scales to be accessed routinely. An alternative to the detailed, all–atom approach consists in turning to somewhat rougher models, that, nonetheless, retain the fundamental characteristics of protein chains. Such is the case of coarse–grained models, in which each amino acid of the protein is represented by a bead located at the vertex of a two– or a three–dimensional lattice [53]. An intermediate description consists of an all–atom representation of the protein in an implicit solvent. It is far from clear, however, whether the delicate interplay of the protein with explicit water molecules is a necessary condition for guaranteeing the correct folding toward the native state.

Whereas MD simulations involving an explicit solvent rarely exceed a few hundreds of ns [54], significantly shorter free energy calculations can be designed advantageously to understand the physical phenomena that drive folding. Among these phenomena, the subtle, temperature–dependent hydrophobic effect [55, 56] remains one of the most investigated to rationalize the collapse of a disordered protein chain into an appropriately folded one. The choice of a pertinent reaction coordinate that characterizes the folding process of a short peptide, let alone a small protein, constitutes a conundrum, unlikely to find a definitive answer in the near future. This intricate problem is rooted in the vast number of degrees of freedom that vary concomitantly as the peptide evolves toward a folded structure. The free energy is, therefore, a function of many variables that cannot be accounted for in a straightforward fashion. Valuable information may, nonetheless, be obtained from simple model systems, for which a non–ambiguous reaction coordinate can be defined. Such is the case of the terminally blocked undecamer of L–leucine organized in an α–helix, the C–terminal residue of which was unfolded from an α–helical conformation to that of a β–strand [57]. Similar calculations have been endeavored with blocked poly–L–alanine of various lengths to examine helix propagation at its N– and C–termini [58]

To highlight the temperature–dependent nature of the hydrophobic effect, the MD simulations were run in the canonical ensemble at 280, 300, 320, 340, 360 and 370 K, using the Nosé–Hoover algorithm implemented in the program COSMOS. Changes in the free energy consecutive to modifications of the last ψ dihedral angle of the homopolypeptide were estimated with the US method. All other torsional angles were restrained softly in a range characteristic of an α–helix. For each temperature, the complete reaction pathway connecting the α–helical state to the β–strand — roughly speaking $-90 \leq \psi \leq +170°$, was broken down in five mutually overlapping windows. The full free energy profiles were subsequently reconstructed employing WHAM. The total simulation length varied from 14 ns at 370 K, to 76 ns at 280 K, on account of the slower relaxation at lower temperatures.

The PMFs shown in Fig. 6 each possess two distinct local minima corresponding to the α–helix and the β–strand, and separated by a maximum of the free energy around 90° . These three conformational states are characterized by different SASAs — *viz.* 134±2, 149±7 and 117±13 Å^2 for the α–helix, the transition state (TS) and the β–strand, respectively. The most striking feature of the PMFs lies in

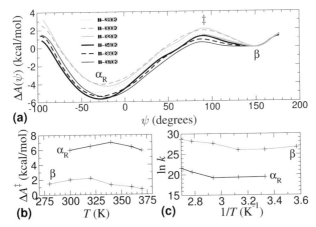

Figure 6. Free energy profile characterizing the unfolding of the undecamer of poly–leucine at different temperatures (**a**). Activation free energies from the α–helix to the transition state, $\ddagger$, and from the β–strand to $\ddagger$ (**b**). Arrhenius plot for the same transitions, inferred from the activated complex theory, *viz.* $\ln k = \ln h/\beta - \beta\Delta A^{\ddagger}$. h is the Planck constant (**c**)

the temperature dependence of the free energy associated to the transition from the α–helix to the TS, $\Delta A_{\alpha_R \to \ddagger}$, and that from the β–strand to the TS, $\Delta A_{\beta \to \ddagger}$. Furthermore, neither $\Delta A_{\alpha_R \to \ddagger}(T)$ nor $\Delta A_{\beta \to \ddagger}(T)$ varies monotonically as the temperature increases. On the contrary, the two profiles exhibit a maximum at a characteristic temperature T_i, suggestive of a non–Arrhenius behavior driven by physical principles that are related to the solvation properties of the hydrophobic homopolypeptide. Considering that the variation of entropy between state Ξ — i.e. either the α–helix or the β–strand, and the TS is $\Delta S^{\ddagger}(T) = -\partial \Delta A^{\ddagger}(T)/\partial T$, it follows that at T_i, the entropy is zero. Since $\Delta A^{\ddagger}(T)$ decreases when $T > T_i$, $\Delta S^{\ddagger}(T)$ is positive in this region, hence, implying that the conformational transition is favored entropically. Moreover, as $\Delta A^{\ddagger}(T) > 0$, the change in internal energy, $\Delta U^{\ddagger}(T) = \Delta A^{\ddagger}(T) + T\Delta S^{\ddagger}(T)$, is also necessarily positive, so that Ξ is the thermodynamically favored state. Symmetrically, when $T < T_i$, $\Delta S^{\ddagger}(T)$ is negative and Ξ is the entropically favored state. On account of the greater SASA for the TS, the transition from Ξ to the TS can be viewed as the transfer of a portion of the side chain from a buried state to a solvent–exposed one, similar in spirit to the translocation of a hydrophobic solute from a non–polar medium to water, that is accompanied by a positive free energy change — the thermodynamic fingerprint of the hydrophobic effect [59]. Put together, the free energy barrier can be ascribed at high temperature to the sole internal energy, to which an increasing entropic contribution is added upon lowering the temperature.

A fundamental issue brought to light in this investigation is the non–Arrhenius behavior of the unfolding kinetics of the hydrophobic peptide, and the underlying temperature dependence of the free energy barrier. This result is in blatant contradiction with the hypothesis commonly adopted in lattice model simulations that the activation free energy is a constant, and further suggests that a temperature dependence should be introduced in these models to capture subtle solvent contributions, like the hydrophobic effect [55].

6 Free Energy Calculations and Membrane Protein Association

To a large extent, our knowledge of how membrane protein domains recognize and associate into functional, three–dimensional entities remains fragmentary. Whereas the structure of membrane proteins can be particularly complex, their TM region is often simple, consisting in general of a bundle of α–helices, or barrels of β–strands. An important result brought to light by deletion experiments indicates that some membrane proteins can retain their biological function upon removal of large fractions of the protein. This is suggestive that rudimentary models, like simple α–helices, can be utilized to understand the recognition and association processes of TM segments into complex membrane proteins. On the road to reach this goal, the "two–stage" model [60] provides an interesting view for rationalizing the folding of membrane proteins. According to this model, elements of the secondary structure — viz. in most cases, α–helices, are first formed and inserted into the lipid bilayer, prior to specific inter–helical interactions that drive the TM segments towards well–ordered, native structures. Capturing the atomic detail of the underlying mechanisms responsible for α–helix recognition and association requires model systems that are supported by robust experimental data to appraise the accuracy of the computations endeavored. Glycophorin A (GpA), a glycoprotein ubiquitous to the human erythro-cyte membrane, represents one such system. It forms non–covalent dimers through the reversible association of its membrane–spanning domain — i.e. residues 62 to 101, albeit only residues 73 to 96 actually adopt an α–helical conformation [61–63]. Inter–helical association has been shown to result from specific interactions involv-ing a heptad of residues, essentially located on one face of each TM segment, as may be seen in Fig. 7.

The reversible association of GpA in a lipid bilayer was modelled using its dimeric, α–helical TM segments immersed in a membrane mimetic formed by a lamella of dodecane placed between two lamellae of water. The ABF method intro-duced in the program NAMD [26] was employed to allow the TM segments to diffuse freely along the reaction coordinate, ξ, chosen to be the distance separating the cen-ters of mass of the two α–helices. Hénin, J.; Pohorille, A.; Chipot, C. Insights into the recognition and association of transmembrane alpha-helices. The free energy of alpha-helix dimerization in glycophorin A. *J. Am. Chem. Soc.* 2005, 127, 8478–8484. Such a free energy calculation is not only challenging methodologically, but it is also of paramount importance from a biophysical standpoint, because it bridges structural data obtained from nuclear magnetic resonance (NMR) [61–63] to thermodynamic data obtained from analytical ultracentrifugation [64, 65] and fluorescence resonance energy transfer (FRET) [66, 67], providing a dynamic view of the recognition and association stages.

Visual inspection of the PMF derived from a 125 ns simulation and describ-ing the reversible association of the α–helices reveals a qualitatively simple profile,

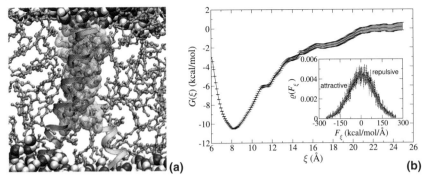

Figure 7. TM domain of GpA formed by a homodimer of α–helices, embedded in a lipid membrane mimetic. The heptad of residues involved in the association of the TM segments are shown as transparent van der Waals spheres. Note the crossing angle between the two α–helices, equal to *ca.* $40°$ (**a**). Free energy profile delineating the reversible association of the TM segments. Insert: Gaussian–like distributions of the force acting along the reaction coordinate, ξ, in the repulsive region — *viz.* $\xi < 8$ Å, and in the attractive region — *viz.* $\xi > 8$ Å (**b**)

featuring a single minimum characteristic of the native, dimeric state. As ξ increases, so does the free energy, progressing by steps that correspond to the successive breaking of inter–helical contacts. Beyond 21 Å, the TM segments are sufficiently separated to assume that they no longer interact. Integration of the PMF in the limit of α–helix association yields the association constant and, hence, the free energy of dimerization, equal to $+11.5\pm0.4$ kcal/mol. Direct and precise comparison of this value with experiment is not possible, because measurements were carried in different environments, namely hydrocarbon *vs.*detergent micelles. It can, nonetheless, be inferred that the value in dodecane probably constitutes an upper bound to the experimental estimates determined in micelles, on account of (i) the greater order imposed by the detergent chains, and, (ii) the hydrophobic fraction of the system that increases with the length of the chain [67].

Deconvolution of the PMF into free energy components illuminates two distinct regimes controlling recognition and association. At large separations, as inter–helical contacts vanish, the helix–helix term becomes progressively negligible, resulting essentially from the interaction of two macro–dipoles. The TM segments are stabilized by favorable helix–solvent contributions. In contrast, at short separations, helix–helix interactions are prominent and govern the change in the free energy near the global minimum. Association proceeds through the transient formation of early, non–native contacts involving residues that act as recognition sites. These contacts are subsequently replaced by contacts in the heptad of residues responsible for association, concomitantly with the tilt of the two α–helices from an upright position to that characteristic of the native dimer.

7 Conclusion

Free energy calculations constitute a tangible link between theory and experiment, by quantifying at the thermodynamic level the physical phenomena modelled by statistical simulations. With twenty years of hindsight gained from methodological development and characterization, a variety of problems of both chemical and biological relevance can now be tackled with confidence. Among the progresses achieved in recent years, a significant step forward has been made in the calculation of free energies along a reaction coordinate, in particular employing the concept of an average force acting along this coordinate [25, 26]. "Alchemical transformations" utilized to mimic site–directed mutagenesis experiments have also benefited from advances in the understanding of the methodology and how the latter should be applied [68]. Efforts to characterize and estimate the error affecting the simulations [10, 11, 69] have equally played an active role in turning free energy calculations into another tool in the arsenal of computational methods available to the modeler. Put together, free energy calculations have come of age to become a predictive approach, instead of remaining at the stage of a mere proof of concept [70]. As has been illustrated in this chapter, they can be applied to numerous problems, ranging from *de novo* drug design to the understanding of biophysical processes in lipid membranes. Free energy calculations, however, cannot yet be considered as "black box", routine jobs. A robust, reliable methodology does not necessarily imply that it can be used blindly. The nature of the problem dictates the choice of the method and the associated protocol — e.g. US *vs.* FEP *vs.* TI *vs.* average force, the number of intermediate λ–states along a reaction coordinate, the pertinent choice of this reaction coordinate, the amount of sampling per individual λ–state, a single *vs.* a dual topology paradigm for "alchemical transformations", or constrained *vs.* unconstrained MD. Furthermore, little effort has been hitherto devoted to the automatization of free energy calculations, through, for instance, a user–friendly definition of the topologies representative of the initial and the final states of a transformation. With the increased access to massively parallel architectures as the price/performance ratio of computer chips continues to fall inexorably, the bottleneck of free energy calculations has shifted from a purely computational aspect to a human one, due to the need of qualified modelers to set these calculations up. This explains why their use in industry, and particularly in the pharmaceutical world, over the past years has remained scarce. Cutting–edge applications of free energy calculations emanate essentially from academic environments, where the focus is not so much on high throughput, but rather on well–delineated, specific problems that often require more human attention than computational power. Yet, it is envisioned that in a reasonably near future, free energy methods will become an unavoidable element of screening pipelines, discriminating between candidates selected from cruder approaches, to retain only the best leads towards a given target.

Acknowledgments

Jérôme Hénin, Surjit Dixit, Olivier Collet, Eric Darve, Andrew Pohorille and Alan E. Mark are gratefully acknowledged for fruitful and inspiring discussions. The author thank the Centre Informatique National de l'Enseignement Supérieur (CINES) and the centre de Calcul Réseaux et Visualisation Haute Performance (CRVHP) for generous provision of CPU time on their SGI Origin 3000 architectures.

References

[1] Kollman, P. A., Free energy calculations: Applications to chemical and biochemical phenomena, *Chem. Rev.* 93, 2395–2417, 1993.

[2] Postma, J. P. M.; Berendsen, H. J. C.; Haak, J. R., Thermodynamics of cavity formation in water: A molecular dynamics study, *Faraday Symp. Chem. Soc.* 17, 55–67, 1982.

[3] Warshel, A., Dynamics of reactions in polar solvents. Semiclassical trajectory studies of electron transfer and proton transfer reactions, *J. Phys. Chem.* 86, 2218–2224, 1982.

[4] Bash, P. A.; Singh, U. C.; Langridge, R.; Kollman, P. A., Free energy calculations by computer simulation, *Science* 236, 564–568, 1987.

[5] Bash, P. A.; Singh, U. C.; Brown, F. K.; Langridge, R.; Kollman, P. A., Calculation of the relative change in binding free energy of a protein–inhibitor complex, *Science* 235, 574–576, 1987.

[6] McQuarrie, D. A., *Statistical mechanics*, Harper and Row: New York, 1976.

[7] Allen, M. P.; Tildesley, D. J., *Computer Simulation of Liquids*, Clarendon Press: Oxford, 1987.

[8] Kirkwood, J. G., Statistical mechanics of fluid mixtures, *J. Chem. Phys.* 3, 300–313, 1935.

[9] Zwanzig, R. W., High–temperature equation of state by a perturbation method. I. Nonpolar gases, *J. Chem. Phys.* 22, 1420–1426, 1954.

[10] Lu, N.; Singh, J. K.; Kofke, D. A.; Woolf, T. B., Appropriate methods to combine forward and reverse free–energy perturbation averages, *J. Chem. Phys.* 118, 2977–2984, 2003.

[11] Lu, N.; Kofke, D. A.; Woolf, T. B., Improving the efficiency and reliability of free energy perturbation calculations using overlap sampling methods, *J. Comput. Chem.* 25, 28–39, 2004.

[12] Mark, A. E. Free Energy Perturbation Calculations. in *Encyclopedia of computational chemistry*, Schleyer, P. v. R.; Allinger, N. L.; Clark, T.; Gasteiger, J.; Kollman, P. A.; Schaefer III, H. F.; Schreiner, P. R., Eds., vol. 2. Wiley and Sons, Chichester, 1998, pp. 1070–1083.

[13] Torrie, G. M.; Valleau, J. P., Nonphysical sampling distributions in Monte Carlo free energy estimation: Umbrella sampling, *J. Comput. Phys.* 23, 187–199, 1977.

[14] Pearlman, D. A.; Kollman, P. A., The overlooked bond–stretching contribution in free energy perturbation calculations, *J. Chem. Phys.* 94, 4532–4545, 1991.

[15] Boresch, S.; Karplus, M., The role of bonded terms in free energy simulations: I. Theoretical analysis, *J. Phys. Chem. A* 103, 103–118, 1999.

[16] Boresch, S.; Karplus, M., The role of bonded terms in free energy simulations: II. Calculation of their influence on free energy differences of solvation, *J. Phys. Chem. A* 103, 119–136, 1999.

[17] Pearlman, D. A., A comparison of alternative approaches to free energy calculations, *J. Phys. Chem.* 98, 1487–1493, 1994.

[18] Beutler, T. C.; Mark, A. E.; van Schaik, R. C.; Gerber, P. R.; van Gunsteren, W. F., Avoiding singularities and neumerical instabilities in free energy calculations based on molecular simulations, *Chem. Phys. Lett.* 222, 529–539, 1994.

[19] Straatsma, T. P.; Berendsen, H. J. C., Free energy of ionic hydration: Analysis of a thermodynamic integration technique to evaluate free energy differences by molecular dynamics simulations, *J. Chem. Phys.* 89, 5876–5886, 1988.

[20] Chandler, D., *Introduction to modern statistical mechanics*, Oxford University Press, 1987.

[21] Kumar, S.; Bouzida, D.; Swendsen, R. H.; Kollman, P. A.; Rosenberg, J. M., The weighted histogram analysis method for free energy calculations on biomolecules. I. The method, *J. Comput. Chem.* 13, 1011–1021, 1992.

[22] Pearlman, D. A., Determining the contributions of constraints in free energy calculations: Development, characterization, amnd recommendations, *J. Chem. Phys.* 98, 8946–8957, 1993.

[23] den Otter, W. K.; Briels, W. J., The calculation of free–energy differences by constrained molecular dynamics simulations, *J. Chem. Phys.* 109, 4139–4146, 1998.

[24] den Otter, W. K., Thermodynamic integration of the free energy along a reaction coordinate in Cartesian coordinates, *J. Chem. Phys.* 112, 7283–7292, 2000.

[25] Darve, E.; Pohorille, A., Calculating free energies using average force, *J. Chem. Phys.* 115, 9169–9183, 2001.

[26] Hénin, J.; Chipot, C., Overcoming free energy barriers using unconstrained molecular dynamics simulations, *J. Chem. Phys.* 121, 2904–2914, 2004.

[27] Jorgensen, W. L.; Ravimohan, C., Monte Carlo simulation of differences in free energies of hydration, *J. Chem. Phys.* 83, 3050–3054, 1985.

[28] Chipot, C.; Kollman, P. A.; Pearlman, D. A., Alternative approaches to potential of mean force calculations: Free energy perturbation versus thermodynamic integration. Case study of some representative nonpolar interactions, *J. Comput. Chem.* 17, 1112–1131, 1996.

[29] Widom, B., Some topics in the theory of fluids, *J. Chem. Phys.* 39, 2808–2812, 1963.

[30] Straatsma, T. P.; Berendsen, H. J. C.; Stam, A. J., Estimation of statistical errors in molecular simulation calculations, *Mol. Phys.* 57, 89–95, 1986.

[31] Chipot, C.; Pohorille, A., Conformational equilibria of terminally blocked single amino acids at the water–hexane interface. A molecular dynamics study, *J. Phys. Chem. B* 102, 281–290, 1998.

[32] Chipot, C.; Millot, C.; Maigret, B.; Kollman, P. A., Molecular dynamics free energy perturbation calculations. Influence of nonbonded parameters on the

free energy of hydration of charged and neutral species, *J. Phys. Chem.* 98, 11362–11372, 1994.

[33] Soriano, P.; Montgomery, C.; Geske, R.; Bradley, A., Targeted disruption of the c–src proto–oncogene leads to osteopetrosis in mice., *Cell* 64, 693–702, 1991.

[34] Lange, G.; Lesuisse, D.; Deprez, P.; Schoot, B.; Loenze, P.; Benard, D.; Marquette, J. P.; Broto, P.; Sarubbi, E.; Mandine, E., Principles governing the binding of a class of non–peptidic inhibitors to the SH2 domain of src studied by X-ray analysis, *J. Med. Chem.* 45, 2915–2922, 2002.

[35] Lange, G.; Lesuisse, D.; Deprez, P.; Schoot, B.; Loenze, P.; Benard, D.; Marquette, J. P.; Broto, P.; Sarubbi, E.; Mandine, E., Requirements for specific binding of low affinity inhibitor fragments to the SH2 domain of ^{pp60}Src are identical to those for high affinity binding of full length inhibitors, *J. Med. Chem.* 46, 5184–5195, 2003.

[36] Kale, L.; Skeel, R.; Bhandarkar, M.; Brunner, R.; Gursoy, A.; Krawetz, N.; Phillips, J.; Shinozaki, A.; Varadarajan, K.; Schulten, K., NAMD2: Greater scalability for parallel molecular dynamics, *J. Comput. Phys.* 151, 283–312, 1999.

[37] Bhandarkar, M.; Brunner, R.; Chipot, C.; Dalke, A.; Dixit, S.; Grayson, P.; Gullingsrud, J.; Gursoy, A.; Humphrey, W.; Hurwitz, D. et al. NAMD *users guide, version 2.5.* Theoretical biophysics group, University of Illinois and Beckman Institute, 405 North Mathews, Urbana, Illinois 61801, September 2003.

[38] Carrupt, P.; Testa, B.; Gaillard, P. Computational approaches to lipophilicity: Methods and applications. in *Reviews in Computational Chemistry*, Lipkowitz, K.; Boyd, D. B., Eds., vol. 11. VCH, New York, 1997, pp. 241–345.

[39] Wohnsland, F.; Faller, B., High–throughput permeability pH profile and high–throughput alkane–water log P with artificial membranes, *J. Med. Chem.* 44, 923–930, 2001.

[40] Bas, D.; Dorison-Duval, D.; Moreau, S.; Bruneau, P.; Chipot, C., Rational determination of transfer free energies of small drugs across the water–oil interface, *J. Med. Chem.* 45, 151–159, 2002.

[41] Rivail, J. L.; Rinaldi, D., A quantum chemical approach to dielectric solvent effects in molecular liquids, *Chem. Phys.* 18, 233–242, 1976.

[42] Chipot, C., Rational determination of charge distributions for free energy calculations, *J. Comput. Chem.* 24, 409–415, 2003.

[43] Pohorille, A.; Wilson, M.A.; New, M.H.; Chipot, C., Concentrations of anesthetics across the water–membrane interface; The Meyer–Overton hypothesis revisited, *Toxicology Lett.* 100, 421–430, 1998.

[44] Takeda, S.; Haga, T.; Takaesu, H.; Mitaku, S., Identification of G protein–coupled receptor genes from the human genome sequence, *FEBS Lett.* 520, 97–101, 2002.

[45] Palczewski, K.; Kumasaka, T.; Hori, T.; Behnke, C. A.; Motoshima, H.; Fox, B. A.; Le Trong, I.; Teller, D. C.; Okada, T.; Stenkamp, R. E.; Yamamoto, M.; Miyano, M., Crystal structure of rhodopsin: A G protein–coupled receptor, *Science* 289, 739–745, 2000.

[46] Archer, E.; Maigret, B.; Escrieut, C.; Pradayrol, L.; Fourmy, D., Rhodopsin crystal: New template yielding realistic models of G–protein–coupled receptors ?, *Trends Pharmacol. Sci.* 24, 36–40, 2003.

[47] Talkad, V. D.; Fortune, K. P.; Pollo, D. A.; Shah, G. N.; Wank, S. A.; Gardner, J. D., Direct demonstration of three different states of the pancreatic cholecystokinin receptor, *Proc. Natl. Acad. Sci. USA* 91, 1868–1872, 1994.

[48] Moroder, L.; Wilschowitz, L.; Gemeiner, M.; Göhring, W.; Knof, S.; Scharf, R.; Thamm, P.; Gardner, J. D.; Solomon, T. E.; Wünsch, E., Zur Synthese von Cholecystokinin–Pankreozymin. Darstellung von [28–Threonin, 31–Norleucin]– und [28–Threonin, 31–Leucin]– Cholecystokinin–Pankreozymin–(25–33)–Nonapeptid, *Z. Physiol. Chem.* 362, 929–942, 1981.

[49] Gigoux, V.; Escrieut, C.; Fehrentz, J. A.; Poirot, S.; Maigret, B.; Moroder, L.; Gully, D.; Martinez, J.; Vaysse, N.; Fourmy, D., Arginine 336 and Asparagine 333 of the human cholecystokinin–A receptor binding site interact with the penultimate aspartic acid and the C–terminal amide of cholecystokinin, *J. Biol. Chem.* 274, 20457–20464, 1999.

[50] Gigoux, V.; Escrieut, C.; Silvente-Poirot, S.; Maigret, B.; Gouilleux, L.; Fehrentz, J. A.; Gully, D.; Moroder, L.; Vaysse, N.; Fourmy, D., Met–195 of the cholecystokinin–A interacts with the sulfated tyrosine of cholecystokinin and is crucial for receptor transition to high affinity state, *J. Biol. Chem.* 273, 14380–14386, 1998.

[51] Gigoux, V.; Maigret, B.; Escrieut, C.; Silvente-Poirot, S.; Bouisson, M.; Fehrentz, J. A.; Moroder, L.; Gully, D.; Martinez, J.; Vaysse, N.; Fourmy, D., Arginine 197 of the cholecystokinin–A receptor binding site interacts with the sulfate of the peptide agonist cholecystokinin, *Protein Sci.* 8, 2347–2354, 1999.

[52] Daggett, V., Long timescale simulations, *Curr. Opin. Struct. Biol.* 10, 160–164, 2000.

[53] Taketomi, H.; Ueda, Y.; Gō, N., Studies on protein folding, unfolding and fluctuations by computer simulation. 1. The effect of specific amino acid sequence represented by specific inter–unit interactions, *Int. J. Pept. Protein Res.* 7, 445–459, 1975.

[54] Duan, Y.; Kollman, P. A., Pathways to a protein folding intermediate observed in a 1–microsecond simulation in aqueous solution, *Science* 282, 740–744, 1998.

[55] Pratt, L. R., Molecular theory of hydrophobic effects: "She is too mean to have her name repeated", *Annu. Rev. Phys. Chem.* 53, 409–436, 2002.

[56] Pratt, L. R.; Pohorille, A., Hydrophobic effects and modeling of biophysical aqueous solution interfaces, *Chem. Rev.* 102, 2671–2692, 2002.

[57] Collet, O.; Chipot, C., Non–Arrhenius behavior in the unfolding of a short, hydrophobic α–helix. Complementarity of molecular dynamics and lattice model simulations, *J. Am. Chem. Soc.* 125, 6573–6580, 2003.

[58] Young, W. S.; Brooks III, C. L., A microscopic view of helix propagation: N and C–terminal helix growth in alanine helices, *J. Mol. Biol.* 259, 560–572, 1996.

[59] Shimizu, S.; Chan, H. S., Temperature dependence of hydrophobic interactions: A mean force perspective, effects of water density, and non–additivity of thermodynamics signature, *J. Am. Chem. Soc.* 113, 4683–4700, 2000.

[60] Popot, J. L.; Engelman, D. M., Membrane protein folding and oligomerization: The two–stage model, *Biochemistry* 29, 4031–4037,1990.

[61] MacKenzie, K. R.; Prestegard, J. H.; Engelman, D. M., A transmembrane helix dimer: Structure and implications, *Science* 276, 131–133, 1997.

[62] MacKenzie, K. R.; Engelman, D. M., Structure–based prediction of the stability of transmembrane helix–helix interactions: The sequence dependence of glycophorin A dimerization, *Proc. Natl. Acad. Sci. USA* 95, 3583–3590, 1998.

[63] Smith, S. O.; Song, D.; Shekar, S.; Groesbeek, M.; Ziliox, M.; Aimoto, S., Structure of the transmembrane dimer interface of glycophorin A in membrane bilayers, *Biochemistry* 40, 6553–6558, 2001.

[64] Fleming, K. G.; Ackerman, A. L.; Engelman, D. M., The effect of point mutations on the free energy of transmembrane α–helix dimerization, *J. Mol. Biol.* 272, 266–275, 1997.

[65] Fleming, K. G., Standardizing the free energy change of transmembrane helix–helix interactions, *J. Mol. Biol.* 323, 2002, 563–571.

[66] Fisher, L. E.; Engelman, D. M.; Sturgis, J. N., Detergents modulate dimerization, but not helicity, of the glycophorin A transmembrane domain, *J. Mol. Biol.* 293, 639–651, 1999.

[67] Fisher, L. E.; Engelman, D. M.; Sturgis, J. N., Effects of detergents on the association of the glycophorin A transmembrane helix, *Biophys. J.* 85, 3097–3105, 2003.

[68] Dixit, S. B.; Chipot, C., Can absolute free energies of association be estimated from molecular mechanical simulations ? The biotin–streptavidin system revisited, *J. Phys. Chem. A* 105, 9795–9799, 2001.

[69] Rodriguez-Gomez, D.; Darve, E.; Pohorille, A., Assessing the efficiency of free energy calculation methods, *J. Chem. Phys.* 120, 3563–3570, 2004.

[70] Simonson, T.; Archontis, G.; Karplus, M., Free energy simulations come of age: Protein–ligand recognition, *Acc. Chem. Res.* 35, 430–437, 2002.

Numerical Methods for Calculating the Potential of Mean Force

Eric Darve

Institute for Computational and Mathematical Engineering, Mechanical Engineering
Department, Stanford University, CA, U.S.A.

Abstract. We review different numerical methods to calculate the potential of mean force
$A(\xi)$ along one or several reaction coordinates using Molecular Dynamics. We will focus
our discussion on thermodynamic integration techniques which attempts to compute $dA/d\xi$
by calculating a mean force acting on ξ. The first technique is based on constraining ξ to a
constant value. We show that the force needed to impose the constraint can be simply related
to the mean force. A recent technique, called Adaptive Biasing Force, is presented, which
attempts to calculate the mean force and remove it from the system. This leads to a uniform
sampling and a diffusive-like motion along the reaction coordinate. A novel implementation
of this technique is derived. The advantages of this new formulation in terms of statistical error
and convergence are discussed.

Key words: potential of mean force, free energy, adaptive biasing force, blue moon,
molecular dynamics, thermodynamic integration

1 Introduction

The calculation of the potential of mean force for molecular systems is the goal of
many molecular dynamics or Monte-Carlo simulations. Applications in the biological field range from protein folding, drug discovery, to the study of biological membranes and ion channels. A typical simulation takes place in a very high-dimensional
space, spanned by the position and momentum of all atomic particles. Despite the
simplicity of the time integrators which are being used, those calculations remain
extremely challenging due to the large differences in time scales. For example, the
fastest time scales are bond stretching and angle bending on the order of femto to
pico second (10^{-15} to 10^{-12} sec). Longer time scales are motion of loops in proteins and collective motion with time scales between pico and nano second (10^{-12}
to 10^{-9} sec). The longest time scales are for folding small proteins (micro second
$= 10^{-6}$ sec) or even seconds for folding very complex proteins. Self assembly of
proteins in membranes to form ion channels is another example.

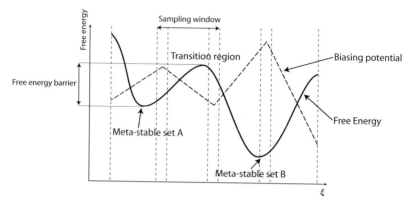

Figure 1. The umbrella sampling method can be used to overcome energy barriers between stable states of the system

To address this computational cost, advances have been made in utilizing computer hardware such has parallel computers (PC clusters, shared memory parallel computers) and more recent computing platforms such as grid computing. In addition to this, a lot of effort has been put into developing new numerical algorithms. Figure 1 represents the typical free energy profile or potential of mean force along a reaction coordinate. The x-axis is the reaction coordinate which could be the distance between the two molecules, a torsion angle along the backbone of a protein, the relative orientation of a α helix with respect to a membrane. The y-axis is the free energy. In general the free energy is related to the probability density function $P(\xi)$ of the reaction coordinate through:

$$A(\xi) = -k_B T \log P(\xi) \tag{1}$$

This equation means that when there is a free energy difference of a few $k_B T$ the probability $P(\xi)$ is reduced considerably, that is those conformations with large $A(\xi)$ are sampled very rarely. This is a very important observation in terms of numerically efficiency. At the transition region for example, the free energy is maximum and typically very few sample points are obtained during the course of the molecular dynamics simulation. In turn this results in very large statistical errors. Those errors can only be reduced by increasing the simulation time, sometimes beyond what is practically feasible.

A basic but powerful method to improve the efficiency of such computations is to split the interval of computation along the reaction coordinate ξ into sub-intervals. In this way, the energy barriers inside each window are usually smaller (at least if the windows are sufficiently small). In addition, in the umbrella sampling method [7, 27, 30, 36, 63, 64], a biasing potential can be used in order to further improve the sampling. If this biasing potential is a function $U_b(x)$, the new Hamiltonian (or energy function) for the system is $H(x) + U_b(x)$. In general the biasing potential needs to be guessed beforehand or can be gradually improved using an iterative refinement process. This process consists in running a first short simulation to estimate the free energy, $A^{(0)}(x)$ and then use this estimate to bias the system using $U_b^{(0)}(x) = -A^{(0)}(x)$. With this first bias, we improve the sampling and obtain a more accurate approximation of the free energy $A^{(1)}(x)$. The biasing potential $U_b^{(i)}(x)$ can be gradually improved in this fashion. Within each window, we obtain a relatively uniform sampling. This leads in general to small statistical errors.

It is clear however that in complex situations it is not possible to make an educated guess of the biasing potential. In particular the position of the transition regions or the height of the barrier may be quite difficult to guess. There may several intermediate states between A and B (see Fig. 1) corresponding to local minima of the free energy. In addition given the dependence of $P(\xi)$ on $A(\xi)$ (see equation 1) even relatively small errors in the biasing potential can lead to regions of space which are not sampled adequately (for example if one maximum in the free energy is not correctly estimated).

Many alternative approaches have been designed, including the adiabatic free energy dynamics (AFED)[53–55] and slow growth. AFED is based on associating the reaction coordinate with a high mass and large temperature. In the adiabatic limit, Rosso[53–55] was able to show that the free energy can be recovered from the probability density function. The use of a large mass and high temperature ensures that the sampling along the reaction coordinate is uniform including near transition regions. This approach requires an explicit definition of generalized coordinates as well as a careful analysis to determine the appropriate mass and temperature to be used for the reaction coordinate. In particular, an appropriate trade-off must be found between satisfying the adiabaticity condition (which requires a large mass) and the efficiency of the simulation (if the mass is too large the reaction coordinate changes extremely slowly).

In the slow growth approach, an external operator is exerting a force on the system such that ξ varies infinitely slowly from A to B. In the limit of infinitely slow speed, a classical result of statistical mechanics is that the work needed to change ξ is equal to the free energy difference ΔA. In practice, this works well if $\dot{\xi}$ is so sufficiently small that the system stays close to equilibrium. An extension of this technique has been proposed by Jarzynski and others[11–14, 20, 28, 31, 32, 38] who provided an exact equation for the free energy difference for finite switching times. If $\dot{\xi}$ is not negligible, significant non equilibrium effects are going to be present. Those lead in general to a heating of the system and larger energies than usual. Despite this deviation from equilibrium statistics, the equation given by Jarzynski is exact. However it has been observed by several authors[52] that even though the equation is exact mathematically, this method leads to relatively large statistical errors which require very long simulation times to be reduced.

Another class of technique is called thermodynamic integration. For a sufficiently smooth function, the free energy difference can always be written as the following integral $A(\xi_1) - A(\xi_0) = \int_{\xi_0}^{\xi_1} dA/d\xi \, d\xi$. The key observation is that it is possible to calculate $dA/d\xi$ by recognizing that it is in fact equal to the following statistical average:

$$\frac{dA}{d\xi} = \left\langle \frac{\partial H}{\partial \xi} \right\rangle_\xi$$

The subscript ξ indicates that the average is computed for a fixed value of ξ, *i.e.* at a given point on our free energy plot (Fig. 1). Such an average corresponds very naturally to a "generalized" force acting on the reaction coordinate ξ. Of course in general ξ is not a particle and therefore no "real" mechanical force is acting on it. But if ξ was indeed a particle coordinate then this expression reduces to the mechanical force acting on this coordinate, *i.e.* $-\partial U/\partial \xi$ where U is the potential energy. Therefore this equation generalizes the notion of force to arbitrary variables which are function of the atomic positions.

Several techniques are available to calculate $\langle \partial H/\partial \xi \rangle$. Ciccotti and others [9, 10, 20–24, 39, 59–61] have developed a technique, called Blue Moon ensemble method, whereby a simulation is performed with ξ fixed at some value. This can be realized by applying an external force, the constraint force, which prevents the value of ξ from changing. From the statistics of this constraint force it is possible to recover the derivative of the free energy $dA/d\xi$. In fact the constraint force is of the form $\lambda \nabla \xi$, that is it is pointing in the direction of the gradient of ξ. We will see later on that $\langle \partial H/\partial \xi \rangle \sim \langle \lambda \rangle$ (the actual equation is more complicated as we will show). This means that the constraint force is a direct measurement of the derivative of the free energy.

An advantage of this technique is that this allows getting as many sample points as needed at each location ξ along the interval of interest. In particular, it is possible to obtain very good statistics even in transition regions which are rarely visited otherwise. This leads in general to an efficient calculation and small statistical errors. Nevertheless, despite its many successes, this method has some difficulties. First, the system needs to be prepared such that ξ has the desired value (at which $dA/d\xi$ needs to be computed). Then an equilibration run needs to be performed such that the bias introduced by the previous procedure is removed. Second it is difficult in general to determine how many points are needed to calculate the integral $\int_{\xi_0}^{\xi_1} \langle \partial H/\partial \xi \rangle \, d\xi$. This is a quadrature problem made difficult by the fact that nothing is known about the function being integrated. On one hand, we want to limit the number of quadrature points used, as calculating $\langle \partial H/\partial \xi \rangle$ is an expensive process. On the other hand, we need to resolve all the fluctuations of $dA/d\xi$ in order to correctly estimate the free energy profile. Third, it may be difficult to sample all the relevant conformations of the system with ξ fixed. This is a more subtle problem illustrated by Fig. 2. Several distinct pathways may exist between A and B. It is usually relatively easy for the molecule to enter one pathway or the other while the system is close to A or B. However, in the middle of the pathway, it may be very difficult to switch to another pathway. This means that if we start a simulation with ξ fixed inside one of the pathway, it is very unlikely that the system will ever cross to explore conformations associated with another pathway. And even if it does so, this procedure will in general lead to large statistical errors as the rate limiting process becomes the transition rate between pathways inside the set $\xi = $ constant.

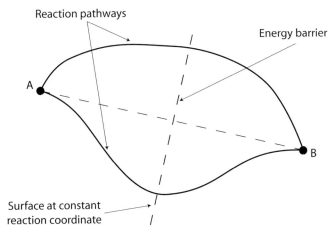

Figure 2. Free energy computation using constraint forces. It is difficult to sample the surface $\xi(x) = \xi$ using a constrained simulation because of the presence of energy barriers separating minima

The adaptive biasing force method (ABF) of Darve and Pohorille [15, 16, 52] is a method developed independently which circumvents those issues. It is based on un-constrained Molecular Dynamics simulations. This is a very efficient approach which begins by establishing a simple formula (see sections below) to calculate $dA/d\xi$ using an un-constrained molecular dynamics simulation (*i.e.* a regular simulation). This derivative represents the mean force acting on ξ. Therefore if we remove this force from the system we obtain a uniform sampling along ξ. This is done by adding an external force or biasing force which is opposite to the one acting on ξ. In fact this biasing force is equal to $dA/d\xi \, \nabla \xi$. What is the resulting motion of the system? There is no average force acting on ξ. However a non zero fluctuating force (with zero mean) remains. Therefore the dynamics of ξ is similar to a diffusive system or a random walk in 1D. Clearly the sampling along ξ is uniform which provides the smallest possible statistical error and an excellent convergence to the exact free energy profile.

This has the advantage of removing the need to a priori guess the biasing potential or to iteratively refine it. Instead the biasing force is estimated from the sampled conformations of the system and continuously updated as the simulation goes. This is therefore an adaptive algorithm. It uses all the statistics obtained so far to improve the sampling of the system. This is the reason why it is superior to the method of biasing potential which requires a series of simulations in order to properly estimate the correct biasing potential. In ABF, only one long simulation is needed to perform the entire calculation.

Importantly, contrary to constrained simulations, the system is allowed to evolve freely and in particular to explore the various pathways connecting A and B. This is one of the reasons why ABF can converge much faster than Blue Moon.

ABF shares some similarities with the technique of Laio et al. [33, 37, 40, 41, 43, 61], called meta-dynamics, where potential energy terms in the form of Gaussian functions are added to the system in order to escape from energy minima and accelerate the sampling of the system. However their approach is not based on an analytical expression for the derivative of the free energy but rather on importance sampling. Contrary to AFED[53–55], ABF is not limited to the adiabatic case.

In this article, we present the basic concepts behind thermodynamic integration. We give an overview of the Blue Moon method which is based on constraining ξ and of ABF. A new equation for ABF is derived which simplifies greatly the implementation process and makes it possible to rapidly implement the technique in an existing molecular dynamics code. We finish with some numerical examples to illustrate some applications of those techniques. Here are the main sections:

- Generalized coordinates. This section contains background material in order to properly introduce different free energy techniques.
- Derivative of the free energy. We give the central formula to practically calculate this derivative in a molecular dynamics code. This is important to facilitate the computer implementation.
- Blue Moon Ensemble. This is the method of constraints.
- Adaptive Biasing Force
- Numerical results

2 Generalized Coordinates and Lagrangian Formulation

Thermodynamic integration and ABF

There are three main techniques to calculate the free energy.

First, free energy perturbation methods attempt to do a direct calculation. Let's say we want to calculate the free energy difference between a system characterized by an Hamiltonian function H_0 and a system characterized by H_1. Free energy perturbation consists in running a simulation with a given Hamiltonian H_0 and calculate along the trajectory the energy difference $\Delta H = H_1 - H_0$. A classical result of statistical mechanics then allows to calculate the free energy difference $\Delta A = A_1 - A_0$ using:

$$\Delta A = -k_B T \log \langle \exp(-\Delta H / k_B T) \rangle$$

Second, some methods are based on computing the probability density function $P(\xi)$ directly from the simulation and then use the relation $A(\xi) = -k_B T \log P(\xi)$. This is approach followed by the umbrella sampling method for example[7, 27, 30, 36, 63, 64].

The third class of methods, thermodynamic integration or TI, calculates $dA/d\xi$ which by integration leads to $A(\xi)$. The derivative $dA/d\xi$ has units of force if ξ is a length and is usually interpreted as the mean force acting on ξ. This mean force can be readily calculated from simulations. In ABF, an external force is added to the system which opposes the mean force on ξ so that the resulting mean force on ξ is zero. This leads to a significant improvement in the convergence and efficiency of the free energy calculation as we will see.

This section will discuss some of the equations used to calculate the derivative of the free energy. In a different form, those results will be used in ABF to both calculate $dA/d\xi$ and bias the system in an adaptive manner.

Generalized coordinates

Molecular dynamics is based on a numerical integration of Newton's equations. The unknowns which are usually considered are the Cartesian coordinates of the atoms. In general those equations can be formulated as:

$$m_i \frac{d^2 x_i}{dt^2} = -\nabla U(x)$$

where x is a vector containing the 3 Cartesian coordinates of all the atoms, m_i are the masses and U is the potential energy. We are interested in computing the free energy A along a reaction coordinate ξ:

$$A(\xi) = -k_B T \, \log \int e^{-\beta H} \, \delta(\xi - \xi(x)) \, dx dp \tag{2}$$

In this integral p are the momenta associated with x, $p_i = m_i \dot{x}_i$.

The dirac delta function $\delta(\xi - \xi(x))$ means that we are effectively integrating over all coordinates x such that $\xi(x) = \xi$. It appears that if we define a set of coordinates of the form $(\xi, q_1, \cdots, q_{N-1})$ and their associated momenta $(p^\xi, p_1^q, \cdots, p_{N-1}^q)$ then this integration would be much simpler:

$$A(\xi) = -k_B T \, \log \int e^{-\beta H} \, dq_1 \cdots dq_{N-1} dp^\xi \cdots dp_{N-1}^q \tag{3}$$

How are those generalized coordinates and momenta defined? In general, we have to formulate our equations using the Lagrangian L. Let's assume that a configuration of our mechanical system can be represented by a set of coordinates q. Then:

$$L(q, \dot{q}) = K(q, \dot{q}) - U(q, \dot{q})$$

where K is the kinetic energy.

For example let's consider the simple 2D case where we have two Cartesian coordinates x and y:

$$L(x, y, \dot{x}, \dot{y}) = \frac{1}{2} m \dot{x}^2 + \frac{1}{2} m \dot{y}^2 - U(x, y)$$

We may express the same Lagrangian using other coordinates such as polar coordinates r and θ. The equation for L would then read:

$$L(r, \theta, \dot{r}, \dot{\theta}) = \frac{1}{2} m \left((\dot{r})^2 + (r\dot{\theta})^2 \right) - U(r, \theta) \tag{4}$$

In general, if we have $q = q(x)$ we can write:

$$\dot{q}_i = \sum_j \frac{\partial q_i}{\partial x_j} \dot{x}_j \overset{\text{def}}{=} [J^{-1}(q)\dot{x}]_i$$

where $J(q)$ is the Jacobian matrix, $[J(q)]_{ij} = \partial x_i/\partial q_j$. The kinetic energy K can be expressed as:

$$\sum_i \frac{1}{2} m_i \dot{x}_i^2 = \frac{1}{2} \dot{x}^t \cdot M \cdot \dot{x}$$

$$= \frac{1}{2} \dot{q}^t \cdot [J^t(q) M J(q)] \cdot \dot{q} \tag{5}$$

where M is a diagonal matrix with $M_{ii} = m_i$ and $\dot{x}^t$ is the transpose of vector $\dot{x}$. We denote Z and M^G the following matrices:

$$Z \overset{\text{def}}{=} J^{-1} M^{-1} (J^{-1})^t \tag{6}$$

$$M^G \overset{\text{def}}{=} Z^{-1} = J^t M J \tag{7}$$

Z_ξ is the first element in the matrix:

$$Z_\xi = \sum_i \frac{1}{m_i} \left(\frac{\partial \xi}{\partial x_i} \right)^2 \tag{8}$$

For our 2D example we obtain for J and Z:

$$J = \begin{pmatrix} \cos(\theta) & -r\sin(\theta) \\ \sin(\theta) & r\cos(\theta) \end{pmatrix}$$

$$Z = \frac{1}{m} \begin{pmatrix} 1 & 0 \\ 0 & \frac{1}{r^2} \end{pmatrix}$$

The kinetic energy is given by:

$$K(\dot{r}, \dot{\theta}) = \frac{1}{2m} \begin{pmatrix} \dot{r} & \dot{\theta} \end{pmatrix} Z^{-1} \begin{pmatrix} \dot{r} \\ \dot{\theta} \end{pmatrix} = \frac{1}{2} m \left((\dot{r})^2 + (r\dot{\theta})^2 \right)$$

This is identical to (4).

For the general case of coordinates q, the generalized momenta are defined by:

$$p_i = \frac{\partial L}{\partial \dot{q}_i}$$

If we use (5) to calculate p_i we obtain:

$$p_i = \frac{\partial}{\partial \dot{q}_i} \left(\frac{1}{2} \dot{q}^t \cdot Z^{-1} \cdot \dot{q} \right)$$

$$= [Z^{-1} \dot{q}]_i$$

The Hamiltonian in generalized coordinate is defined as:

$$H(q, p) = K(q, p) + U(q, p)$$
$$= \frac{1}{2} p^t \, Z \, p + U(q, p) \tag{9}$$

The equations of motion are given by:

$$\frac{dq_i}{dt} = [Zp]_i$$
$$\frac{dp_i}{dt} = -\frac{1}{2} p^t \frac{\partial Z}{\partial q_i} p - \frac{\partial U}{\partial q_i}$$

Let's take our 2D example to see how those equations are applied. The momenta are defined by:

$$p_r = m\dot{r}$$
$$p_\theta = mr^2\dot{\theta}$$

p_θ is in fact the angular momentum of the particle. The equations of motion for the momenta are:

$$\frac{dp_r}{dt} = \frac{1}{m}\frac{1}{r^3} p_\theta^2 - \frac{\partial U}{\partial r} = mr\dot{\theta}^2 - \frac{\partial U}{\partial r}$$
$$\frac{dp_\theta}{dt} = -\frac{\partial U}{\partial \theta}$$

It can be verified that $-\frac{\partial U}{\partial \theta}$ is the torque acting on the particle, and the second equation is the equation for the angular momentum.

Using the formalism developed above for generalized coordinates, we can derive an expression for the derivative of the free energy. Let's differentiate (3) with respect to ξ:

$$\frac{dA}{d\xi} = \frac{\int \frac{\partial H}{\partial \xi} e^{-\beta H} \, dq_1 \cdots dq_{N-1} dp^\xi \cdots dp_{N-1}^q}{\int e^{-\beta H} \, dq_1 \cdots dq_{N-1} dp^\xi \cdots dp_{N-1}^q}$$
$$= \left\langle \frac{\partial H}{\partial \xi} \right\rangle_\xi \tag{10}$$

It is important to understand that the partial derivative with respect to ξ requires a definition of the coordinates q since the partial derivative has to be computed by varying ξ, q being constant. However it is clear that even though generalized coordinates appear explicitly in this expression, the result of the averaging $\langle \partial H/\partial \xi \rangle_\xi$ is independent of the particular choice of q. We will see later on that it is possible to derive expressions which can be numerically evaluated without the need to define generalized coordinates. For now, we show that the derivative of the free energy can be written as:

$$\left\langle \frac{\partial H}{\partial \xi} \right\rangle_\xi = \left\langle \frac{\partial U}{\partial \xi} - k_B T \frac{\partial \log |J|}{\partial \xi} \right\rangle_\xi \tag{11}$$

This derivative can be seen as resulting from two contributions: the mechanical forces acting along ξ and the variations of the volume element associated with the generalized coordinates which is equal to $|J|$. The term $-1/\beta\, \partial \log |J|/\partial \xi$ is an entropic contribution.

Let's assume that $\xi = r$ and we are using polar coordinates (r, θ). In this case, $|J| = r$. The formula therefore reads:

$$\left\langle \frac{\partial H}{\partial r} \right\rangle_r = \left\langle \frac{\partial U}{\partial r} - \frac{1}{\beta r} \right\rangle_r$$

As we explained earlier, the difficulty in this formulation is that generalized coordinates appear explicitly in the form of the Jacobian $|J|$ which may be difficult to calculate in many cases. It is therefore desirable to find an expression where the explicit knowledge of the generalized coordinates is not required. This is done in the next section.

Proof. The partial derivative of H with respect to ξ can be computed using (9):

$$\frac{\partial H}{\partial \xi} = \frac{1}{2} p^t \frac{\partial Z}{\partial \xi} p + \frac{\partial U}{\partial \xi} \tag{12}$$

We can obtain a simple expression for $\langle \partial H/\partial \xi \rangle_\xi$ by integrating over the momenta p. To do so first consider the following integral:

$$\int \left(p^t \cdot \frac{\partial Z}{\partial \xi} \cdot p \right) \exp\left(-\frac{\beta}{2} p^T \cdot Z \cdot p \right) dp \tag{13}$$

It is possible to show that for any matrix A and B:

$$\int u^t \cdot B \cdot u \, \exp(-u^t \cdot A \cdot u) = \frac{\mathrm{Tr}(A^{-1}B)}{2} \int \exp(-u^t \cdot A \cdot u) \tag{14}$$

Applying this result to (13), we get:

$$\int \left(p^t \cdot \frac{\partial Z}{\partial \xi} \cdot p \right) \exp\left(-\frac{\beta}{2} p^T \cdot Z \cdot p \right) dp = $$
$$\frac{\mathrm{Tr}(\frac{2}{\beta} Z^{-1} \frac{\partial Z}{\partial \xi})}{2} \int \exp\left(-\frac{\beta}{2} p^T \cdot Z \cdot p \right) dp$$

For any matrix $A(t)$ we also have the following identity:

$$\frac{d}{dt} \log |A(t)| = \mathrm{Tr}\left(A^{-1} \frac{dA}{dt} \right) \tag{15}$$

where $|A(t)|$ is the determinant of $A(t)$. Applying this result we obtain:

$$\int \left(p^t \cdot \frac{\partial Z}{\partial \xi} \cdot p \right) \exp \left(-\frac{\beta}{2} p^T \cdot Z \cdot p \right) dp =$$

$$= k_B T \frac{\partial \log |Z|}{\partial \xi} \int \exp \left(-\frac{\beta}{2} p^T \cdot Z \cdot p \right) dp \qquad (16)$$

Using (6) which defines matrix Z:

$$\log |Z| = \log(|J|^{-1} |M|^{-1} |J|^{-1})$$
$$= -2 \log |J| - \log |M| \qquad (17)$$

Using (16), (17) and (12) for $\partial H / \partial \xi$, we finally have:

$$\left\langle \frac{\partial H}{\partial \xi} \right\rangle_\xi = \left\langle \frac{\partial U}{\partial \xi} - k_B T \frac{\partial \log |J|}{\partial \xi} \right\rangle_\xi \qquad (18)$$

3 Derivative of the Free Energy

We will now prove that for an arbitrary vector field w such that $w \cdot \nabla \xi \neq 0$:

$$\frac{dA}{d\xi} = \left\langle \nabla U \cdot \frac{w}{w \cdot \nabla \xi} - k_B T \nabla \cdot \frac{w}{w \cdot \nabla \xi} \right\rangle_\xi \qquad (19)$$

The key advantage of this equation is that it does not involve generalized coordinates explicitly and is therefore very convenient from a computational standpoint.

Proof. As usual in thermodynamic integration, we try to compute the derivative of A rather than A itself. By differentiating directly (2) rather than introducing generalized coordinates we obtain:

$$\frac{dA}{d\xi} = -k_B T \frac{\int e^{-\beta U} \delta'(\xi - \xi(x)) \, dx}{\int e^{-\beta U} \delta(\xi - \xi(x)) \, dx} \qquad (20)$$

In order to simplify this expression and get rid of the δ' we are going to do an integration by parts. In the previous expression the derivative is with respect to ξ. To obtain derivatives with respect to x we use the chain rule of differentiation:

$$\frac{d \, \delta(f(x))}{dx} = f'(x) \, \delta'(f(x))$$

Applying this equation to our case, we obtain:

$$\frac{\partial \, \delta(\xi - \xi(x))}{\partial x_i} = -\frac{\partial \xi}{\partial x_i} \delta'(\xi - \xi(x)) \qquad (21)$$

Let's assume now that we have a vector field w such that $w \cdot \nabla \xi \neq 0$. Examples are:

$$\begin{cases} w_i = \dfrac{\partial \xi}{\partial x_i}, \text{ or} \\[2ex] w_i = \dfrac{1}{m_i}\dfrac{\partial \xi}{\partial x_i} \end{cases}$$

(21) leads to:

$$(w \cdot \nabla)\, \delta(\xi - \xi(x)) = -(w \cdot \nabla \xi)\, \delta'(\xi - \xi(x)) ,$$

from which, after dividing through by $-w \cdot \nabla \xi$:

$$\delta'(\xi - \xi(x)) = -\left(\frac{w}{w \cdot \nabla \xi}\right) \cdot \nabla \, \delta(\xi - \xi(x))$$

Inserting this result in (20), we obtain for the numerator:

$$\int e^{-\beta U}\, \delta'(\xi - \xi(x))\, dx = -\int e^{-\beta U}\left(\frac{w}{w \cdot \nabla \xi}\right) \cdot \nabla \, \delta(\xi - \xi(x))\, dx$$

We can now integrate by parts:

$$\int e^{-\beta U}\, \delta'(\xi - \xi(x))\, dx = -\int \nabla \left(e^{-\beta U}\frac{w}{w \cdot \nabla \xi}\right) \delta(\xi - \xi(x))\, dx$$

The divergence can be computed using the product rule to obtain:

$$\nabla \left(e^{-\beta U}\frac{w}{w \cdot \nabla \xi}\right) = e^{-\beta U}\left(-\beta \frac{\nabla U \cdot w}{w \cdot \nabla \xi} + \nabla \cdot \frac{w}{w \cdot \nabla \xi}\right)$$

For example if $w = \nabla \xi$ we obtain the following expression:

$$e^{-\beta U}\left(-\beta \frac{\nabla U \cdot \nabla \xi}{|\nabla \xi|^2} + \nabla \cdot \frac{\nabla \xi}{|\nabla \xi|^2}\right)$$

We have a new expression for $dA/d\xi$ which is valid for any choice of w satisfying the condition mentioned above:

$$\frac{dA}{d\xi} = -k_B T \frac{\int e^{-\beta U}\left(-\beta \frac{\nabla U \cdot w}{w \cdot \nabla \xi} + \nabla \cdot \frac{w}{w \cdot \nabla \xi}\right) \delta(\xi - \xi(x))\, dx}{\int e^{-\beta U}\, \delta(\xi - \xi(x))\, dx}$$

Using the notation $\langle \rangle$ we get (19):

$$\frac{dA}{d\xi} = \left\langle \nabla U \cdot \frac{w}{w \cdot \nabla \xi} - k_B T \nabla \cdot \frac{w}{w \cdot \nabla \xi} \right\rangle_\xi$$

Discussion of (19)

To give an example with a specific choice of w, consider $w = \nabla \xi$. Then:

$$\frac{dA}{d\xi} = \left\langle \frac{1}{|\nabla \xi|^2} \left(\nabla U \cdot \nabla \xi - k_B T \left(\nabla^2 \xi - 2 \frac{\nabla \xi \cdot \mathcal{H}(\xi) \cdot \nabla \xi}{|\nabla \xi|^2} \right) \right) \right\rangle_\xi \qquad (22)$$

where $\mathcal{H}(\xi)$ is the Hessian matrix of ξ:

$$\mathcal{H}(\xi)_{ij} = \frac{\partial^2 \xi}{\partial x_i \partial x_j} \; ,$$

and $\nabla^2 \xi$ is the Laplacian of ξ:

$$\nabla^2 \xi = \sum_i \frac{\partial^2 \xi}{\partial x_i^2} \; .$$

It may be surprising that the choice of w is arbitrary. This may become clearer when one realizes that there is a direct connection between the choice of generalized coordinates in (11) and the choice of w. Just as the choice of $(q_1, \cdots, q_{N-1})$ is arbitrary, so is the choice of w.

The connection can be made by defining:

$$w_i = \frac{\partial x_i}{\partial \xi}$$

With this choice, the condition $w \cdot \nabla \xi \neq 0$ is trivially satisfied since

$$w \cdot \nabla \xi = \sum_i \frac{\partial x_i}{\partial \xi} \frac{\partial \xi}{\partial x_i} = 1$$

Our new (19) can be seen to be identical to (11). Consider first $\nabla U \cdot (w/(w \cdot \nabla \xi))$ in (19):

$$\nabla U \cdot \frac{w}{w \cdot \nabla \xi} = \sum_i \frac{\partial U}{\partial x_i} \frac{\partial x_i}{\partial \xi} = \frac{\partial U}{\partial \xi}$$

This is the same term as in (11).

The second term is equal to:

$$\frac{\partial \log |J|}{\partial \xi} = \mathrm{Tr} \left(J^{-1} \frac{\partial J}{\partial \xi} \right)$$

$$= \sum_{ij} \frac{\partial q_i}{\partial x_j} \frac{\partial}{\partial \xi} \left[\frac{\partial x_j}{\partial q_i} \right]$$

$$= \sum_{ij} \frac{\partial q_i}{\partial x_j} \frac{\partial}{\partial q_i} \left[\frac{\partial x_j}{\partial \xi} \right]$$

$$= \sum_j \frac{\partial}{\partial x_j} \left(\frac{\partial x_j}{\partial \xi} \right) = \nabla w$$

Thus we proved that:

$$\frac{\partial U}{\partial \xi} - k_B T \frac{\partial \log |J|}{\partial \xi} = \nabla U \cdot \frac{w}{w \cdot \nabla \xi} - k_B T \quad \nabla \cdot \frac{w}{w \cdot \nabla \xi}$$

for $w_i = \partial x_i / \partial \xi$. There is a direct link between a choice of generalized coordinates and the choice of w. The significant practical difference of course is that choosing w can be done without explicitly defining the set of generalized coordinates, for example with $w_i = \partial \xi / \partial x_i$ or $w_i = 1/m_i \, \partial \xi / \partial x_i$.

If we choose for example, $\xi = r$ and $w = \nabla \xi$ we recover:

$$\nabla U \cdot \frac{w}{w \cdot \nabla \xi} - k_B T \nabla \cdot \frac{w}{w \cdot \nabla \xi} = \frac{\partial U}{\partial r} - \frac{1}{\beta r}$$

4 Potential of Mean Constraint Force

Having found a more convenient expression to calculate $dA/d\xi$ there remains to develop an algorithm which overcomes the sampling difficulty associated with high energy barriers. There are basically two approaches. The first one, called Blue Moon ensemble method, consists in imposing a constraint force on the system such that ξ is constant. We will show that the average of this constraint force is related to $dA/d\xi$ through a simple equation. The procedure is then as follows. One discretizes the interval of interest $[\xi_{min}, \xi_{max}]$ into a number of quadrature points. At each point, we run a constraint simulation from which $dA/d\xi$ can be computed. A quadrature can then be used to estimate $\Delta A = \int_{\xi_0}^{\xi_1} dA/d\xi \, d\xi$.

The second one, ABF, consists in computing this average and removing it from the system. This leads to a diffusive like motion along ξ and a much better convergence of the calculation. We will describe ABF in Sect. 5.

4.1 Constrained Simulation

In the Blue Moon method, a force of the form $\lambda \nabla \xi$ is applied at each step such that ξ remains constant throughout the simulation. To see how λ can be related to $dA/d\xi$, we recall that the free energy A is the effective or average potential acting on ξ. From physical intuition, it should be true that $-dA/d\xi$ is the average of the force acting on ξ. In a constraint simulation, this force is equal to $-\lambda$. Therefore we can expect to have $dA/d\xi \sim \langle \lambda \rangle$. We now make this statement more rigorous.

In constrained simulations, the Hamiltonian H is supplemented by a Lagrange multiplier:

$$H^\lambda \stackrel{\text{def}}{=} H + \lambda(\xi - \xi(x)) = K + U + \lambda(\xi - \xi(x)) \tag{23}$$

The additional term $\lambda(\xi - \xi(x))$ is needed to enforce the constraint $\xi(x) = \text{constant}$. This corresponds to an additional force equal to $-\nabla(\lambda(\xi - \xi(x)))$. λ is going to chosen such that $\xi(x) = \xi$. Therefore the force can be expressed more simply as

$\lambda \nabla \xi$. The interpretation is then quite natural. In order to enforce the constraint we are applying a force parallel to $\nabla \xi$ which opposes the mechanical force acting on ξ.

The condition $\xi(x) = $ constant implies in particular that $\ddot{\xi} = 0$. With this condition it is possible to derive an expression for λ as a function of positions and velocities:

$$
\begin{aligned}
\ddot{\xi} &= \frac{d}{dt}\dot{\xi} \\
&= \frac{d}{dt}(\nabla \xi \cdot \dot{x}) \\
&= \nabla \xi \cdot \ddot{x} + \frac{d\nabla \xi}{dt} \cdot \dot{x} \\
&= \sum_i \frac{1}{m_i}\frac{\partial \xi}{\partial x_i}\left(-\frac{\partial U}{\partial x_i} + \lambda \frac{\partial \xi}{\partial x_i}\right) + \dot{x}^t \cdot \mathcal{H} \cdot \dot{x}
\end{aligned}
$$

where we assumed that $\xi - \xi(x) = 0$. Solving for $\ddot{\xi} = 0$, we obtain:

$$
\lambda = Z_\xi^{-1}\left(\sum_i \frac{1}{m_i}\frac{\partial \xi}{\partial x_i}\frac{\partial U}{\partial x_i} - \dot{x}^t \cdot \mathcal{H} \cdot \dot{x}\right) \tag{24}
$$

See (8) for the definition of Z_ξ. Therefore λ is in general a function of x and $\dot{x}$. With this expression for λ, we see that if we start a simulation with $\xi = \xi(x)$ and $\dot{\xi} = 0$ then $\xi = \xi(x)$ at all times. Note that, different methods are employed numerically (see SHAKE or RATTLE[1]) to avoid a "drift" of ξ. Those methods are not based on a direct application of (24).

4.2 Fixman Potential

One important difference between constrained and un-constrained simulations is that the sampling of p_ξ is different. Recall that

$$
p_\xi = (Z^{-1})_{\xi\xi}\dot{\xi} + \sum_{i \leq N-1}(Z^{-1})_{\xi i}\frac{dq_i}{dt}
$$

In constrained simulations, $\dot{\xi} = 0$ so that p_ξ is not an *independent* variable but rather a function of q and p_q. Let's consider an arbitrary function $f(x)$ and the following average:

$$
\frac{\int f(x)e^{-\beta H}\,\delta(\xi(x) - \xi)\,dxdp}{\int e^{-\beta H}\,\delta(\xi(x) - \xi)\,dxdp} \tag{25}
$$

The Hamiltonian function of a system which is constrained with $\xi(x) = \xi$ and $\dot{\xi} = 0$ is given by:

$$
H_\xi = \frac{1}{2}(p^q)^t\,(M_q^G)^{-1}\,p^q + U(q) \tag{26}
$$

where the matrix M_q^G is a submatrix of M^G associated with q:

$$[M_q^G]_{ij} = \sum_k m_k \frac{\partial x_k}{\partial q_i} \frac{\partial x_k}{\partial q_j} \qquad (27)$$

What is the difference between sampling according to H and to H_ξ? The main difference is that when using H, p_ξ is sampled according to the correct distribution, whereas, when using H_ξ, p_ξ is a function of q and p and is not sampled as an independent variable. However averages using H can be easily connected to averages using H_ξ. Indeed for H the contribution of p_ξ is equal to:

$$\int e^{-\beta/2 \, Z_\xi p_\xi^2} \, dp_\xi \propto Z_\xi^{-1/2}$$

Therefore we have the following equality for the average of f in (25):

$$\frac{\int f(x)e^{-\beta H} \, \delta(\xi(x) - \xi) \, dxdp}{\int e^{-\beta H} \, \delta(\xi(x) - \xi) \, dxdp} = \frac{\int f(x)Z_\xi^{-1/2}e^{-\beta H_\xi} \, dqdp^q}{\int Z_\xi^{-1/2}e^{-\beta H_\xi} \, dqdp^q}$$

This equation is often written as:

$$\langle f \rangle_\xi = \frac{\left\langle f \, Z_\xi^{-1/2} \right\rangle_{\xi,\dot\xi}}{\left\langle Z_\xi^{-1/2} \right\rangle_{\xi,\dot\xi}} \qquad (28)$$

where $\langle\rangle_{\xi,\dot\xi}$ denotes an average with ξ constant and $\dot\xi = 0$.
    Considering the weight factor

$$Z_\xi^{-1/2}e^{-\beta H_\xi} = e^{-\beta(H_\xi + \frac{1}{2\beta} \log Z_\xi)}$$

we see that the average of f can be computed with a modified Hamiltonian

$$H_\xi^F(q, p^q) = H_\xi(q, p^q) + \frac{1}{2\beta} \log Z_\xi(q) \qquad (29)$$

The second term in this potential is the so-called Fixman potential. With this potential we simply have:

$$\langle f \rangle_\xi = \langle f \rangle_{\xi,\dot\xi}^F$$

where the F denotes an average at $\xi(x) = \xi$, $\dot\xi = 0$ and with Hamiltonian H_ξ^F.

4.3 Potential of Mean Constraint Force

We are now going to establish a key result which is a connection between the free energy $A(\xi)$ and constrained simulations. Specifically, there is a very simple relation between $A(\xi)$ and H_ξ^F.

$A(\xi)$ is defined in terms of a partition function:

$$A(\xi) = -k_B T \log \int e^{-\beta H} dq_1 \cdots dq_{N-1} dp^{\xi} \cdots dp^q_{N-1}$$

We've seen that the Hamiltonian H_ξ^F is a function of q and p^q only. Let's try to calculate the integral over p^ξ to get rid of this variable. The Hamiltonian H can be written as:

$$H = \frac{1}{2}\left((p^\xi)^t Z_\xi p^\xi + 2(p^\xi) Z_{\xi q} p^q + (p^q)^t Z_q p_q\right) + U(\xi, q)$$

The integral over p^ξ can be calculated analytically and is equal to:

$$\int e^{-\frac{\beta}{2}((p^\xi)^t Z_\xi p^\xi + 2(p^\xi) Z_{\xi q} p^q)} dp^\xi = \sqrt{\frac{\pi}{2\beta}} (Z_\xi)^{-1/2} e^{\frac{1}{2}(p^q)^t Z_{q\xi}(Z_\xi)^{-1} Z_{\xi q} p^q}$$

Some additional algebra then shows that:

$$A(\xi) = -k_B T \log \int e^{-\beta H_\xi^F} dq dp^q - k_B T \log \sqrt{\frac{\pi}{2\beta}}$$

(we used the fact that the inverse of M_q^G is equal to $Z_q - Z_{q\xi}(Z_\xi)^{-1} Z_{\xi q}$).

A calculation similar to the one performed to obtain (10) finally gives:

$$\frac{dA}{d\xi} = \left\langle \frac{\partial H_\xi^F}{\partial \xi} \right\rangle_{\xi, \dot\xi}^F \tag{30}$$

This equation is a very important result and shows that constrained simulations can be used to calculate $dA/d\xi$. A possible algorithm would consist in running a constrained simulation with the Hamiltonian H_ξ^F (which contains the extra Fixman potential $1/(2\beta) \log Z_\xi(q)$), calculate the rate of change of H_ξ^F with ξ at each step. Finally by averaging this rate of change the derivative of A can be computed.

Computing $\partial H_\xi^F/\partial \xi$ will in general prove to be cumbersome. To facilitate the implementation of the method, we will derive an equation with the constraint force, used to impose the constraint $\xi(x) = \xi$ throughout the simulation. We use a constrained Lagrangian with the Fixman potential introduced in the previous section (see (23)):

$$L^F = L - \frac{1}{2\beta} \log Z_\xi - \lambda^F (\xi - \xi(x))$$

It is easier to use a Lagrangian formulation since the constraint $\dot\xi = 0$ can be naturally expressed in this setting.

Now let's find an expression for $\langle \partial H_\xi^F/\partial \xi \rangle_{\xi, \dot\xi}^F$ involving λ^F. Using the Lagrangian equations with L^F, we have:

$$\frac{d}{dt}\left(\frac{\partial L^F}{\partial \dot\xi}\right) \stackrel{\text{def}}{=} \left(\frac{\partial L^F}{\partial \xi}\right) = \frac{\partial L}{\partial \xi} - \frac{1}{2\beta}\frac{\partial}{\partial \xi} \log Z_\xi + \lambda^F$$

Since $\dot{\xi} = 0$, we furthermore have:

$$\frac{\partial L}{\partial \xi} = \frac{\partial}{\partial \xi} \left(\frac{1}{2} \begin{pmatrix} \dot{\xi} \\ \dot{q} \end{pmatrix}^t M^G \begin{pmatrix} \dot{\xi} \\ \dot{q} \end{pmatrix} - U \right)$$

$$= \frac{\partial}{\partial \xi} \left(\frac{1}{2} \dot{q}^t M_q^G \dot{q} - U \right) = \frac{\partial L_\xi}{\partial \xi}$$

where L_ξ is the Lagrangian of the constrained system (q, p^q) (see (26)). We therefore get the following result for λ^F:

$$\lambda^F = \frac{d}{dt} \left(\frac{\partial L^F}{\partial \dot{\xi}} \right) - \frac{\partial L_\xi}{\partial \xi} + \frac{1}{2\beta} \frac{\partial}{\partial \xi} \log Z_\xi$$

Since our expression for $dA/d\xi$ involves H_ξ^F (equation (30)), we need replace the Lagrangian L_ξ by H_ξ^F.

For this, we show that in general for a set $(q_1, \cdots, q_N, p_1, \cdots, p_N)$:

$$\frac{\partial H}{\partial q_1}\bigg|_{q_i, i \neq 1, p} = -\frac{\partial L}{\partial q_1}\bigg|_{q_i, i \neq 1, \dot{q}} \tag{31}$$

To see this, we use the definition of H and L:

$$H(q, p) = \frac{1}{2} p^t Z p + U$$

$$L(q, p) = \frac{1}{2} \dot{q}^t Z^{-1} \dot{q} - U$$

with $Zp = \dot{q}$. Since

$$\frac{\partial Z^{-1}}{\partial q_1} = -Z^{-1} \frac{\partial Z}{\partial q_1} Z^{-1}$$

we see that $\partial H / \partial q_1 = -\partial L / \partial q_1$.

With (31), we have:

$$\lambda^F = \frac{d}{dt} \left(\frac{\partial L^F}{\partial \dot{\xi}} \right) + \frac{\partial H_\xi}{\partial \xi} + \frac{1}{2\beta} \frac{\partial}{\partial \xi} \log Z_\xi$$

$$= \frac{d}{dt} \left(\frac{\partial L^F}{\partial \dot{\xi}} \right) + \frac{\partial H_\xi^F}{\partial \xi} \tag{32}$$

using the definition (29) of H_ξ^F.

Using (30) for $dA/d\xi$, we now have an expression of the derivative which involves the Lagrange multiplier λ^F and the Lagrangian L^F:

$$\frac{\partial A}{\partial \xi} = \left\langle \lambda^F - \frac{d}{dt} \frac{\partial L^F}{\partial \dot{\xi}} \right\rangle_{\xi, \dot{\xi}}^F \tag{33}$$

For Hamiltonian systems, it is generally true that the phase-space average of the time derivative of any function of q and p is equal to zero. Therefore, (32) says that even though $\partial H_\xi^F/\partial\xi$ is different from λ^F point wise, their averages are equal.

From (33), we get the following result for $dA/d\xi$:

$$\frac{dA}{d\xi} = \left\langle \lambda^F \right\rangle_{\xi,\dot\xi}^F \tag{34}$$

4.4 Derivative of the Free Energy without the Fixman Potential

Den Otter [20–23], Sprik [60, 61] and Darve[16] used a different expression where the Fixman potential is not used. In that case we need to add correction terms to account for the different sampling of p_ξ. First, the weight $Z_\xi^{-1/2}$ in (28) must be re-introduced. Second, the Lagrange multiplier is going to be slightly different since the Fixman potential $1/(2\beta) \log Z_\xi$ is not used. Considering (24) for λ as a function of x and $\dot x$, we see that the Lagrange multiplier λ^F needs to be replaced by:

$$\lambda + \frac{1}{2\beta Z_\xi} \sum_i \frac{1}{m_i} \frac{\partial\xi}{\partial x_i} \frac{\partial \log Z_\xi}{\partial x_i}$$

The final expression therefore reads:

$$\frac{dA}{d\xi} = \frac{\left\langle Z_\xi^{-1/2} \left(\lambda + \frac{1}{2\beta Z_\xi} \sum_i \frac{1}{m_i} \frac{\partial\xi}{\partial x_i} \frac{\partial \log Z_\xi}{\partial x_i} \right) \right\rangle_{\xi,\dot\xi}}{\left\langle Z_\xi^{-1/2} \right\rangle_{\xi,\dot\xi}} \tag{35}$$

Examples. Let's consider $\xi = r$ in 2D. In this case:

$$\nabla\xi = u_r, \quad Z_\xi = \frac{1}{m}$$

where $u_r = \mathbf{r}/r$. Thus:

$$\frac{\partial A}{\partial r} = \langle \lambda \rangle_{r,\dot r}$$

For $\xi = r^2$:

$$\nabla\xi = 2r\, u_r, \quad Z_\xi = \frac{4r^2}{m}, \quad \nabla \log Z_\xi = \frac{2}{r} u_r$$

The expression for $dA/d\xi$ then reads:

$$\frac{dA}{d\xi} = \langle \lambda \rangle_{\xi,\dot\xi} + \frac{1}{2\beta\xi}$$

Let's take a final example with $\xi = \cos(\theta)$, $0 \le \theta \le \pi$, in 3D where θ is the angle with the z-axis. Then:

$$\nabla \xi = -\sin\theta \frac{u_\theta}{r}, \quad Z_\xi = \frac{\sin^2\theta}{mr^2},$$

$$\nabla \log Z_\xi = \frac{2\cot\theta}{r} u_\theta - \frac{2}{r} u_r$$

where u_θ is a unit vector defined by:

$$u_\theta = u_r \times \frac{u_r \times u_z}{|u_r \times u_z|}$$

The vector u_z is a unit vector pointing in the z direction. We finally get:

$$\frac{dA}{d\xi} = \frac{\langle r\lambda \rangle_{\xi,\dot\xi}}{\langle r \rangle_{\xi,\dot\xi}} - \frac{\xi}{\beta(1-\xi^2)} \tag{36}$$

4.5 A More Compact Expression

The formula that we gave for PCMF, equation (35), has the drawback of requiring the calculation of the second derivative of ξ with respect to x. Even though this is not a theoretical limitation, from a practical standpoint it would be convenient to have an expression involving first derivatives of ξ only. This can be obtained by introducing the constrained Hamiltonian H_ξ and doing the following expansion:

$$A(\xi) = -k_B T \, \log \int e^{-\beta H} \, \delta(\xi - \xi(x)) \, dxdp$$

$$= -k_B T \, \log \int e^{-\beta H_\xi} \, dqdp^q$$

$$- k_B T \, \log \frac{\int e^{-\beta H} \, \delta(\xi - \xi(x)) \, dxdp}{\int e^{-\beta H_\xi} \, dqdp^q} \tag{37}$$

See (26) for the definition of H_ξ.

The first term can be computed in a straightforward way using the Lagrange multiplier λ. A proof similar to the one given in Sect. 4.3 (see (34)) shows that:

$$\frac{d}{d\xi} \left(-k_B T \, \log \int e^{-\beta H_\xi} \, dqdp^q \right) = \langle \lambda \rangle_{\xi,\dot\xi}$$

The Fixman potential is not needed here since we are averaging with respect to H_ξ, i.e. $\xi(x) = \xi$ and $\dot\xi = 0$.

The second term is (37) can be expressed in terms of Z_ξ:

$$\frac{\int e^{-\beta H} \, \delta(\xi - \xi(x)) \, dxdp}{\int e^{-\beta H_\xi} \, dqdp^q} = \sqrt{\frac{\pi}{2\beta}} \frac{\int Z_\xi^{-1/2} e^{-\beta H_\xi} \, dqdp^q}{\int e^{-\beta H_\xi} \, dqdp^q}$$

$$= \sqrt{\frac{\pi}{2\beta}} \left\langle Z_\xi^{-1/2} \right\rangle_{\xi,\dot\xi}$$

If we are interested (as is almost always the case) in computing free energy differences, the factor $\sqrt{\pi/2\beta}$ can be ignored. The factor $Z_\xi^{-1/2}$ represents a contribution to the free energy, in addition to the "mechanical" forces given by λ, due to entropic effects.

The new expression for the derivative of the free energy is:

$$A(\xi) = \int \langle \lambda \rangle_{\xi,\dot{\xi}} \, d\xi - k_B T \, \log \left\langle Z_\xi^{-1/2} \right\rangle_{\xi,\dot{\xi}} \tag{38}$$

This expression involves only the computation of the gradient of ξ. A comparison with the previous (35) which uses the same constrained simulation but a different equation for $dA/d\xi$ shows the simplicity of this new expression. A similar expression was derived by Schlitter et al. [56–58, 62].

Let's take the three previous examples of Sect. 4.4.

In 2D, take $\xi = r$. In this case Z_ξ is a constant and therefore can be omitted from (38). We directly get $A(\xi) = \int \langle \lambda \rangle_{\xi,\dot{\xi}} \, d\xi$.

For $\xi = r^2$, $Z_\xi = 4r^2/m$, we obtain:

$$A(\xi) = \int \langle \lambda \rangle_{\xi,\dot{\xi}} \, d\xi + \frac{\log 4\xi/m}{2\beta}$$

However if we add a constant to $A(\xi)$ we get:

$$A(\xi) = \int \langle \lambda \rangle_{\xi,\dot{\xi}} \, d\xi + \frac{\log \xi}{2\beta}$$

This is the same result as before (Sect. 4.4).

The last example is $\xi = \cos(\theta)$. In this case $Z_\xi = \sin^2 \theta/mr^2$ and we get:

$$A(\xi) = \int \langle \lambda \rangle_{\xi,\dot{\xi}} \, d\xi + \frac{1}{2\beta} \log(1 - \xi^2) - k_B T \log\langle r \rangle_{\xi,\dot{\xi}}$$

This equation is now different from equation (36). The derivative of $\log(1 - \xi^2)/2\beta$ gives an identical term $-\xi/\beta(1 - \xi^2)$, but the weight r appears as a separate term $-k_B T \log\langle r \rangle$ instead of $\langle r\lambda \rangle/\langle r \rangle$.

5 Adaptive Biasing Force

We now describe a different approach which is simpler than Blue Moon and very efficient. It does not require running a constrained simulation and can be performed entirely with a single Molecular Dynamics run.

One reason for inefficiencies of constraint methods is that they may prevent an efficient sampling of the set $\xi(x) = \xi$. This is illustrated by Fig. (2). It is very common that many pathways separated by high energy barriers exist to go from A to B. In constrained simulation, the system can get trapped in one the pathways. In the most serious cases, this leads to quasi non ergodic effect where only a part of the set $\xi(x) = \xi$ is effectively explored. In less serious cases, the convergence is quite slow.

Another approach which does not suffer from such problems is the Adaptive Biasing Force (ABF) method. This method is based on computing the mean force on ξ and then removing this force in order to improve the sampling. This leads to a uniform sampling along ξ. The dynamics of ξ corresponds to a random walk with zero mean force. Only the fluctuating part of the instantaneous force on ξ remains. This method is quite simple to implement and leads to a very small statistical error and excellent convergence.

First, we provide an expression to compute the derivative of the free energy for unconstrained simulations. Second, we will discuss the calculation of the biasing force and the algorithmic implementation of the method.

5.1 Derivative of the Free Energy

Darve et al. [15, 16, 52] derived the following formula for the derivative of the free energy:

$$\frac{\partial A}{\partial \xi} = - \left\langle m_\xi \ddot{\xi} - \frac{2m_\xi^2}{\beta} \sum_{ij} \frac{1}{m_i m_j} \frac{\partial \xi}{\partial x_i} \frac{\partial \xi}{\partial x_j} \frac{\partial^2 \xi}{\partial x_i \partial x_j} \right\rangle$$

where

$$m_\xi = \frac{1}{Z_\xi}$$

This expression involves first and second order derivatives of ξ with respect to x. While this equation has been successfully used for many computations, a more convenient expression can be derived which involves derivatives with respect to time, which are easy to calculate using Molecular Dynamics, and first order derivatives with respect to space. No second derivatives are required. This significantly simplifies the implementation. This equation is:

$$\frac{dA}{d\xi} = - \left\langle \frac{d}{dt} \left(m_\xi \frac{d\xi}{dt} \right) \right\rangle_\xi \tag{39}$$

where $m_\xi = Z_\xi^{-1}$. For a single reaction coordinate $Z_\xi = \sum_r \frac{1}{m_r} (\partial \xi / \partial x_r)^2$.

This formula can be generalized to several reaction coordinates with little difficulty. In that case, m_ξ is a matrix defined as the inverse of matrix Z_ξ:

$$[Z_\xi]_{kl} = \sum_r \frac{1}{m_r} \frac{\partial \xi_k}{x_r} \frac{\partial \xi_l}{x_r}$$

In Darve[15] for example, an equation is derived involving the derivative of m_ξ with respect to Cartesian coordinates x_i (equation (36) for example). This is not required for our present equation (39).

Examples. Using the 2D examples from the previous section, for $\xi = r$, (39) becomes:

$$\frac{dA}{d\xi} = -\left\langle \frac{d}{dt}\left(m\frac{d\xi}{dt}\right)\right\rangle_\xi$$

For $\xi = r^2$, we obtain:

$$\frac{dA}{d\xi} = -\left\langle \frac{d}{dt}\left(\frac{m}{4\xi}\frac{d\xi}{dt}\right)\right\rangle_\xi$$

Finally for $\xi = \cos\theta$, we obtain:

$$\frac{dA}{d\xi} = -\left\langle \frac{d}{dt}\left(\frac{mr^2}{1-\xi^2}\frac{d\xi}{dt}\right)\right\rangle_\xi$$

We now prove this result and show that it is a consequence of the fundamental equation (19) (unconstrained formulation with configurational derivatives) with the choice

$$w_i = \frac{1}{m_i}\frac{\partial\xi}{\partial x_i}\;.$$

Proof. The derivative $d\xi/dt$ can be written as $d\xi/dt = \nabla\xi\cdot\dot{x}$. With the product rule for derivatives, we get:

$$\frac{d}{dt}\left(Z_\xi^{-1}\frac{d\xi}{dt}\right) = \frac{d}{dt}\left(Z_\xi^{-1}\nabla\xi\right)\cdot\dot{x} + \left(Z_\xi^{-1}\nabla\xi\right)\ddot{x}$$

$$= (\dot{x})^t\cdot\nabla\left(Z_\xi^{-1}\nabla\xi\right)\cdot\dot{x} + \left(Z_\xi^{-1}\nabla\xi\right)\ddot{x} \tag{40}$$

In the last line, we used again $d/dt = (\dot{x})^t\cdot\nabla$. The average over momenta with ξ fixed can be computed analytically. If we use $m_i\ddot{x}_i = -\partial U/\partial x_i$, we obtain:

$$-\left\langle \frac{d}{dt}\left(Z_\xi^{-1}\frac{d\xi}{dt}\right)\right\rangle_\xi =$$

$$= -\left\langle (\dot{x})^t\cdot\nabla\left(Z_\xi^{-1}\nabla\xi\right)\cdot\dot{x} + \left(Z_\xi^{-1}\nabla\xi\right)\ddot{x}\right\rangle_\xi$$

$$= -\left\langle k_BT\,\mathrm{Tr}\left[M^{-1}\nabla\left(Z_\xi^{-1}\nabla\xi\right)\right] - Z_\xi^{-1}\sum_i\frac{1}{m_i}\frac{\partial\xi}{\partial x_i}\frac{\partial U}{\partial x_i}\right\rangle_\xi$$

$$= \left\langle \sum_i [Z_\xi^{-1}\frac{1}{m_i}\frac{\partial\xi}{\partial x_i}\frac{\partial U}{\partial x_i} - k_BT\frac{\partial}{\partial x_i}(Z_\xi^{-1}\frac{1}{m_i}\frac{\partial\xi}{\partial x_i})]\right\rangle_\xi$$

It is identical to the general equation (19) with the choice $w_i = 1/m_i\,\partial\xi/\partial x_i$.

5.2 Implementation of ABF

The Adaptive Biasing Force (ABF) algorithm[15, 16, 52] computes the free energy using (39). In addition, an external force is applied to the system in order to get a uniform sampling along ξ. Assume that we bin the interval of interest for ξ and that we have collected $n^k(N)$ samples in bin k after N steps in an MD simulation. We use those samples to compute a running average of the force acting along ξ:

$$F_\xi^k(N) = \frac{1}{n^k(N)} \sum_{l=1}^{n^k(N)} \frac{d}{dt}\left(m_\xi \frac{d\xi}{dt}\right)(x_l^k) \qquad (41)$$

where x_l^k corresponds to sample l in bin k. We will explain in the next section how the time derivatives can be computed. The external force applied to the system is chosen equal to $-F_\xi^k$.

In general when very few samples are available the force $F_\xi^k(N)$ will not be an accurate approximation of $dA/d\xi$. Large variations in $F_\xi^k(N)$ may lead to non equilibrium effects and systematic bias of the calculation. However it is possible to mitigate this effect by adding a ramp function which reduces the variations from one step to the next of the external force applied in a given bin. The external force applied to the system can be chosen equal to:

$$-R(n^k(N))F_\xi^k(N) \qquad (42)$$

$$R(n) = \begin{cases} n/N_0 & \text{if } n \le N_0 \\ 1 & \text{if } n > N_0. \end{cases} \qquad (43)$$

It has been found in numerical tests that N_0 should be chosen relatively small, $N_0 = 100$ for example. In general it is important to have a good balance between improving rapidly the sampling of the system, which requires a small N_0, and not introducing large non equilibrium effects, which requires a large N_0. However, the effect of those non-equilibrium perturbations occur only at the beginning of the simulation and disappear rapidly as the number of sample points in the bin increases. For N samples, this initial systematic error decays like $1/N$. The statistical error, however, decays much more slowly like $1/\sqrt{N}$. This means that initial sampling errors quickly become negligible compared to statistical errors and therefore won't negatively affect the accuracy of the computation. Numerical tests have shown that choosing a small N_0 is preferred as it rapidly provides a good sampling along the reaction coordinate ξ.

5.3 Numerical Calculation of the Time Derivatives

In this section we describe the implementation of ABF using equation (39). The main issue is the computation of the time derivative. If we calculate the time derivative at half time step $t + \Delta t/2$ then we can approximate:

$$\frac{d}{dt}\left(Z_\xi^{-1}\frac{d\xi}{dt}\right) = \frac{(Z_\xi^{-1}\dot{\xi})(t + \Delta t) - (Z_\xi^{-1}\dot{\xi})(t)}{\Delta t} + O(\Delta t^2)$$

In general a Molecular Dynamics simulation which uses Velocity Verlet has a global accuracy of $O(\Delta t^2)$ therefore it is not necessary to calculate the time derivative with greater accuracy. $Z_\xi^{-1}\dot{\xi}$ can be computed using:

$$Z_\xi^{-1}\dot\xi = Z_\xi^{-1}\nabla\xi \cdot v , \quad \text{where } v = \dot x$$

The first term $Z_\xi^{-1}\nabla\xi$ is a function of x only. Using velocity Verlet, x is computed with local accuracy $O(\Delta t^4)$; v is computed with accuracy $O(\Delta t^2)$ if the following approximation is used:

$$v(t + \Delta t/2) \stackrel{\text{def}}{=} \frac{x(t + \Delta t) - x(t)}{\Delta t} + O(\Delta t^2)$$

This approximation is not sufficient since it would lead to an error in $O(\Delta t)$ for $d(Z_\xi^{-1}\dot\xi)/dt$.

Before introducing a more accurate approximation for v, we recall the basic velocity Verlet procedure:

$$v(t + \Delta t/2) \stackrel{\text{def}}{=} v(t - \Delta t/2) + \Delta t\, a(t) \tag{44}$$

$$x(t + \Delta t) = x(t) + \Delta t\, v(t + \Delta t/2) \tag{45}$$

where $a(t) = d^2x/dt^2$. It can be shown that the following expression provides an estimate of $v(t)$ with accuracy $O(\Delta t^4)$:

$$v(t) \stackrel{\text{def}}{=} \frac{v(t + \Delta t/2) + v(t - \Delta t/2)}{2} - \frac{\Delta t}{12}(a(t + \Delta t) - a(t - \Delta t)) + O(\Delta t^4)$$

$$= \frac{v(t + \Delta t/2)}{2} - \frac{\Delta t}{12}a(t + \Delta t) + \frac{v(t - \Delta t/2)}{2} + \frac{\Delta t}{12}a(t - \Delta t) + O(\Delta t^4)$$

$$\stackrel{\text{def}}{=} \frac{1}{2}v^+(t + \Delta t/2) + \frac{1}{2}v^-(t - \Delta t/2) + O(\Delta t^4)$$

Using this approximation we can now calculate the time derivative at $t + \Delta t/2$, as:

$$\frac{d}{dt}\left(Z_\xi^{-1}\frac{d\xi}{dt}\right) = \frac{(Z_\xi^{-1}\dot\xi)(t + \Delta t) - (Z_\xi^{-1}\dot\xi)(t)}{\Delta t} + O(\Delta t^3)$$

$$= \frac{1}{\Delta t}\Big(Z_\xi^{-1}(t + \Delta t)\nabla\xi(t + \Delta t) \cdot v(t + \Delta t)$$

$$- Z_\xi^{-1}(t)\nabla\xi(t)\cdot v(t)\Big) + O(\Delta t^3) \tag{46}$$

$$= \frac{1}{2\Delta t}(p_\xi^+(t + \Delta t) - p_\xi^+(t) + p_\xi^-(t + \Delta t) - p_\xi^-(t)) \tag{47}$$

where

$$p_\xi^+(t) \stackrel{\text{def}}{=} \frac{Z_\xi^{-1}(t)\nabla\xi(t)\cdot v^+(t + \Delta t/2)}{2\Delta t}$$

$$= \frac{Z_\xi^{-1}(t)\nabla\xi(t)\cdot \left[v(t + \Delta t/2) - \frac{\Delta t}{6}a(t + \Delta t)\right]}{2\Delta t},$$

$$p_\xi^-(t) \stackrel{\text{def}}{=} \frac{Z_\xi^{-1}(t)\nabla\xi(t)\cdot v^-(t - \Delta t/2)}{2\Delta t}$$

$$= \frac{Z_\xi^{-1}(t)\nabla\xi(t)\cdot \left[v(t - \Delta t/2) + \frac{\Delta t}{6}a(t - \Delta t)\right]}{2\Delta t}$$

In this expression Z_ξ, $\nabla\xi$ and a are functions of x and can computed at t and $t + \Delta t$. The velocity at half step is directly provided by the velocity Verlet algorithm. This means that to calculate $d(Z_\xi^{-1}\dot{\xi})/dt$ at time $t + \Delta t/2$ we need to collect data from time steps $t - \Delta t$, t, $t + \Delta t$, and $t + 2\Delta t$.

In ABF, an external force $\lambda\nabla\xi$ is applied to the system to improve the sampling along ξ. Since this force is added, the calculation of the derivative of $-Z_\xi^{-1}\dot{\xi}$ must be modified. Consider equation (40). The first term does not need any correction since x and $\dot{x}$ are sampled according to the correct distribution. However $\ddot{x}$ includes the ABF force whose contribution needs to be removed to compute the free energy derivative. The correction is therefore equal to:

$$(Z_\xi^{-1}\nabla\xi) \, M^{-1} \, (\lambda\nabla\xi) = \lambda$$

Since we approximate $-d(Z_\xi^{-1}\dot{\xi})/dt$ at $t + \Delta t/2$ we need to add the following correction:

$$\frac{1}{2}(\lambda(t) + \lambda(t + \Delta t))$$

Algorithm

The algorithm is summarized in Algorithm 4 at the end of this article.

6 Numerical Results

6.1 Method

The first example involves calculating the potential of mean force for the rotation of the C-C bond in 1,2-dichloroethane (DCE) dissolved in water. In the second example, the potential of mean force for the transfer of fluoromethane (FMet) across a water-hexane interface is obtained. Both test cases have been studied before [2, 15, 16, 48].

The first system consisted of one DCE molecule surrounded by 343 water molecules, all placed in a cubic box whose edge length was 21.73 Å. This yielded a water density approximately equal to 1 g/cm^3. The second system contained one FMet molecule and a lamella of 486 water molecules in contact with a lamella of 83 hexane molecules. This system was enclosed in a box, with $x, y-$dimensions equal to 24×24 Å^2 and $z-$dimension, perpendicular to the water-hexane interface, equal to 150 Å. Thus, the system contained one liquid-liquid interface and two liquid-vapor interfaces. The same geometry was used in a series of previous studies on the transfer of different solutes across a water-hexane interface [8, 50]. In both cases, periodic boundary conditions were applied in the three spatial directions.

Water-water interactions were described by the TIP4P model [34]. The models of DCE and FMet were described in detail previously [2, 49]. Interactions between different components of the system (the solute, water and hexane) were defined from the standard combination rules [35]. All intermolecular interactions were truncated smoothly with a cubic spline function between 8.0 and 8.5 Å. Cutoff distances were

---

Algorithm 4 ABF algorithm

Loop over time steps $i = 1, \cdots, n$:

  a1 $\leftarrow M^{-1} \times F(r0)$ /* Force computation at t   */

  a1 $\leftarrow$ ABF(a1, r0, v)   /* Add ABF force and update FE statistics   */

  v $\leftarrow v + dt \times a1$ (44)   /* Velocity update $t - \Delta t/2 \rightarrow t + \Delta t/2$ */

  r0 $\leftarrow r0 + dt \times v$ (45)   /* Position update $t \rightarrow t + \Delta t$ */

End of loop

$F(x)$ is the vector of mechanical forces: $F = -\nabla U$. "$\leftarrow$" is the assignment operator. The following ABF routine needs to be implemented:

a1 = function ABF(a1,r0,v)

/* * a1: acceleration at time t *

  * r0: position at time t *

  * v: velocity at time $t - \Delta t/2$ */

save ZD0, pxi0, pxi1, a0   /* saved between function calls   */

k $\leftarrow$ bin corresponding to $\xi(r0)$

Rk $\leftarrow$ ramp $R(n^k(N))$ in bin k   (43)   /* $R(n^k(N))$ */

la $\leftarrow -$ Rk$\times$F(k) / n(k)   (42)   /* $\lambda(t) = -R(n^k(N))F^k_\xi(N)$ */

a1 $\leftarrow$ a1 + la $\times M^{-1} \times \nabla \xi(r0)$ /* Add ABF   */

pxip $\leftarrow$ ZD0 $\cdot$ (v/2 $-$ (dt/12) $\times$ a1)   (47)   /* $p^+_\xi(t - \Delta t)/(2\Delta t)$ */

ZD0 $\leftarrow$ 1/dt $\times (Z^{-1}_\xi(r0) \nabla \xi(r0))$ (46)   /* $1/\Delta t \; Z^{-1}_\xi(t) \nabla \xi(t)$ */

pxim $\leftarrow$ ZD0 $\cdot$ (v/2 + (dt/12) $\times$ a0)   (47)   /* $p^-_\xi(t)/(2\Delta t)$ */

a0 $\leftarrow$ a1

pxi0 $\leftarrow$ pxip + pxi0   (47)   /* $[d(Z^{-1}_\xi \dot\xi)/dt](t - 3\Delta t/2)$ */

If $i \geq 4$:

  k0 $\leftarrow$ bin at step $t - 3\Delta t/2$.

  n(k0) $\leftarrow$ n(k0)+1 (increment $n^{k0}(N)$).

  F(k0) $\leftarrow$ F(k0) $-$ pxi0 (add new sample $-d(m_\xi \dot\xi)/dt$ to bin k_0). See equation (41).

Endif

pxi0 $\leftarrow -$ pxip + pxim + pxi1 $-$ la/2 (47) /*$(-p^+_\xi(t - \Delta t)+p^-_\xi(t)-p^-_\xi(t - \Delta t))/(2\Delta t)$

                                 $-(\lambda(t - \Delta t) + \lambda(t))/2$ */

pxi1 $\leftarrow -$ pxim $-$ la/2   (47) /* $-p^*_\xi(t)/(2\Delta t) - \lambda(t)/2$

End function ABF

"$\cdot$" is the dot product. At $t = 0$, the variables n, F, pxi0, pxi1, a0, ZD0 must be initialized to 0. There are more sophisticated versions of this algorithm including some where extrapolation schemes are used to estimate $dA/d\xi$ in bins where few sample points have been gathered. For example, if we don't have enough statistics in bin k, more accurate estimates in surrounding bins can be used to extrapolate $dA/d\xi$ in k.

---

measured between carbon atoms of DCE and FMet, oxygen atoms of water and united atoms of hexane.

    The equations of motion were integrated using the velocity Verlet algorithm with a 1 fs time step for DCE and 2 fs time step for FMet. The temperature was kept constant at 300 K using the Martyna et al. implementation [42] of the Nosé-Hoover algorithm [29, 45]. This algorithm allows for generating configurations from a canonical ensemble. Bond lengths and bond angles of water and hexane molecules were kept fixed using RATTLE [1].

For DCE in water, the Cl-C-C-Cl torsional angle ξ was taken as the reaction coordinate. For the transfer of FMet across the water-hexane interface, ξ was defined as the z component of the distance between the centers of mass of the solute and the hexane lamella.

The sets of calculations performed for both systems are summarized in Table 1. The total number of steps was roughly $2 \cdot 10^6$ for the DCE test and $6 \cdot 10^6$ for the FMet test.

Table 1. Summary of ABF calculations performed for the test cases

Test	$[\xi_a, \xi_b]$	Δt	# windows	bin size	simulation time / bin	total # steps
DCE	$[0, 180]$ (deg)	1 (fs)	5	5 (deg)	0.4 (ns)	$2 \cdot 10^6$
FMet	$[-15, 10]$ (Å)	2 (fs)	5	0.2 (Å)	2.5 (ns)	$6.25 \cdot 10^6$

In the DCE test, we used 5 overlapping windows (see Table 2) over the interval $[0°, 180°]$. The MD trajectory in each window was 0.4 ns long. Force statistics were collected in bins 5° wide. In the FMet test, we used 5 non-overlapping windows (details in Table 2) over the interval $[-15, 10]$ Å. The MD trajectory in each window was 2.5 ns long. Force statistics were collected in bins 0.2 Å wide.

Table 2. Setup for windows used in ABF

Test	Units	Window 1	Window 2	Window 3	Window 4	Window 5
DCE	deg	$[0, 46]$	$[26, 82]$	$[62, 118]$	$[107, 154]$	$[134, 180]$
FMet	Å	$[-15, -10]$	$[-10, -5]$	$[-5, 0]$	$[0, 5]$	$[5, 10]$

6.2 Dichloromethane

In Fig. 3 we show the free energy profile for the rotation of DCE around the Cl-C-C-Cl torsional angle. This profile, obtained using ABF, is in excellent agreement with the previously calculated reference curve [15]. The free energies for the *gauche* and *trans* conformations in water are nearly the same. In contrast, in the gas phase the *trans* conformation is favored by 1.1 kcal/mol [2, 48]. This means that, compared to the *trans* rotamer, the *gauche* conformation is stabilized in the aqueous environment by 1.4 kcal/mol. This can be explained by favorable interactions between the permanent dipole of DCE in the *gauche* state and the surrounding water. These interactions

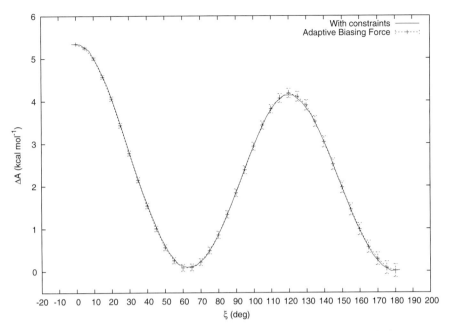

Figure 3. Free energy computation using ABF and a constrained simulation

are absent if DCE is in the *trans* conformation because, by symmetry, this state has no dipole moment. The free energy barrier was found to be equal to 4.2 kcal/mol, approximately 0.7 kcal/mole higher than the same barrier in the gas phase.

6.3 Fluoromethane

The free energy profile for the transfer of Fluoromethane (FMet) across a water-hexane interface obtained using ABF is shown in Fig. 4. The free energy exhibits a minimum at the interface, which is approximately 2 kcal/mol deep (compared to the free energy in bulk water). The existence of this minimum is due to the lower density at the interface between weekly interacting liquids, such as water and oil, compared to the densities in the bulk solvents. As a result, the probability of finding a cavity sufficiently large to accommodate the solute increases and the corresponding free energy cost of inserting a small, nonpolar or weakly polar solute decreases [46]. Similar free energy profiles were found for a wide range of other solutes [8, 47, 50]. The free energy difference between FMet in water and hexane is approximately equal to 0.5 kcal/mol, which corresponds well to the measured partition coefficient between these two liquids [46].

The maximum error is 0.12 kcal/mol. The profile is also in a very good agreement with the profile obtained for the same process using the particle insertion method [46], which is highly accurate for small solutes.

Despite its apparent simplicity, this calculation is relatively difficult to perform. Rodriguez-Gomez et al. [52] demonstrated that ABF performs much better than slow

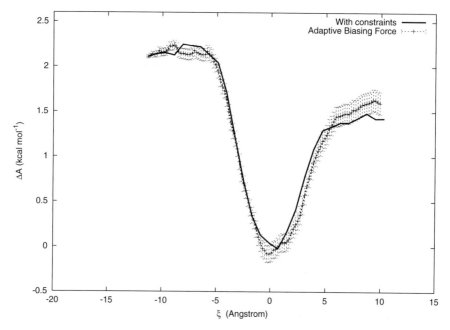

Figure 4. Free energy computation using ABF and a constrained simulation

growth (where the free energy is obtained from the work needed to change ξ adiabatically from -12Å to 10 Å), and than fast growth (non-equilibrium method of Jarzynski and Crooks[11–14, 20, 28, 31, 32, 38]).

6.4 Alanine Dipeptide

We applied ABF to study the isomerization of alanine dipeptide in water. This is one of the simplest peptides.The system is shown on Fig. 5. We consider the angles (ϕ, ψ). We chose this example to illustrate ABF with multiple reaction coordinates, for which the algorithm generalizes in a straightforward manner.

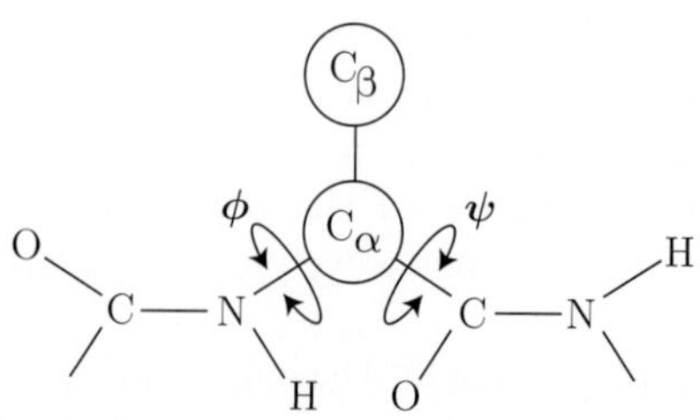

Figure 5. Schematic representation of alanine dipeptide. We compute the free energy as a function of ϕ and ψ. The multi-dimensional calculation is a direct extension of the single reaction coordinate case

One can observe that the calculation cost increases rapidly with the number of dimensions. If we assume for example that we need 10 quadrature points in each dimension to calculate $A(\xi)$ and that we need 10^5 sample conformations at each quadrature point to accurately calculate $\partial^d A/\partial\xi_1 \cdots \partial\xi_d$, then the computational cost scales like $\mathcal{O}(10^5 10^d)$ where d is the number of reaction coordinates. Therefore the cost increases quite rapidly with d.

The system we considered here is composed of the peptide with explicit water molecules. Periodic boundary conditions are used. 480 explicit water molecules with the TIP4P[34] force field were used. The box size is $[24.41, 24.41, 24.41]$ Å. The computed derivatives $(\partial A/\partial\phi, \partial A/\partial\psi)$ are shown on Fig. 6. The following rectangular windows were used (in degrees): for ϕ [180,245], [235,305], [295,360], and for ψ [−120, −55], [−65, 5], [−5, 65], [55, 125], [115, 180]. This corresponds to $\Delta\phi = 60°$ and $\Delta\psi = 60°$ with an overlap of $10°$ between windows. Note that the overlap is not a requirement of the method since we are calculating the derivative of the free energy. Those are point-wise values and do not need to be matched between windows. Those windows can be used to assess heuristically the accuracy of the calculation. We ran a 1 ns trajectory using ABF in each window. The time step was 1 fs.

In principle, the mean force is (up to the sign) the gradient of the free energy: $\nabla A = -\langle (F_\phi, F_\psi)\rangle$. However in practice because of statistical errors the estimated $\langle (F_\phi, F_\psi)\rangle$ is not the gradient of a function (it is not a conservative force). In order to reconstruct $A(\phi, \psi)$ a mean square procedure or the line integration method of Wu et al. [65] can be used.

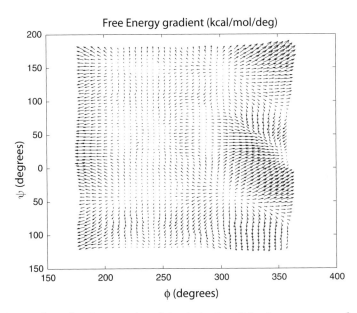

Figure 6. Two dimensional vector plot of the derivative of the free energy as a function of (ϕ, ψ). This derivative is the opposite of the mean force acting on (ϕ, ψ)

Using least-square, we approximate $A(\phi, \psi)$ using:

$$A(\phi, \psi) \approx \sum_k \alpha_k B_k(\phi, \psi)$$

where $B_k(\phi, \psi)$ is a suitable basis set, for example a piecewise linear approximation. The coefficients α_k are found by least square minimization. Let's denote (ϕ_l, ψ_l) the set of points where we calculated $F_\phi^l = -\langle F_\phi \rangle$ and $F_\psi^l = -\langle F_\psi \rangle$. The least square minimization finds the coefficients α_k which minimize:

$$\min_{\alpha_k} \sum_l \left(\sum_k \alpha_k \frac{\partial B_k(\phi_l, \psi_l)}{\partial \phi} - F_\phi^l \right)^2 + \left(\sum_k \alpha_k \frac{\partial B_k(\phi_l, \psi_l)}{\partial \psi} - F_\psi^l \right)^2$$

The line-integration method, which was used here, is a simpler approach where A is estimated using:

$$A(\phi, \psi) = A(\phi_0, \psi_0) - \frac{1}{2} \left(\int_{(\phi_0, \psi_0)}^{(\phi, \psi_0)} \langle F_\phi \rangle \, d\phi + \int_{(\phi, \psi_0)}^{(\phi, \psi)} \langle F_\psi \rangle \, d\psi \right.$$

$$\left. + \int_{(\phi_0, \psi_0)}^{(\phi_0, \psi)} \langle F_\psi \rangle \, d\psi + \int_{(\phi_0, \psi)}^{(\phi, \psi)} \langle F_\phi \rangle \, d\phi \right)$$

This corresponds to estimating $A(\phi, \psi)$ using the average of two different integration paths from (ϕ_0, ψ_0) to (ϕ, ψ).

The result for the reconstructed free energy is shown on Fig. 7. The error was estimated by running another set of simulations using the ABF estimate as a biasing potential and computing the error $E(\phi, \psi)$ using:

$$E(\phi, \psi) = \left| k_B T \ln \frac{P(\phi, \psi)}{P(\phi_0, \psi_0)} \right|$$

where $(\phi_0, \psi_0) = (270, 30)$. It was found to be around 0.2 kcal/mol everywhere except in high free energy region where the error was found to be around 0.6 kcal/mol. A longer simulation would be required to further reduce this error.

7 Conclusion

We discussed various methods to calculate the potential of mean force. The basic techniques are based on umbrella sampling and free energy perturbation. We described another approach based on thermodynamic integration which attempts to calculate the derivative of A by computing a mean force acting on ξ. Historically, this was first done by running a constrained simulation and averaging the constraint force $\lambda \nabla \xi$. This is the force which needs to be applied to the system in order to keep ξ constant. A more recent technique was developed, the adaptive biasing force (ABF),

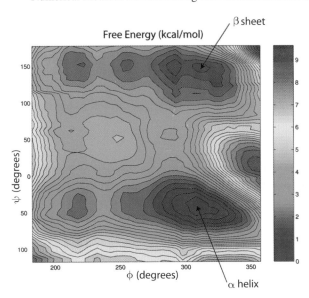

Figure 7. Two dimensional contour plot of the free energy as a function of ϕ (x axis) and ψ (y axis). The units are in kcal/mol. Note that the ϕ axis run from 180 to 360° while the ψ axis run from -120 to 180°. The two most important minima correspond to the values found in an α helix and a β sheet. Regions with very large free energies (up to around 9 kcal/mol) were sampled adequately by ABF

which improves sampling by removing the mean force on ξ. The resulting system randomly samples conformations along ξ with uniform probability. This leads to a very good convergence. ABF is an adaptive method which constantly modifies the external force as additional samples are gathered and more accurate averages are obtained. In the limit of a large number of sample points we converge to an Hamiltonian system and a uniform sampling along ξ. ABF is a very efficient method when sampling of the conformational space is rendered difficult by the presence of energy barriers. It works well with a single reaction coordinates or many.

We are currently working on applying ABF to Monte-Carlo simulations which would allow for example doing transition path sampling simulations [3–6, 17–19, 25, 26, 44]. In addition, we are considering extending ABF to alchemical transformations[51] where one molecule is transformed into another. Those types of simulations are very popular for drug discovery analysis, to compare mutant and wild types of proteins, etc.

Acknowledgments

I would like to thank Andrew Pohorille (NASA Ames Research Center) for valuable discussions and sharing his ideas. David Rodriguez-Gomez (NASA Ames Research

Center) performed the simulations for Alanine Dipeptide using ABF. The work presented in this article was performed in part using a Joint Research Interchange Grant between NASA and Stanford University. Some of the computer time was provided by NASA's Columbia parallel computer. We thank the reviewers of this manuscript for helpful remarks and references.

References

[1] H. C. Andersen. RATTLE: a 'velocity' version of the SHAKE algorithm for molecular dynamics calculations. *J Comp Phys*, 52:24–34, 1983.

[2] I. Benjamin and A. Pohorille. Isomerization reaction dynamics and equilibrium at the liquid-vapor interface of water — A molecular dynamics study. *J Chem Phys*, 98:236–242, 1993.

[3] P. G. Bolhuis, D. Chandler, C. Dellago, and P. L. Geissler. Transition path sampling: throwing ropes over rough mountain passes, in the dark. *Ann Rev Phys Chem*, 53:291 – 318, 2002.

[4] P. G. Bolhuis, C. Dellago, and D. Chandler. Sampling ensembles of deterministic transition pathways. *Faraday Discussions*, 110:421–436, 1998.

[5] P. G. Bolhuis, C. Dellago, and D. Chandler. Reaction coordinates of biomolecular isomerization. *Proc. Natl. Acad. Sci. U.S.A.*, 97(11):5877–5882, 2000.

[6] P. G. Bolhuis, C. Dellago, P. L. Geissler, and D. Chandler. Transition path sampling: throwing ropes over mountains in the dark. *J. Phys.: Condens. Matter*, 12(8A):A147–A152, 2000.

[7] C. Chipot. Rational determination of charge distributions for free energy calculations. *J Comp Chem*, 24(4):409–15, 2003.

[8] C. Chipot, M.A. Wilson, and A. Pohorille. Interactions of anesthetics with the water-hexane interface. A molecular dynamics study. *J Phys Chem B*, 101:782–791, 1997.

[9] G. Ciccotti and M. Ferrario. Rare events by constrained molecular dynamics. *J Mol Liquids*, 89(1/3):1–18, 2000.

[10] G. Ciccotti and M. Ferrario. Blue moon approach to rare events. *Mol Sim*, 30(11-12):787–793, 2004.

[11] G. E. Crooks. Nonequilibrium measurements of free energy differences for microscopically reversible Markovian systems. *J Stat Phys*, 90:1481–1487, 1998.

[12] G. E. Crooks. Entropy production fluctuation theorem and the nonequilibrium work relation for free energy differences. *Phys Rev E*, 60:2721–2726, 1999.

[13] G. E. Crooks. *Excursions in Statistical Dynamics*. PhD thesis, U.C. Berkeley, Department of Chemistry, 1999.

[14] G. E. Crooks. Path-ensemble averages in systems driven far from equilibrium. *Phys Rev E*, 61:2361–2366, 2000.

[15] E. Darve and A. Pohorille. Calculating free energies using average force. *J Chem Phys*, 115:9169–9183, 2001.

[16] E. Darve, M.A. Wilson, and A. Pohorille. Calculating free energies using a scaled-force molecular dynamics algorithm. *Mol. Sim.*, 28:113–144, 2002.

[17] C. Dellago, P. G. Bolhuis, and D. Chandler. Efficient transition path sampling: application to Lennard-Jones cluster rearrangements. *J Chem Phys*, 108(22):9236–45, 1998.

[18] C. Dellago, P. G. Bolhuis, and D. Chandler. On the calculation of reaction rate constants in the transition path ensemble. *J Chem Phys*, 110(14):6617–25, 1999.

[19] C. Dellago, P. G. Bolhuis, F. S. Csajka, and D. Chandler. Transition path sampling and the calculation of rate constants. *J Chem Phys*, 108(5):1964–77, 1998.

[20] W. K. den Otter. The calculation of free-energy differences by constrained molecular-dynamics simulations. *J Chem Phys*, 112:7283–7292, 2000.

[21] W. K. den Otter. Thermodynamic integration of the free energy along a reaction coordinate in Cartesian coordinates. *J Chem Phys*, 112(17):7283–7292, 2000.

[22] W. K. den Otter and W. J. Briels. The calculation of free-energy differences by constrained molecular-dynamics simulations. *J Chem Phys*, 109(11):4139–4146, 1998.

[23] W. K. den Otter and W. J. Briels. Free energy from molecular dynamics with multiple constraints. *Mol Phys*, 98:773–781, 2000.

[24] J. M. Depaepe, J. P. Ryckaert, E. Paci, and G. Ciccotti. Sampling of molecular conformations by molecular dynamics techniques. *Mol Phys*, 79(3):515–22, 1993.

[25] P. L. Geissler, C. Dellago, and D. Chandler. Chemical dynamics of the protonated water trimer analyzed by transition path sampling. *Phys Chem Chem Phys*, 1(6):1317–1322, 1999.

[26] P. L. Geissler, C. Dellago, and D. Chandler. Kinetic pathways of ion pair dissociation in water. *J Phys Chem B*, 103(18):3706–3710, 1999.

[27] M. F. Hagan, A. R. Dinner, D. Chandler, and A. K. Chakraborty. Atomistic understanding of kinetic pathways for single base-pair binding and unbinding in DNA. *Proc. Natl. Acad. Sci. U.S.A.*, 100(24):13922–13927, 2003.

[28] D. A. Hendrix and C. Jarzynski. A "fast growth" method of computing free energy differences. *J Chem Phys*, 114:5974–5981, 2001.

[29] W. G. Hoover. Canonical dynamics: equilibrium phase space distributions. *Phys Rev A*, 31:1695–1697, 1985.

[30] D. M. Huang, P. L. Geissler, and D. Chandler. Scaling of hydrophobic solvation free energies. *J Phys Chem B*, 105(28):6704–6709, 2001.

[31] G. Hummer. Fast-growth thermodynamic integration: error and efficiency analysis. *J Chem Phys*, 114:7330–7337, 2001.

[32] G. Hummer and A. Szabo. Free energy reconstruction from nonequilibrium single-molecule pulling experiments. *Proc. Natl. Acad. Sci. U.S.A.*, 98:3658–3661, 2001.

[33] M. Iannuzzi, A. Laio, and M. Parrinello. Efficient exploration of reactive potential energy surfaces using Car-Parrinello molecular dynamics. *Phys Rev Lett*, 90(23):238302–4, 2003.

[34] W. L. Jorgensen, J. Chandrasekhar, J. D. Madura, R. W. Impey, and M. L. Klein. Comparison of simple potential functions for simulating liquid water. *J Chem Phys*, 79:926–935, 1983.

[35] W. L. Jorgensen, J. D. Madura, and C. J. Swenson. Optimized potential energy functions for liquid hydrocarbons. *J Am Chem Soc*, 106:6638–6646, 1984.

[36] S. Kumar, D. Bouzida, R.H. Swendsen, P.A. Kollman, and J.M. Rosenberg. The weighted histogram analysis method for free-energy calculations on biomolecules. l. The method. *J Comp Chem*, 13(8):1011–1021, 1992.

[37] A. Laio and M. Parrinello. Escaping free-energy minima. *Proc. Natl. Acad. Sci. U.S.A.*, 99(20):12562–12566, 2002.

[38] J. Liphardt, S. Dumont, S. B. Smith, I. Tinoco, and C. Bustamante. Equilibrium information from nonequilibrium measurements in an experimental test of Jarzynski's equality. *Science*, 296:1832–1835, 2002.

[39] L. Maragliano, M. Ferrario, and G. Ciccotti. Effective binding force calculation in dimeric proteins. *Mol Sim*, 30(11-12):807–816, 2004.

[40] R. Martonak, A. Laio, and M. Parrinello. Predicting crystal structures: the Parrinello-Rahman method revisited. *Phys Rev Lett*, 90(7):075503–1, 2003.

[41] R. Martonak, A. Laio, and M. Parrinello. Predicting crystal structures: the Parrinello-Rahman method revisited. *Phys Rev Lett*, 90(7):075503/1–075503/4, 2003.

[42] G. J. Martyna, M. L. Klein, and M. Tuckerman. Nosé-Hoover chains: The canonical ensemble via continuous dynamics. *J Chem Phys*, 97:2635–2643, 1992.

[43] C. Micheletti, A. Laio, and M. Parrinello. Reconstructing the density of states by history-dependent metadynamics. *Phys Rev Lett*, 92(17):170601–4, 2004.

[44] P. Nielaba, M. Mareschal, and G. Ciccotti, editors. *Bridging the time scale gap with transition path sampling*, 2002.

[45] S. Nosé. A molecular dynamics method for simulations in the canonical ensemble. *Mol Phys*, 52:255–268, 1984.

[46] A. Pohorille, C. Chipot, M. New, and M.A. Wilson. Molecular modeling of protocellular functions. In L. Hunter and T.E. Klein, editors, *Pac Symp Biocomp '96*, pages 550–569. World Scientific, 1996.

[47] A. Pohorille, M.H. New, K. Schweighofer, and M.A. Wilson. Computer simulations of small molecules in membranes. In D.W. Deamer, editor, *Membrane Permeability: 100 years since Ernst Overton*, Current Topics in Membranes, pages 49–76. Academic Press, 1999.

[48] A. Pohorille and M.A. Wilson. Isomerization reactions at aqueous interfaces. In J. Jortner, R.D. Levine, and B. Pullman, editors, *Reaction Dynamics in Clusters and Condensed Phases — The Jerusalem Symposia on Quantum Chemistry and Biochemistry*, volume 26, page 207. Kluwer, 1993.

[49] A. Pohorille and M.A. Wilson. Excess chemical potential of small solutes across water-membrane and water-hexane interfaces. *J Chem Phys*, 104:3760–3773, 1996.

[50] A. Pohorille, M.A. Wilson, and C. Chipot. Interaction of alcohols and anesthetics with the water-hexane interface: a molecular dynamics study. *Prog. Colloid Polym. Sci.*, 103:29–40, 1997.

[51] R. J. Radmer and P. A. Kollman. Free energy calculation methods: a theoretical and empirical comparison of numerical errors and a new method for qualitative estimates of free energy changes. *J Comp Chem*, 18(7):902–19, 1997.

[52] D. Rodriguez-Gomez, E. Darve, and A. Pohorille. Assessing the efficiency of free energy calculation methods. *J Chem Phys*, 120(8):3563–78, 2004.

[53] L. Rosso, J.B. Abrams, and M.E. Tuckerman. Mapping the backbone dihedral free-energy surfaces in small peptides in solution using adiabatic free-energy dynamics. *J Phys Chem B*, 109(9):4162–4167, 2005.

[54] L. Rosso, P. Minary, Z.W. Zhu, and M.E. Tuckerman. On the use of the adiabatic molecular dynamics technique in the calculation of free energy profiles. *J Chem Phys*, 116(11):4389–4402, 2002.

[55] L. Rosso and M.E. Tuckerman. An adiabatic molecular dynamics method for the calculation of free energy profiles. *Mol Sim*, 28(1-2):91–112, 2002.

[56] J. Schlitter and M. Klahn. The free energy of a reaction coordinate at multiple constraints: a concise formulation. *Mol Phys*, 101(23-24):3439–3443, 2003.

[57] J. Schlitter and M. Klahn. A new concise expression for the free energy of a reaction coordinate. *J Chem Phys*, 118(5):2057–60, 2003.

[58] J. Schlitter, W. Swegat, and T. Mulders. Distance-type reaction coordinates for modelling activated processes. *J Mol Model*, 7(6):171–177, 2001.

[59] A. Sergi, G. Ciccotti, M. Falconi, A. Desideri, and M. Ferrario. Effective binding force calculation in a dimeric protein by molecular dynamics simulation. *J Chem Phys*, 116(14):6329–38, 2002.

[60] M. Sprik and G. Ciccotti. Free energy from constrained molecular dynamics. *J Chem Phys*, 109(18):7737–7744, 1998.

[61] A. Stirling, M. Iannuzzi, A. Laio, and M. Parrinello. Azulene-to-naphthalene rearrangement: the Car-Parrinello metadynamics method explores various reaction mechanisms. *Europ J Chem Phys Phys Chem*, 5(10):1558–1568, 2004.

[62] W. Swegat, J. Schlitter, P. Kruger, and A. Wollmer. MD simulation of protein-ligand interaction: formation and dissociation of an insulin-phenol complex. *Biophys J*, 84(3):1493–506, 2003.

[63] T. V. Tolpekina, W. K. den Otter, and W. J. Briels. Influence of a captured solvent molecule on the isomerization rates of calixarenes. *J Phys Chem B*, 107(51):14476–14485, 2003.

[64] P. Virnau and M. Muller. Calculation of free energy through successive umbrella sampling. *J Chem Phys*, 120(23):10925–30, 2004.

[65] Z. Wu and L. Li. A line-integration based method for depth recovery from surface normals. *Computer Vision, Graphics, and Image processing*, 43:53–66, 1988.

Replica-Exchange-Based Free-Energy Methods

Christopher J. Woods[1], Michael A. King[2], and Jonathan W. Essex[1]

[1] School of Chemistry, University of Southampton, Southampton SO17 1BJ, UK
[2] Celltech Group plc, 208 Bath Road, Slough SL1 3WE, UK

Abstract. The calculation of relative free-energies that involve large reorganizations of the environment is one of the great challenges of condensed-phase simulation. To meet this challenge, we have developed a new free-energy technique that combines the advantages of the Hamiltonian replica-exchange method with thermodynamic integration. We have tested and compared this technique with other free-energy methods on a range of systems including the binding of ligands to p38 kinase. The use of replica-exchange moves leads to dramatic improvements in configurational sampling, and a reduction of simulation errors. Subtle solvation effects are explicitly taken into account in the free-energy results. This is achieved at no extra computational cost, relative to standard free-energy methods.

Key words: Free Energy, Replica Exchange, Thermodynamic Integration

1 Introduction

The calculation of relative free energies is still one of the great challenges of condensed phase simulation. Of particular difficulty is the calculation of relative free energies that involve substantial reorganization of the environment, as in the case of the binding of different ligands to a protein. This leads to restricted sampling, and the system can become locked in local minima. Recently, Replica-Exchange[1, 2] (RE) methods have been proposed that improve sampling. They achieve this by running multiple replicas of the system at different temperatures, pressures, or with different Hamiltonians, and by periodically swapping the coordinates of selected pairs. The use of these techniques has been shown to improve the quality of conformational sampling of protein and polypeptide systems[1–7].

2 Free Energy Calculations

One method to calculate the relative free energy of two systems, here labelled A and B, is to use Finite Difference Thermodynamic Integration (FDTI)

[8–12]. This method uses a λ-coordinate to connect the two systems together. The λ-coordinate is used to scale the forcefield for the simulation such that at $\lambda = 0.0$ the forcefield represents system A, at $\lambda = 1.0$ the forcefield represents system B, and at λ values in-between the forcefield represents a non-physical hybrid of the two systems. Simulation trajectories are generated at several points, or windows, across λ. These trajectories could be generated via either Molecular Dynamics or Monte Carlo. These trajectories are used to estimate the gradient of the free energy, G with respect to λ at each λ-window, $\left(\frac{\partial G}{\partial \lambda}\right)_\lambda$. The relative free energy between the systems A and B, ΔG_{AB}, is equal to the integral of the free energy gradients across λ. The numerical integration of the free energy gradients may be performed using the trapezium rule[13] or via Gaussian quadrature[10, 11].

The gradient of the free energy with respect to λ at each window can be approximated using the Zwanzig equation[14];

$$G_{\lambda+\Delta\lambda} - G_\lambda = -k_B T \ln \left\langle \exp\left(\frac{-(E_{\lambda+\Delta\lambda}(x) - E_\lambda(x))}{k_B T}\right)\right\rangle_\lambda , \qquad (1)$$

where G_λ is equal to the free energy at λ, k_B is Boltzmann's constant, T is the simulation temperature, $E_\lambda(x)$ is the energy of the system at λ for configuration x, $E_{\lambda+\Delta\lambda}(x)$ is the energy of the system at $\lambda + \Delta\lambda$ for the same configuration x, and $\langle...\rangle_\lambda$ represents an average over all of the configurations contained in the trajectory generated at λ. If $\Delta\lambda$ is sufficiently small, then the finite difference approximation may be used, and

$$\left(\frac{\partial G}{\partial \lambda}\right)_\lambda \approx \frac{G_{\lambda+\Delta\lambda} - G_\lambda}{\Delta\lambda} . \qquad (2)$$

2.1 Replica Exchange Thermodynamic Integration

FDTI estimates the free energy gradient at each λ-window via an average over a single simulation trajectory generated at that λ-window. This can lead to random sampling errors in the calculated average, as the simulation may have been trapped within a local minimum, and thus this trajectory may not have sampled a sufficient range of configurations. One way of solving this problem would be to run multiple simulations at each λ-window and to average the results together, though this would not be a satisfactory solution. A better approach is to use the relatively new Replica Exchange method[1, 2]. A Replica Exchange simulation consists of multiple replicas of the system running in parallel. Each replica has a different property, and the coordinates of different replicas are periodically tested and swapped. It has been recognised that Replica Exchange can be applied to the calculation of relative free energies by using replicas with different λ-values.[1] Replica Exchange can thus be applied to a standard FDTI simulation by recognising that neighbouring λ-windows can be tested and swapped via the Replica Exchange methodology. This would allow the free energy average at each λ-window to be composed from coordinates generated by multiple trajectories. This combination of Replica Exchange with FDTI is called Replica Exchange Thermodynamic Integration (RETI)[16–18]

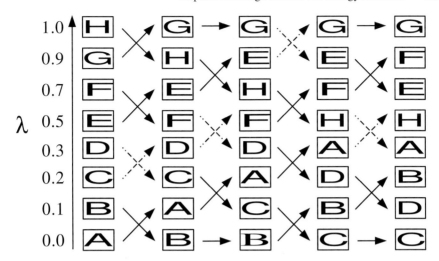

1. Create replicas of the system (A to H) and distribute over λ. Perform normal NVT or NPT sampling for each replica in parallel.

2. Perform a *Replica Exchange* test between neighbouring replicas. Swap the coordinates of pairs that pass. Do nothing with pairs that fail. Then continue the NVT or NPT sampling.

3. Keep repeating this test/sampling cycle until the trajectories have converged. At this point, the set of coordinates for each λ-value, e.g. (B,A,C,B,D) for λ = 0.1, form a correct ensemble distribution.

Figure 1. Protocol for a RETI simulation. This is identical to the protocol of an FDTI simulation, except for the addition of a periodic λ-swap test between neighbouring λ-windows. It should be noted that the implementation of the λ-swap test is trivial, and should have a negligible effect on the computational expense of the simulation

and the method is shown diagrammatically in Fig. 1. The λ-swap move between configurations i at $\lambda = A$ and j at $\lambda = B$ is tested via;

$$\exp\left[\frac{1}{k_BT}\left(E_B(j) - E_B(i) - E_A(j) + E_A(i)\right)\right] \geq rand(0,1) , \qquad (3)$$

where $E_B(j)$ is the energy of configuration j at $\lambda = B$, $E_A(j)$ is the energy of configuration j at $\lambda = A$, $E_A(i)$ is the energy of configuration i at $\lambda = A$, $E_B(i)$ is the energy of configuration i and $\lambda = B$, and $rand(0,1)$ is a uniform random number generated between 0.0 and 1.0.

3 Hydration of Water and Methane

The RETI method was first tested and compared to FDTI by application to the calculation of the relative hydration free energies of water and methane. The details of these experiments have been described previously[16, 17], but in summary, the system consisted of a periodic box of 1679 TIP4P water molecules, simulated in the NPT ensemble at a pressure of 1 atm and a temperature of 298 K. At the centre of the box was placed a single TIP4P water molecule that was perturbed into an OPLS united atom methane molecule using a λ-coordinate. At $\lambda = 0.0$ the system represented water in water, while at $\lambda = 1.0$ the system represented methane in water. The relative hydration free energy of water and methane was calculated using both RETI and FDTI simulations. To allow a fair comparison, the conditions used for each of these simulations were identical. Monte Carlo (MC) sampling was used to generate the trajectories for each simulation. To investigate the size of random sampling error, the FDTI and RETI simulations were repeated four times, using a different random number seed so that different trajectories would be generated. Each simulation used twenty one windows spaced evenly across λ, and the free energy gradients were approximated using $\Delta\lambda = 0.001$. Coordinate λ-swap moves were attempted every 50 K (thousand) MC moves during the RETI simulations, and both simulations sampled 10 M (million) moves per λ-window, with the first 3 M moves discarded as equilibration.

The calculated relative hydration free energies are shown in Fig. 2 and compared to the experimental value[19]. The free energy results calculated via the RETI method are significantly better than those produced via FDTI. The four RETI results are very consistent, with a spread of less than 0.2 kcal mol^{-1}. The addition of the RETI λ-swap move has reduced random sampling error, as is made clear in a plot of the radial distribution function between the water-methane and the solvating waters as a function of λ (Fig. 3).

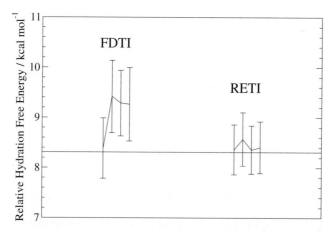

Figure 2. The relative hydration free energy of water and methane as calculated from four FDTI simulations and four RETI simulations. The experimental value of 8.31 kcal mol^{-1} is shown as a horizontal line. The four RETI free energies are very consistent, have smaller error bars, and are all very close to the experimental value

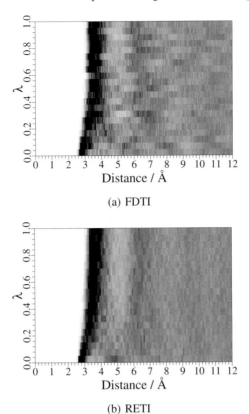

(a) FDTI

(b) RETI

Figure 3. The radial distribution function (RDF) between the perturbing water-methane solute and the oxygens of the solvating water molecules, as a function of λ (Water at $\lambda = 0.0$, methane at $\lambda = 1.0$). The scale runs from white (0.0), through grey(1.0) to black (2.0). The RDFs produced from a FDTI and RETI simulation are shown. Note how the locations of the solvation peaks change smoothly with respect to λ for RETI, and the increase in detail compared to the plot produced from the FDTI simulation

4 Halide Binding to a Calix[4]Pyrrole Derivative

The second test system for the RETI method was the calculation of the relative binding free energies of halides to a calix[4]pyrrole derivative. The details of this test have been described previously[16, 18], so again only a summary of the results will be presented here. For this system, RETI and FDTI were used to calculate the relative binding free energies of fluoride, chloride and bromide ions to a calix[4]pyrrole derivative that had been shown to be a specific binder for fluoride ions[15]. The relative binding free energies were calculated using both bound leg simulations, where one ion was perturbed into another while bound to the calix[4]pyrrole host, and a free leg, where the ions were perturbed while free in solvent. Both the bound and free legs were run in a periodic box of DMSO solvent molecules (1714 for the bound leg, 640 for the free leg), contaminated with 30 randomly positioned water molecules. To investigate the effect of random sampling error, the contaminating water molecules were repositioned randomly for each λ-window of each FDTI and RETI simulation. 21 λ-windows were evenly distributed across the λ-coordinate for each perturbation, and 10 M steps of Monte Carlo simulation were performed within each window, with free energy averages collected over the last 7 M steps. The full results for these simulations have been presented previously[16, 18], and they show that the relative binding free energies predicted via RETI were of a higher quality than those prediced via FDTI. As these results have been already presented, only a particularly interesting aspect of those results will be presented here. The λ-swap move of RETI allowed the free energy gradient at each λ-window to be formed as an average over configurations generated by multiple trajectories. This is in contrast to standard FDTI, where the free energy average is only formed over the configurations generated via a single trajectory. This is demonstrated clearly by Fig. 4, which plots the shortest distance between the halide ion and the closest water molecule as a function of λ. If a contaminating water molecule diffused to within 2-3 Å of the halide then it could form a strong hydrogen bonding interaction with the halide. Because of the relatively small number of contaminating water molecules, a single trajectory may

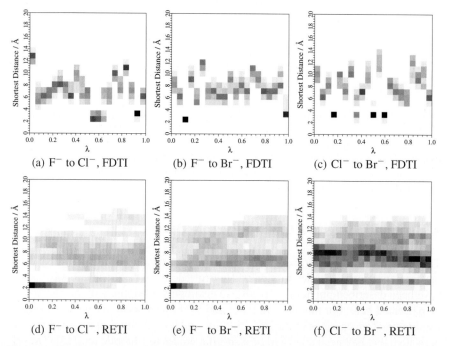

Figure 4. Shortest distance between the central halide and water oxygen atom, as a function of λ. Distances were calculated every 50 K MC steps and are histogrammed every 0.5 Å. The color scale runs from white (zero density), through grey (medium density) to black (high density)

not have sampled any configurations that involved this important hydrogen bonding interaction. Because FDTI only uses a single trajectory to form the free energy average at each λ-window, this meant that some of the averages were generated from trajectories that sampled no hydrogen-bonding to the halide, while others were generated from trajectories that were mostly sampling hydrogen-bonding configurations. In contrast, the λ-swap move of RETI enabled hydrogen-bonding and non-hydrogen-bonding configurations to be swapped across the entire λ-coordinate, thus enabling all of the free energy gradients to include configurations sampled from both types of trajectory. In addition, the λ-swap move correctly weighted the hydrogen-bonding configurations, increasing their probability for λ values that represented fluoride, and decreasing their probability for chloride and bromide.

5 Conclusion

Replica Exchange Thermodynamic Integration (RETI) is a novel free energy method that was developed to overcome the problems of random sampling error in free energy calculations. RETI has been applied successfully to the calculation of the relative hydration free energy of water and methane[17], the relative binding free energies of halides to a calix[4]pyrrole derivative[18] and to the relative binding of ligands to p38 kinase[16] (not presented here). All these applications showed that the addition of the λ-swap move has resulted in improved sampling and significantly improved free energy results. The λ-swap move was trivial to implement, and had a negligible impact on the computational cost of the simulation.

Acknowledgements

We would like to thank the BBSRC, EPSRC and Celltech Group plc for funding this research, and to Prof. W. L. Jorgensen for the provision of the MCPRO program and the associated source code[20].

References

[1] Sugita, Y., Kitao, A., Okamoto, Y.: Multidimensional replica-exchange method for free-energy calculations. J. Chem. Phys., **113**, 6042–6051 (2000)

[2] Fukunishi, H., Watanabe, O., Takada, S.: On the Hamiltonian replica exchange method for efficient sampling of biomolecular systems: Application to protein structure prediction. J. Chem. Phys., **116**, 9058–9067 (2002)

[3] Hansmann, U. H. E.: Parallel tempering algorithm for conformational studies of biological molecules. Chem. Phys. Lett., **281**, 140–150 (1997)

[4] Sugita, Y., Okamoto, Y.: Replica-exchange molecular dynamics method for protein folding, Chem. Phys. Lett., **314**, 141–151 (1999)

[5] Sanbonmatsu, K. Y., Garcia, A. E.: Structure of Met-enkephalin in explicit aqueous solution using replica exchange molecular dynamics, Proteins, **46**, 225–234 (2002)

[6] Zhou, R. H., Berne, B. J.: Can a continuum solvent model reproduce the free energy landscape of a beta-hairpin folding in water? Proc. Natl. Acad. Sci., **99**, 12777–12782 (2002)

[7] Verkhivker, G. M., Rejto, P. A., Bouzida, D., Arthurs, S., Colson, A. B., Freer, S. T., Gehlhaar, D. K., Larson, V., Luty, B. A., Marrone, T., Rose, P. W.: Parallel simulated tempering dynamics of ligand-protein binding with ensembles of protein conformations. Chem. Phys. Lett., **337**, 181–189 (2001)

[8] Mezei, M.: The finite-difference thermodynamic integration, tested on calculating the hydration free-energy difference between acetone and dimethylamine in water. J. Chem. Phys., **86**, 7084–7088 (1987)

[9] Guimaraes, C. R. W., Alencastro, R. B.: Evaluating the relative free energy of hydration of new thrombin inhibitor candidates using the finite difference thermodynamic integration (FDTI) method. Int. J. Quantum. Chem., **85**, 713–726 (2001)

[10] Guimaraes, C. R. W., Alencastro, R. B.: Thrombin inhibition by novel benzamidine derivatives: A free-energy perturbation study. J. Med. Chem., **45**, 4995–5004 (2002)

[11] Guimaraes, C. R. W., Alencastro, R. B.: Thermodynamic analysis of thrombin inhibition by benzamidine and p-methylbenzamidine via free-energy perturbations: Inspection of intraperturbed-group contributions using the finite difference thermodynamic integration (FDTI) algorithm. J. Phys. Chem. B, **106**, 466–476 (2002)

[12] Kamath, S., Coutinho, E., Desai, P.: Calculation of relative binding free energy difference of DHFR inhibitors by a Finite Difference Thermodynamic Integration (FDTI) approach. J. Biomol. Struct. Dyn., **16**, 1239–1244 (1999)

[13] Pearlman, D. A., Charifson, P. S.: Are free energy calculations useful in practice? A comparison with rapid scoring functions for the p38 MAP kinase protein system. J. Med. Chem., **44**, 3417–3423 (2001)

[14] Zwanzig, R. W.: High-Temperature Equation of State by a Perturbation Method. I. Nonpolar Gases. J. Chem. Phys., **22**, 1420–1426 (1954)

[15] Woods, C. J., Camiolo, S., Light, M. E., Coles, S. J., Hursthouse, M. B., King, M. A., Gale, P. A., Essex, J. W.: Fluoride-selective binding in a new deep cavity calix[4]pyrrole: Experiment and theory. J. Am. Chem. Soc., **124**, 8644–8652 (2002)

[16] Woods, C. J.: The Development of Free Energy Methods for Protein-Ligand Complexes. PhD Thesis, University of Southampton, Southampton, UK (2003)

[17] Woods, C. J., King, M. A., Essex, J. W.: The Development of Replica-Exchange-Based Free-Energy Methods. J. Phys. Chem. B, **107**, 13703–13710 (2003)

[18] Woods, C. J., King, M. A., Essex, J. W.: Enhanced Configurational Sampling in Binding Free-Energy Calculations. J. Phys. Chem. B, **107**, 13711–13718 (2003)

[19] Zhu, T. H., Li, J. B., Hawkins, G. D., Cramer, C. J., Truhlar, D. G.: Density functional solvation model based on CM2 atomic charges. J. Chem. Phys., **109**, 9117–9133 (1998)
[20] Jorgensen, W. L.: MCPRO 1.5, Yale University, New Haven, CT (1996)

Part V

Fast Electrostatics and Enhanced Solvation Models

Implicit Solvent Electrostatics
in Biomolecular Simulation

Nathan A. Baker[1], Donald Bashford[2] and David A. Case[3]

[1] Dept. of Biochemistry and Molecular Biophysics, Washington University, St. Louis,
   MO 63110, USA
[2] Dept. of Molecular Biotechnology, St. Jude Childrens Research Hospital, Memphis,
   TN 38105, USA
[3] Dept. of Molecular Biology, The Scripps Research Institute, La Jolla, CA 92037, USA

Abstract. We give an overview of how implicit solvent models are currently used in protein simulations. The emphasis is on numerical algorithms and approximations: since even folded proteins sample many distinct configurations, it is of considerable importance to be both accurate and efficient in estimating the energetic consequences of this dynamical behavior. Particular attention is paid to calculations of pH-dependent behavior, as a paradigm for the analysis of electrostatic interactions in complex systems.

Key words: Electrostatics, biomolecular simulation, implicit solvent, continuum solvent, Poisson-Boltzmann, Generalized Born

1 Introduction

Computer simulations have become an important method for understanding structure, dynamics, and function of proteins. Models of inter- and intra-molecular energetics are important components of such simulations. These energetic properties are determined by a combination of short- and long-range interactions. Short-range energetics include a variety of contributions such as van der Waals, bonding, angular, and torsional interactions. On the other hand, long-range energetics are predominately governed by electrostatic interactions. Due to their slow decay over distance, electrostatics cannot be neglected in biomolecular modeling as these forces are important at all length scales. Therefore, models which accelerate the evaluation of these interactions can provide important benefits to molecular simulation.

Due to the importance of electrostatics in biomolecular systems, a variety of computational methods have been developed for better understanding these interactions (see [1–6] and references therein). Popular methods for understanding electrostatic interactions in biomolecular systems can be classified as "explicit" methods which treat the solvent in full atomic detail and "implicit" methods which model solvent influences in a pre-averaged continuum fashion. Explicit solvent methods, by defini-tion, offer a more detailed description of biomolecular solvation; however, they also require integration over numerous solvent degrees of freedom. These extra degrees of freedom dramatically increase computational requirements and can limit the ability to use explicit solvent methods to generate converged estimates of thermodynamic and kinetic observables from a biomolecular simulation.

The sampling issues associated with explicit solvent treatments have necessitated the development of robust implicit solvent approaches. As their name implies, im-plicit solvent techniques are derived by pre-averaging over the solvent and counterion coordinates. The result of this averaging is a linear and local continuum dielectric model for solvent response and a mean field charge continuum for the counterion distribution. Despite some artifacts in the implicit solvent approach (e.g., see Roux [3] or Holm et al. [7] and references therein), these methods have enjoyed widespread use over recent years due to their significant reduction of complexity in solvated bio-molecular systems.

In this chapter, we will focus on implicit solvent methods for biomolecules in water, but the logical first proving ground for such applications is small molecules in water. In the most straightforward approach, the solute is described as a quantum-mechanical system in a region of dielectric dielectric constant 1 (where water cannot penetrate), surrounded by the high-dielectric solvent. This demands self-consistency between the electronic structure and the reaction potential arising from the partial orientation of solute dipoles around the solute. The dependence of the reaction po-tential on the solute charge distribution and hence, its electronic structure, gives rise to the self-consistency requirement. The earliest applications of this idea date back to Onsager [8] who used a simplified spherical model. Application to more com-plex molecular shapes has been pioneered by Tomasi and co-workers [9, 10], and comprehensive reviews of this approach are available [10, 11]. The terms, polarized continuum model or self-consistent reaction field are sometimes used to describe such calculations.

The application of continuum model to the *entire* solvent region (including water molecules in direct contact with the solute) seems like a severe approximation, but actually is known to give quite a good account of solvation free energies and pK_a behavior in small, fairly rigid molecules [11–16]. This is undoubtedly due in part to a careful parameterization of the boundary between the solvent and solute, so that the average energetic consequences of even first-shell waters are incorporated into the continuum model. Models for this dividing surface can be made that appear to be transferable, and not overly dependent on the detailed chemical nature of the solute. (This effective boundary varies with temperature in a way that is not easy to model; for this reason, continuum models are much less successful in predicting quantities like solvation enthalpies and entropies.)

A simpler, molecular mechanics, model describes the electrostatic properties of the solute with a set of fixed partial charges, usually centered on the atoms. The essential energetic idea, illustrated in Fig. 1, is to break the process of bringing a molecule from vacuum to solvent into three hypothetical steps: reduction to zero of the molecule's charges in vacuum; solvation of the purely non-polar molecule; and

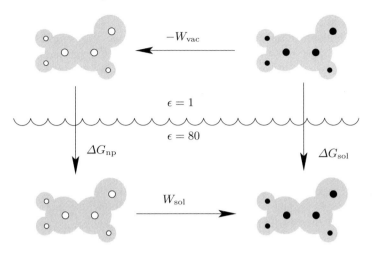

Figure 1. Thermodynamic cycle for computing solvation free energies. Filled circles represent atoms having partial charges, whereas empty circles represent a hypothetical molecule in which the partial charges have been set to zero

restoration of the original partial charges in the solvent environment. The overall solvation free energy is then,

$$\Delta G_{\mathrm{sol}} = \Delta G_{\mathrm{np}} - W_{\mathrm{vac}} + W_{\mathrm{sol}} , \tag{1}$$

where W_{vac} and W_{sol} are the work of charging in the vacuum and solvent environment, respectively, and ΔG_{np} is the free energy of solvation of the hypothetical molecule with all partial charges set to zero. For some small molecules, the ΔG_{np} term can be estimated as the solvation energy of a sterically equivalent alkane. More often, empirical formulas relating ΔG_{np} to surface area and/or volume are used [17–20]. Our focus here, however, will be on the electrostatic, or "polarization" component ΔG_{pol} defined by,

$$\Delta G_{\mathrm{pol}} = W_{\mathrm{sol}} - W_{\mathrm{vac}} . \tag{2}$$

Implicit solvent models have been used to study biomolecular function for over 80 years, starting with work by Born on ion solvation [21], Linderström-Lang [22] and Tanford and Kirkwood [23] on protein titration, and Flanagan et al on hemoglobin dimer assembly [24]. However, the applicability of these models was dramatically improved in the 1980's with increasing computer power and the availability of new numerical implicit solvent models which accurately described biomolecular geometries. Numerical methods for biomolecular implicit solvent electrostatics were first introduced by Warwicker and Watson [25], who described the numerical solution of the Poisson-Boltzmann (PB) equation to obtain the electrostatic potential in the active site of an enzyme.

    In addition to PB approaches, which we will discuss in the next section, further speed advantages may be had by adopting additional approximations. Currently, the most popular of these fall under the category of "generalized Born" models, and these will be discussed in Sect. 3. It is worth noting, however, that computational efficiency, although an important consideration, is not the only, or even the primary reason, for an interest in implicit solvent models. Here are some additional considerations that make many investigators willing to put up with the inevitable loss of physical realism that arises from replacing explicit solvent with a continuum:

1. There is no need for the lengthy equilibration of water that is typically necessary in explicit water simulations; implicit solvent models correspond to instantaneous solvent dielectric response.
2. Continuum simulations generally give improved sampling, due to the absence of viscosity associated with the explicit water environment; hence, the macromolecule can more quickly explore the available conformational space.
3. There are no artifacts of periodic boundary conditions; the continuum model corresponds to solvation in an infinite volume of solvent.
4. New (and simpler) ways to estimate free energies become feasible; since solvent degrees of freedom are taken into account implicitly, estimating free energies of solvated structures is much more straightforward than with explicit water models [26–28].
5. Implicit models provide a higher degree of algorithm flexibility. For instance, a Monte-Carlo move involving a solvent exposed side-chain would require nontrivial re-arrangement of the nearby water molecules if they were treated explicitly. With an implicit solvent model this complication does not arise.
6. Most ideas about "landscape characterization" (that is, analysis of local energy minima and the pathways between them) make good physical sense only with an implicit solvent model. Trying to find a minimum energy structure of a system that includes a large number of explicit solvent molecules is both difficult and generally pointless: the enormous number of potential minima that differ only in the arrangement of water molecules might easily overwhelm an attempt to understand the important nature of the protein solute.

We now turn to an analysis of computational algorithms that are prominent in models employing continuum solvent ideas.

2 Poisson-Boltzmann Methods

2.1 Derivation

The Poisson-Boltzmann (PB) equation is derived from a continuum model of the solvent and aqueous counterion medium around the biomolecules of interest [1, 3–5, 7]. There are numerous derivation of the PB equation based on statistical mechanics treatments (see Holm et al [7] for a review). However, the simplest derivation begins with Poisson's equation [29–31]:

$$-\nabla \cdot \epsilon(\mathbf{x})\nabla\phi(\mathbf{x}) = 4\pi\rho(\mathbf{x}) \text{ for } \mathbf{x} \in \Omega \tag{3}$$

$$\phi(\mathbf{x}) = \phi_0(\mathbf{x}) \text{ for } \mathbf{x} \in \partial\Omega \tag{4}$$

which describes the electrostatic potential $\phi(\mathbf{x})$ in an inhomogeneous dielectric medium of relative permittivity $\epsilon(\mathbf{x})$ due to a charge distribution $\rho(\mathbf{x})$. The equation is usually solved in a domain Ω with a Dirichlet condition $\phi_0(\mathbf{x})$ on the domain boundary $\partial\Omega$ that describes the asymptotic behavior of the solution.

For a system without mobile aqueous counterions, the charge distribution includes only "fixed" biomolecular charges, usually modeled as point charges centered on the atoms:

$$\rho_f(\mathbf{x}) = \sum_{i=1}^{M} Q_i \delta\left(\mathbf{x} - \mathbf{x}_i\right) , \tag{5}$$

where Q_i are the charge magnitudes and $\mathbf{x}_i$ are the charge positions. For a system with mobile counterions, the charge distribution also includes a term which describes the mean-field distribution of m different species of mobile ions:

$$\rho_m(\mathbf{x}) = e_c \sum_{j=1}^{m} \overline{n}_j z_j \exp\left[-e_c z_j \phi(\mathbf{x})/kT - V_j(\mathbf{x})/kT\right] \tag{6}$$

where $\overline{n}_j$ is the number density of ions of species j (in the absence of an external field), z_j is the valence of ions of species j, e_c is the elementary charge, and V_j is the steric interaction between the biomolecule and ions of species j which prevents overlap between the biomolecular and mobile counterion charge distributions. It is assumed that the macromolecule is at infinite dilution and that ϕ differs from zero only in a region near the macromolecule so that the $\overline{n}_j$ are adequate normalization constants.

Substitution of these two charge densities into the Poisson equation gives the "full" or nonlinear PB equation:

$$-\nabla \cdot \epsilon(\mathbf{x})\nabla\phi(\mathbf{x}) - 4\pi e_c \sum_{j=1}^{m} \overline{n}_j z_j \exp\left[-e_c z_j \phi(\mathbf{x})/kT - V_j(\mathbf{x})/kT\right]$$

$$= 4\pi \sum_{i=1}^{M} Q_i \delta\left(\mathbf{x} - \mathbf{x}_i\right) \text{ for } \mathbf{x} \in \Omega \tag{7}$$

$$\phi(\mathbf{x}) = \phi_0(\mathbf{x}) \text{ for } \mathbf{x} \in \partial\Omega . \tag{8}$$

This equation can be simplified somewhat for a 1:1 electrolyte such as NaCl where $q_1 = 1$, $q_2 = -1$. Assuming the steric interactions with the biomolecule are the same for all ion species ($V_1 = V_2 = V$), 7 reduces to

$$-\nabla \cdot \epsilon(\mathbf{x})\nabla\phi(\mathbf{x}) + 8\pi e_c \overline{n} e^{-V(\mathbf{x})/kT} \sinh[e_c\phi(\mathbf{x})/kT]$$

$$= 4\pi \sum_{i=1}^{M} Q_i \delta\left(\mathbf{x} - \mathbf{x}_i\right) \text{ for } \mathbf{x} \in \Omega \tag{9}$$

This is the canonical form of the nonlinear PB equation, and it is useful to examine how the biomolecular structure enters into each term. The dielectric function $\epsilon(\mathbf{x})$ is related to the shape of the biomolecule; assuming lower values of $2 - 20$ in the biomolecular interior and higher values proportional to bulk solvent outside the biomolecule. This dielectric function has been represented by a number of models, including discontinuous descriptions [9, 25] based on unions of atomic spheres, or a more complex molecular surface definition [32, 33] as well as smoother spline-based [34] and Gaussian [35] description. Not surprisingly, the results of PB calculations have been shown to be sensitive to the choice of surface definition; different surfaces appear to require different parameter sets [36–38]. The ion accessibility parameter $\bar{\kappa}^2(\mathbf{x})$ (see 2.12) usually varies from a value proportional to the bulk ionic strength outside the biomolecule to a value of zero at distances between the biomolecule and ions that are less than the ionic radii. Like the dielectric function, this accessibility function $\bar{\kappa}^2(\mathbf{x})$ is often implemented as either a smooth or discontinuous function. Finally, the biomolecular data enters explicitly into the fixed charge distribution on the right hand side of the equation via the partial atomic charge magnitudes $\{Q_i\}$ and locations $\{\mathbf{x}_i\}$.

For many systems where the mean-field linear dielectric approximations implicit in the PB equation are valid [7], the nonlinear PB equation can be reduced to a linear equation by assuming $e_c\phi(\mathbf{x})/kT \ll 1$ in the ion-accessible solvent region. In this case, the exponential functions (whether in the sinh of 9 or the exp of 7) can be truncated at first order in the Taylor series. By also assuming all steric factors are the same $V_j = V$, this linearization yields the linearized PB equation:

$$-\nabla \cdot \epsilon(\mathbf{x})\nabla\phi(\mathbf{x}) + \bar{\kappa}^2(\mathbf{x})\phi(\mathbf{x}) = 4\pi \sum_{i=1}^{M} Q_i \delta\left(\mathbf{x} - \mathbf{x}_i\right) \text{ for } \mathbf{x} \in \Omega \quad (10)$$

$$\phi(\mathbf{x}) = \phi_0(\mathbf{x}) \text{ for } \mathbf{x} \in \partial\Omega . \quad (11)$$

where

$$\bar{\kappa}^2(\mathbf{x}) = e^{-V(\mathbf{x})/kT} \frac{4\pi e_c^2}{kT} \sum_s \bar{n}_s z_s^2 = e^{-V(\mathbf{x})/kT} \frac{8\pi e_c^2 I}{kT} , \quad (12)$$

and I is the ionic strength $(1/2)\sum_s \bar{n}_s z_s^2$. Note that the above differs somewhat from the usual expression for κ^2 in Debye–Hueckel theory; in regions where $V(\mathbf{x}) = 0$, and the dielectric constant is that of the solvent, they are related by $\kappa^2 = \bar{\kappa}^2/\epsilon$.

Some of the primary uses of the PB equations are the calculation of electrostatic energies and forces for use in biomolecular simulations. The most intuitive way to calculate energy is through a charging process. The incremental work, δG of adding an increment of charge density, $\delta\rho$ to an already-existing charge distribution that is producing a potential $\phi(\mathbf{x}; \rho)$ is $\int \delta\rho(\mathbf{x})\phi(\mathbf{x}; \rho) \, d\mathbf{x}$. If the potential is linear in the charge, as in the Poisson or Linearized Poisson–Boltzmann equations, the integration from zero to the full charge distribution gives,

$$G[\phi] = \frac{1}{2} \int_\Omega \rho_f \phi d\mathbf{x} . \quad (13)$$

This formula suggests a practical procedure for calculating molecular energetics: solve the relevant linear equation for ϕ, and then multiply times the charge distribution through 13 or (more commonly) the analogous sum over point charges. The application of this idea is discussed further in Sect. 4.2.

A deeper approach which is useful for analyzing the mathematical problem of determining the potential, developing approximation schemes, and for the calculation of quantities such as forces is based on the fact that the PB equation is related to an energy functional in ϕ. The PB equation then arises from the requirement that ϕ minimize this functional while satisfying the boundary conditions. In this sense, the PB equation directly defines a free energy functional [39–41]. For the nonlinear PB equation in 9, this functional is:

$$G[\phi] = \int_{\Omega} \left[\rho_f \phi - \frac{\epsilon}{8\pi} (\nabla \phi)^2 - 2kT\bar{n}e^{-V/kT} \left(\cosh \left(\frac{e_c\phi}{kT} \right) - 1 \right) \right] d\mathbf{x} . \quad (14)$$

This functional involves several physically-intuitive terms, including (in order) the charge-potential interaction energy, the dielectric polarization energy, and the energy required to assemble the counterion distribution. Like the nonlinear PB equation, this can be linearized for small $\phi(\mathbf{x})$ to give the linearized PB equation free energy:

$$G[\phi] = \int_{\Omega} \left[\rho_f \phi - \frac{\epsilon}{8\pi} (\nabla \phi)^2 - \frac{\bar{\kappa}^2}{2} \phi^2 \right] d\mathbf{x} . \quad (15)$$

One can confirm that this expression is consistent with the work-of-charging expression, 13 through integration by parts and substitution of the original PB equation (10).

The free energy expressions in (14) and (10) can be differentiated with respect to atomic displacements to give electrostatic forces [34, 42]. The saddle-point approximation used to derive the PB equation implies $\delta G[\phi]/\delta \phi = 0$, therefore the force on atom i derived from the nonlinear PB equation is given by differentiation with respect to the atomic coordinate $\mathbf{y}_i$

$$\mathbf{F}_i[\phi] = - \int_{\Omega} \left[\phi \left(\frac{\partial \rho_f}{\partial \mathbf{y}_i} \right) - \frac{(\nabla \phi)^2}{8\pi} \left(\frac{\partial \epsilon}{\partial \mathbf{y}_i} \right) \right.$$

$$\left. + 2\bar{n}e^{-V/kT} \left(\cosh(e_c\phi/kT) - 1 \right) \left(\frac{\partial V}{\partial \mathbf{y}_i} \right) \right] d\mathbf{x} . \quad (16)$$

The corresponding force equation for the linearized PB is

$$\mathbf{F}_i[\phi] = - \int_{\Omega} \left[\phi \left(\frac{\partial \rho_f}{\partial \mathbf{y}_i} \right) - \frac{(\nabla \phi)^2}{8\pi} \left(\frac{\partial \epsilon}{\partial \mathbf{y}_i} \right) - \frac{(\phi)^2}{2} \left(\frac{\partial \bar{\kappa}^2}{\partial \mathbf{y}_i} \right) \right] d\mathbf{x} . \quad (17)$$

The papers cited above present a number of different numerical methods for evaluating the force and energy integrals based on stability considerations and different dielectric and ion accessibility surface definitions.

As mentioned above, it is important to note that both the nonlinear and linearized PB equations are approximate theories which should not be applied blindly to biomolecular systems — particularly those with high charge densities. The PB equation is based on a mean field approximation of the statistical mechanics of the counterion system; as such it neglects counterion correlations and fluctuations which become important at high ion concentrations and valencies [7]. Furthermore, the Poisson equation is based on the assumption of linear and local polarization of the solvent with respect to an applied field [43]. This assumption can break down under high fields or in highly-ordered systems of water. In short, the PB equation (and other implicit solvent models described in this chapter) works best for describing the electrostatic properties of biomolecules with low linear charge density in solutions of monovalent ions at low concentration.

2.2 Numerical Methods

Analytical solutions of the PB equations are not available for realistic biomolecular geometries. Therefore, this equation is generally solved numerically by a variety of computational methods. There are a few methods which try to solve the PB equations using atom-centered basis functions [44, 45]. However, the most common numerical techniques are based on discretization of the problem domain into simplices which support locally-defined basis functions. These methods include finite difference, finite element, and boundary element methods (see Fig.2). By projecting the solution into this discrete basis, these methods transform the continuous partial differential equation into a set of (possibly nonlinear) coupled algebraic equations. In particular, Newton methods [46, 47] are typically used used to transform the nonlinear equations into linear problems which can be solved using a variety of matrix algebra methods [48].

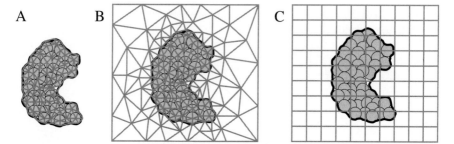

Figure 2. Discretization schemes used in solution of the PB equation: A. boundary element, B. finite element, C. finite difference

Finite element (FE) methods [48, 49] use an adaptive discretization of the problem domain, often into triangles (2D) or tetrahedra (3D). These methods offer a significant degree of adaptivity, allowing the problem unknowns to be distributed to minimize the error in the calculated solution (see Fig. 2A,B). When implemented correctly, FE methods can leverage very efficient algebraic multigrid solvers and thereby allow optimal numerical performance: the solution of an equation with N unknowns in $\mathcal{O}N$ time [50]. FE methods have been successfully applied to both the linear and nonlinear versions of the PB equation [51–55]. Furthermore, using parallel refinement methods developed by Bank and Holst [56], FE methods have been ported to massively parallel supercomputers and thereby enabled the solution of the PB equation for large microtubule structures [57].

Boundary element (BE) methods (see Fig. 2A) represent a particular implementation of finite element methods for surface representations of Poisson's equation [58, 59] and related implementations for linearized [60, 61] and nonlinear [62] versions of the PB equation. These methods rely on a discontinuous description of the solute-solvent dielectric interface; for the Poisson equation, this interface serves as the surface upon which a 2D finite element discretization is performed. For the PB equation, a number of BE methods are available. The linearized PB equation can be solved by extending the surface mesh a short distance from the biomolecule [60, 61]. The nonlinear version of the PB equation is solved by a boundary element formulation in conjunction with a coarse volume discretization in a manner similar to the finite difference Green function-based solution of the PB equation proposed by Levitt and co-workers [63]. The advantage of all BE methods is a much smaller numerical system to be solved (due to the lower dimensionality of the surface discretization). However, BE methods generate a much denser matrix problem than FE or finite difference methods and must be accelerated by wavelet or multigrid techniques [61, 64].

Finite difference (FD) methods rely on a general tensor product (typically Cartesian) mesh discretization of the system (see Fig. 2C). These discretizations include traditional finite difference [65, 66] and finite volume or multigrid [46, 47, 67] methods. All of these methods use various orders of Taylor expansion to generate "stencils" which represent the derivative operators in the PB equation. When applied to the PB equation, these stencils form a series of sparse algebraic equations which are solved with a variety of techniques. In particular, large FD problems benefit from a multigrid solution [46, 47] of the PB equation which allows $\mathcal{O}(N)$ solution of equations involving N grid points. Although FD methods do not permit adaptive refinement in the manner of BE and FE discretizations, they can employ a unique method known as electrostatic focusing [68, 69] to emulate limited adaptivity. As illustrated in Fig. 3B, focusing allows users to choose coarse FD grids for global calculations which are then used to define the boundary conditions on finer grids at regions of interest (binding and active sites, titratable residues, etc.). The result of successive levels of focusing are highly accurate local solutions to the PB equation with reduced levels of computational effort (compared to a global solution at high resolution). Focusing methods methods have also been parallelized [67] following the parallel FE ideas of Bank and Holst [56] and later implemented in a similar manner by Balls and Colella [70]. See Fig. 3 for a schematic of the FD and FE implementations of these parallel approaches. The resulting methods were used to solve the PB equation for a number of large biomolecular systems, including the ribosome and a million-atom microtubule [67].

Table 1 presents a partial list of the more popular PB solvers available at the time of publication.

Table 1. Popular PB equation software packages

Software	Description	Availability
APBS[67]	Solves the PB equation in parallel using the PMG FD multigrid package[46, 47] and the FEtk finite element software[50]	`http://agave.wustl.edu/apbs/`
Delphi[71]	Solves the PB equation sequentially with a highly optimized FD Gauss-Seidel solver	`http://trantor.bioc.columbia.edu/delphi/`
GRASP[72]	Solves the PB equation sequentially to obtain solutions for visualization	`http://trantor.bioc.columbia.edu/grasp/`
MEAD[73]	Solves the PB equation sequentially with an FD successive over-relaxation (SOR) solver	`http://www.scripps.edu/mb/bashford/`
ZAP[35]	Solves the PB equation very rapidly with a FD discretization and a smooth dielectric function	`http://www.eyes-open.com/products/toolkits/zap.html`
UHBD[74]	Multipurpose program with emphasis on Brownian dynamics; solves PB equation sequentially with an FD preconditioned conjugate gradient solver	`http://mccammon.ucsd.edu/uhbd.html`
MacroDox	Multipurpose program with emphasis on Brownian dynamics; solves PB equation sequentially with an FD SOR solver	`http://prin.chem.tntech.edu/`
Jaguar[53, 54]	Multipurpose program with emphasis on quantum chemistry; solves PB equation sequentially with an FE solver	`http://www.schrodinger.com/Products/jaguar.html`
CHARMM[75]	Multipurpose program with emphasis on molecular dynamics; solves PB equation sequentially with a FD multigrid solver and can be linked to APBS	`http://yuri.harvard.edu`
Amber[76]	Multipurpose program with emphasis on molecular dynamics; solves PB equation sequentially with a preconditioned conjugate gradient solver	`http://amber.scripps.edu`

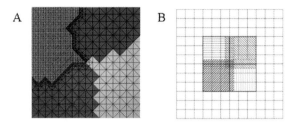

Figure 3. Parallel finite element and finite difference approaches to the PB equation. A. Depiction of parallel adaptive refinement in a finite element solution of the PB equation [56, 57]. The global domain is subdivided among the available processors; the problem is solved on the entire (coarse) domain. However, subsequent adaptive refinement is constrained to the appropriate subdomain specific to each processor. B. Demonstration of sequential [68] and parallel [67] focusing. In sequential focusing, a coarse grid (finer black lines) is used to obtain the global solution and set the boundary conditions on a finer grid (boundary denoted by heavy black line) surrounding the region of interest. Parallel focusing proceeds along similar lines except that the finer grid is further subdivided into "focusing subdomains" which are assigned to specific processors. The focusing proceeds as usual, except the final subdomain assigned to each processor is much smaller than with sequential focusing

3 Generalized Born and Related Approximations

The underlying physical picture on which the generalized Born approximation is based is the same as for the Poisson-Boltzmann calculations discussed above. In the case of simple spherical ion of radius a and charge eQ, the potentials can be found analytically and the result is the well-known Born formula [21],

$$\Delta G_{pol} = -\frac{Q^2}{2a}\left(1 - \frac{1}{\epsilon_{sol}}\right) \tag{18}$$

If we imagine a "molecule" consisting of charges $Q_1 \ldots Q_N$ embedded in spheres of radii, $a_1 \ldots a_N$, and if the separation r_{ij} between any two spheres is sufficiently large in comparison to the radii, then the solvation free energy can be given by a sum of individual Born terms, and pairwise Coulombic terms:

$$\Delta G_{pol} \simeq \sum_i^N -\frac{Q_i^2}{2a_i}\left(1 - \frac{1}{\epsilon_{sol}}\right) + \frac{1}{2}\sum_i^N\sum_{j\neq i}^N \frac{Q_iQ_j}{r_{ij}}\left(\frac{1}{\epsilon_{sol}} - 1\right) \tag{19}$$

where the factor $(1/\epsilon_{sol} - 1)$ appears in the pairwise terms because the Coulombic interactions are re-scaled by the change of dielectric constant upon going from vacuum to solvent.

The project of generalized Born theory can be thought of as an effort to find a relatively simple analytical formula, resembling 19, which for real molecular geometries will capture as much as possible, the physics of the Poisson equation. The linearity of the Poisson equation (or the linearized Poisson–Boltzmann equation) assures that ΔG_{pol} will indeed be quadratic in the charges, as both (13) and (19) assume.

However, in calculations of ΔG_{pol} based on direct solution of the Poisson equation, the effect of the dielectric constant is not generally restricted to the form of a pre-factor, $(1/\epsilon_{sol} - 1)$, nor is it a general result that the interior dielectric constant, ϵ_{in} has no effect. With these caveats in mind, we seek a function f^{GB}, to be used as follows:

$$\Delta G_{pol} \simeq - \left(1 - \frac{1}{\epsilon_{sol}}\right) \frac{1}{2} \sum_{ij} \frac{Q_i Q_j}{f_{ij}^{GB}} \qquad (20)$$

Here the self $(i = j)$ f^{GB} terms can be thought of as "effective Born radii," while in the off-diagonal terms, it becomes an effective interaction distance. The most common form chosen [77] is

$$f_{ij}^{GB}(r_{ij}) = \left[r_{ij}^2 + R_i R_j \exp(-r_{ij}^2/4R_i R_j)\right]^{1/2}, \qquad (21)$$

in which the R_i are the effective Born radii of the atoms, which generally depend not only on a_i, the radius of atom i, but on on the radii and relative positions of all other atoms. In principle, R_i could be chosen so that if one were to solve the Poisson equation for a charge Q_i placed at the position of atom i, and no other charges, and a dielectric boundary determined by all of the molecule's atoms and their radii, then the self energy of charge i in its reaction field would be equal to $-(Q^2/2R_i)(1 - 1/\epsilon_{sol})$. These values of R_i have been called "perfect" radii, and are known to give a reasonably good approximation to Poisson theory when used in conjunction with 21 [78]. Obviously, this procedure would have no practical advantage over a direct calculation of ΔG_{pol} using a numerical solution of the Poisson equation. To find a more rapidly calculable approximation for the effective Born radii, we turn to a formulation of electrostatics in terms of integration over energy density.

3.1 Derivation in Terms of Energy Densities

In the classical electrostatics of a linearly polarizable media [79] the work required to assemble a charge distribution can be formulated either in terms of a product of the charge distribution with the electric potential (as in 13), or in terms of the scalar product of the electric field $\mathbf{E}$ and the electric displacement $\mathbf{D}$:

$$G = \frac{1}{2} \int_{\Omega} \rho_f(\mathbf{x})\phi(\mathbf{x})d\mathbf{x} \qquad (22)$$

$$= \frac{1}{8\pi} \int_{\Omega} \mathbf{E} \cdot \mathbf{D} d\mathbf{x} \qquad (23)$$

We now introduce the essential approximation used in most forms of generalized Born theory: that the electric displacement is Coulombic in form, and remains so even as the exterior dielectric is altered from 1 to ϵ_{sol} in the solvation process. In other words, the displacement due to the charge of atom i (which for convenience is here presumed to lie on the origin) is,

$$\mathbf{D}_i \approx \frac{Q_i \mathbf{r}}{r^3} . \tag{24}$$

This is termed, the Coulomb field approximation. In the spherically symmetric case (as in the Born formula), this approximation is exact, but in more complex geometries, there may be substantial deviations. The work of placing a charge Q_i at the origin within a "molecule" whose interior dielectric constant is ϵ_{in}, surrounded by a medium of dielectric constant ϵ_{ex} and in which no other charges have yet been placed is then,

$$G_i = \frac{1}{8\pi} \int (\mathbf{D}/\epsilon) \cdot \mathbf{D} d\mathbf{x} \approx \frac{1}{8\pi} \int_{in} \frac{Q_i^2}{r^4 \epsilon_{in}} d\mathbf{x} + \frac{1}{8\pi} \int_{ex} \frac{Q_i^2}{r^4 \epsilon_{ex}} d\mathbf{x} . \tag{25}$$

The electrostatic component of the solvation energy is found by taking the difference in W_i when ϵ_{ex} is changed from 1.0 to ϵ_{sol},

$$\Delta G_{pol,i} = \frac{1}{8\pi} \left(\frac{1}{\epsilon_{sol}} - 1 \right) \int_{ex} \frac{Q_i}{r^4} d\mathbf{x} \tag{26}$$

where the contribution due to the interior region has canceled in the subtraction. Comparing 26 to the Born formula (18) or to (20) or (19), we conclude that the effective Born radius should be,

$$R_i^{-1} = \frac{1}{4\pi} \int_{ex} \frac{1}{r^4} d\mathbf{x} \tag{27}$$

It is convenient to re-write this in terms of integration over the interior region, excluding a radius a_i around the origin,

$$R_i^{-1} = a_i^{-1} - \frac{1}{4\pi} \int_{in, r>a_i} \frac{1}{r^4} d\mathbf{x} . \tag{28}$$

Note that in the case of a monatomic ion, where the molecular boundary is simply the sphere of radius a_i, this equation becomes $R_i = a_i$ and the Born formula is recovered exactly.

The integrals in (27) or (28) can be calculated numerically by a variety of quadrature schemes [77, 80–82], which have most of the usual tradeoffs between accuracy and computational efficiency. Many investigators have chosen to accept additional approximation in order to obtain pairwise analytical formulas for the effective radii. If the molecule consisted of a set of non-overlapping spheres of radius a_j at positions, $\mathbf{r}_{ij}$ relative to atom i, then 28 could be written as a sum of integrals over spherical volumes (see Fig. 4),

$$R_i^{-1} = a_i^{-1} - \frac{1}{4\pi} \sum_j \int_{|\mathbf{r} - \mathbf{r}_{ij}| < a_j} \frac{1}{r^4} d\mathbf{x} . \tag{29}$$

The integrals over spheres can then be calculated analytically [83] leading to,

$$R_i^{-1} = a_i^{-1} - \sum_j \frac{a_j}{2(r_{ij}^2 - a_j^2)} - \frac{1}{4r_{ij}} \log \frac{r_{ij} - a_j}{r_{ij} + a_j} . \tag{30}$$

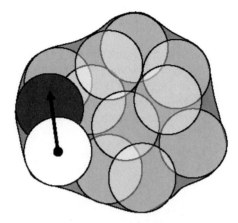

Figure 4. Representation of a molecule as a set of overlapping spheres. The integral needed in 29 can be approximately written as a sum over all of the spheres except for the white one, as in 29

An analytical expression is also available for the case of atom j overlapping with the central atom, i, provided j does not overlap any other atom j' [83]. Although in practice, the atoms j do overlap one another to some extent, these overlaps can be neglected to a first approximation, and empirical corrections can be introduced to compensate for neglect of overlap. This is referred to as the pairwise descreening approximation [84]. Hawkins et al. [84, 85] have introduced such a scheme based on re-scaling the van der Waals radii by factors S_j. The expression for the generalized Born radii takes the form,

$$R_i^{-1} = a_i^{-1} - \sum_j H(r_{ij}, S_j a_j) , \qquad (31)$$

where H is a rather complex expression which, apart from rescaling, is essentially 30 if i and j do not overlap, and having different functional forms in overlapping cases [85].

3.2 Limitations and Variations of the GB Model

The crux of the generalized Born model is the Coulomb field approximation, 24, and this is also the main source of its deviation from solvation energies calculated using solutions of the Poisson equation. In spherically symmetric cases, such as the case analyzed in the original Born theory, $\mathbf{D}$ is given exactly by the Coulomb field, and GB solvation goes over into Poisson-equation solvation. Several studies [81–83] have analyzed deviations from the Coulomb-field approximation for the case of charges at arbitrary positions within a spherical dielectric boundary, a case for which analytical solutions of the the Poisson equation are available [86]. The Coulomb field approximation leads to significant over-estimation of self-energies, and under-estimation of the screening of charge–charge interactions. Some additional quantitative sense of the limitations of the Coulomb field approximation can be gained by considering the case of a charge near a planar dielectric boundary, a situation analyzed in [87]. In the usual case where $\epsilon_{sol} \gg \epsilon_{in}$, the magnitude of ΔG_{pol} is underestimated by a factor of almost 2 compared to the exact expression. This suggests that for charges buried somewhat below the surface of large macromolecules, the solvation energy may be underestimated, and thus the effective radii overestimated due to the Coulomb-field approximation.

It is not yet clear to what extent these limitations in the generalized Born approach might be ameliorated by clever parameterization, or by explicit attempts to go beyond the Coulomb field approximation. There are a number of "flavors" of GB theory in current use; these have recently been reviewed [88, 89], and we only give a brief account of these here.

1. Schaefer and Karplus [90] presented a general formalism for decomposing energy functions based on integration of the Coulomb-field energy density into pairwise atomic terms. The integration over the solute interior in 28 can be rewritten as an integral over all space with the integrand multiplied by a step function $P(\mathbf{r})$ whose value is 1 in the molecular interior and zero elsewhere. This function can then be written as a sum of atomic terms,

$$P(\mathbf{r}) = \sum_j P_j(\mathbf{r}) \,, \tag{32}$$

   where, for example, the P_j might be step functions corresponding to Vornoi volumes of the atoms, but in principle, any function satisfying 32 is admissible. Schaefer and Karplus [90] proposed a Gaussian form for the P_k, normalized according to the "effective volume" of each atom, which is a parameter that characterizes its contribution to the total solvent-inaccessible volume of the solute. Based on this idea, they developed an analytical, continuous and differentiable pairwise-atomic expression for the electrostatic energy of the solute, which they term ACE (Analytical treatment of Continuum Electrostatics). The model compares reasonably well with the results of numerical solution of the Poisson equation for a test set of small molecules and several proteins, and is suitable for use in molecular mechanics calculations.

2. Recently, Onufriev et al. have developed a different approach to the problem of overlapping spheres versus realistic molecular surfaces in generalized Born theory [91, 92]. They noted that pairwise GB theories using functional forms such as 31 have been parameterized to perform quite well for the calculation of small molecule solvation free energies, but do not do as well at reproducing Poisson calculations for macromolecules, particularly the energy terms pertaining to deeply buried atoms. The dielectric shape represented by overlapping spheres still includes numerous small voids in the molecular interior (shown in red in Fig. 4). This is essentially the difference between defining the dielectric interface as the van der Waals surface, which is the assumption of conventional GB theory or as the molecular (or solvent-excluded) surface [93], which is the usual definition in Poisson theory, and which is illustrated as the outer extent of color in Fig. 4. For small molecules, the distinction makes little difference, but for

macromolecules, use of the van der Waals surface leads to internal cavities of solvent dielectric that are unrealistic because they are too small for a solvent-sized probe to fit into. To compensate for this "missed volume," Onufriev et al. introduce *ad hoc* correction factors that effectively increase the volumes of deeply buried atomic spheres. Solvation energy contributions from atoms near the surface are not so strongly affected. This means that the parameterizations that have been developed for small molecules with considerable effort can be carried over into macromolecular calculations with little or no change. This is now the standard generalized Born model in the Amber and NAB program packages.

3. Ghosh et al.[94] have proposed an alternative approach, in which the Coulomb field is still used in place of the correct field, and Green's theorem is used to convert the volume integral in 28 to a surface integral. There are potential computational advantages in the surface integral approach, especially for large systems, where the number of points required to discretize the surface should grow less rapidly than in a volume representation. In practice, empirical short-range and long-range corrections are added to improve agreement with numerical Poisson theory.

4. Other investigators have returned to numerical quadrature to evaluate 28 over the proper molecular volume[81, 82]. These methods (which may also include terms designed to correct deficiencies in the Coulomb field approximation) are generally slower than pairwise methods, but can also more faithfully mimic numerical Poisson results[89].

4 Applications

PB and GB methods have been used in numerous biomedical and biophysical applications; see [1, 3–6, 69] for more information. We have only space here to illustrate a few of these.

4.1 Electrostatic Potential Analysis

One of the most recognizable products of computational electrostatic calculations are the colorful images produced by coloring biomolecular surfaces by electrostatic potentials or by rendering electrostatic isocontours around a biomolecule. As mentioned above, Table 1 presents several popular software packages for performing PB calculations. This form of analysis has proven useful for giving some insight into the role of electrostatics in biomolecular function, although over-analysis with these qualitative methods can be inappropriate.

In addition to visual analysis, electrostatic potentials have been used in a more quantitative manner to identify active and ligand-binding sites [95–98], to predict protein-protein [99–106] and protein-membrane [107–109] membrane interfaces at and around the biomolecular surfaces.

4.2 Energetic Analysis

Energy analysis is one of the primary quantitative applications of PB methods. This analysis is driven by the functionals 14 and 15 for the nonlinear and linearized PB equations, respectively. Numerical solutions of the PB equation using FE and FD methods (with a few exceptions, see [63]) typically include self-energies which represent the interaction of an atomic charge distribution with itself. These self-energies are artifacts; the true self-energy of a point charge is infinite. Furthermore, these self-energies are extremely sensitive to the discretization and can give rise to large variations in the total energy as the mesh position and spacing is changed. For these reasons, energies obtained from a single PB calculation are meaningless; they must be performed as a series of calculations to calculate a free energy difference. If performed correctly, these differences should be free from self-energy artifacts.

In general, self-energies are removed through solvation energy calculations. These calculations determine the change in energy for transferring the protein charges from a homogeneous dielectric (the same as the protein interior) to an inhomogeneous dielectric with the appropriate permittivity values:

$$\Delta_{\mathrm{solv}} G = G_{\mathrm{sys}} - G_{\mathrm{ref}} \tag{33}$$

where G_{sys} is the energy of the system with the inhomogeneous dielectric and G_{ref} is the energy of the system with the homogeneous dielectric. Both calculations use exactly the same conformations of the molecule and the same discretization. This procedure removes the self-energies because the discontinuities due to the rapidly-varying Green function near the atomic centers are canceled by the two calculations with the same (interior) dielectric coefficient, discretization, and biomolecular conformation.

Total energies can be easily obtained from solvation energies by adding the interaction energies between the biomolecular charge distributions using a uniform dielectric equal to that of the protein interior. As these charge interactions are usually available as analytical expressions, they can be easily calculated *without* the self-interaction terms. For example, Coulomb's law energies describe the interaction between point charges and can be calculated according to

$$G = \sum_i^M \sum_{j>i}^M \frac{Q_i Q_j}{\epsilon_p r_{ij}} \, , \tag{34}$$

where ϵ_p is the dielectric of the protein interior. When this Coulomb's law expression is added to the solvation energy as calculated with point charges, the total electrostatic energy is obtained. As an application of this procedure, consider the binding energy calculation depicted in Fig. 5. In this case, both the protein and ligand molecules change conformation upon binding; removal of self-energies is very important to the accuracy of the calculation.

The electrostatic energies and forces calculated in this fashion have a number of other applications. Recent work has focused on the use of PB in several "high-throughput" applications such as Brownian [110–112] and Langevin [76, 113–115]

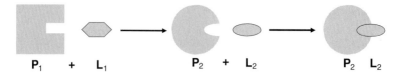

Figure 5. A schematic of ligand binding to a biomolecule with conformational change

dynamics, free energy calculations [116–118], and pK_a analysis [119–121]. These types of applications require repeated solution of the PB equation millions of times during the course of a simulation. As such, current research in PB methodology has worked to accelerate solvers for such demanding applications. Due to this "need for speed", FD methods have gained the most widespread use and demonstrated the best speed and efficiency of the methods described in the previous section. The lack of adaptivity in FD methods actually contributes to their efficiency by providing very simple problem setup and calculation of observables such as forces and energies. It should be noted that adaptive finite element methods are often the optimal choice for very large and strongly nonlinear problems, including the PB equation under certain circumstances. For the high-throughput methods mentioned above, adaptive methods do not *currently* provide the necessary level of efficiency. However, it seems likely that, with development of rapid discretization and error estimation methods, these approaches will eventually become the methods of choice for rapid solution of the PB equation.

4.3 Studying Protonation Equilibria

Solution acidity is an important thermodynamic variable that can affect biochemical function in a way that is often as profound as that of temperature or of the concentration of other allosteric effectors such as cofactors or phosphates. Many *in vitro* experiments mimic cellular compartments by regulating pH closely, commonly with buffering agents. The experimental study of titration behavior and the response of biomolecules to changes in pH has a long history, and there is a large amount known about the thermodynamics of proton binding [122, 123]. Structural correlations are less well-developed, but are becoming of increasing interest as methods for monitoring site-specific proton binding (particular by NMR) become more routine.

There is also a long history of theoretical and computational approaches to study of behavior of proteins and nucleic acids as a function of solution acidity [124–127]. This is known to be a difficult problem, since almost all biomolecules have multiple sites that can bind or release protons, and these are coupled to one another in complex ways. The computational study of the this problem has recently been reviewed [126], and we can only give a brief overview of the field here.

In principle, the most rigorous way to estimate an individual pK_a value for a protein side-chain would involve a free energy simulation connecting the protonated and de-protonated forms of the molecule:

$$pK_a = -\log_{10} K_a = \frac{\Delta G}{kT \ln 10} \tag{35}$$

In a molecular mechanics approach (where covalent bonds cannot be broken) this in practice would involve parallel, explicit solvent simulations on the protein of interest and on a model compound with the same functionality and with a known pK_a. The computed pK_a difference can then be added to the known model compound value to estimate the macromolecular result. This model effectively assumes that energetic contributions outside the molecular mechanics model (such as the strength of the O–H chemical bond) are the same in the protein as in the model compound. Calculations of this general kind were introduced nearly 80 years ago by Linderström-Lang [22] who modeled the protein as a sphere with the charges of the ionizable groups spread uniformly over its surface. As it become understood that proteins were not fluid globules but contained more specific structures, Tanford and Kirkwood [128, 129] developed a model of protein ionizable sites as point charges uniformly spaced at a short fixed distance beneath the surface of a sphere. As actual protein structures became known, the Tanford–Kirkwood model was adjusted to use charge placements derived from the actual structures [130], and empirical corrections for differential solvent exposure were introduced [131, 132].

The modern models, which have nearly displaced the sphere-based models, can be thought of as the natural extension of Tanford and Kirkwood's ideas to more detailed and realistic molecular surface shapes and charge distributions (see Fig. 6). The fundamental assumptions are as follows: (1) The free energy of ionization can be divided into an internal part that includes bond breaking and other electronic structure changes and is confined to a relatively small number of atoms within the ionizing functional group, and an external part that includes interactions with the larger surroundings. (2) The internal part is the same for both the ionizing group in the protein and the corresponding model compound, but the external part may differ. (3) Since the steric changes in protonation/deprotonation are subtle, and similar in both the protein and model compound, the steric contribution to the difference of the external part can be neglected. (4) The remaining difference in the external part is purely electrostatic. (5) A implicit solvent model is adequate to describe this electrostatic difference. As a practical matter, other simplifying approximations are often introduced as well, such as neglect of conformational change and simplification of charge models; but these approximations can be lifted without changing the basic ideas.

4.4 Multiple Interacting Sites

Our discussion so far has considered the thermodynamics of protonation of a single titrating site, where it makes sense to define a pK_a value as in 35, and the fraction of sites that have a proton bound can be trivially calculated once K_a is known. Proteins, of course, have multiple titrating sites, and one must consider the relative energetics of different protonation possibilities, and then the statistics of thermal ensembles over the possible states. A rough "pK_a" can then be identified as the pH value at which the populations of the protonated and deprotonated forms

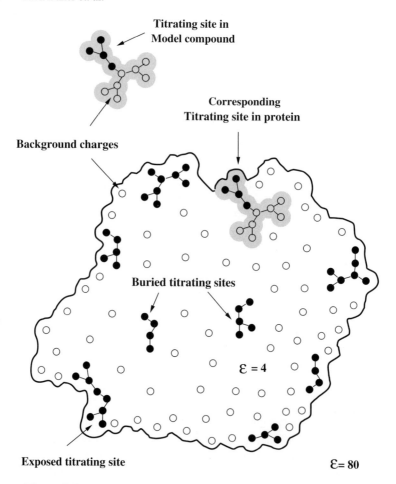

Figure 6. Continuum solvent model for computing pKa values in proteins

are equal. In extreme cases, however, this may not be well-defined or unique, and it is really the properties of the entire titration curve that are of greatest interest [122, 127, 133, 134]. Generally, proton binding is slightly anti-cooperative (because repulsive charge-charge interactions grow as more protons bind), but significant excursions from "normal" behavior can occur, either as the result of strong coupled interactions between nearby sites, or as a result of pH-dependent changes, such as global unfolding at low or high pH.

By "protonation state" of the protein, we mean a specification of which sites are protonated and which are deprotonated. If the protein has N sites, each with two possible protonation states, there are 2^N possible protonation states of the protein. Let them be denoted by the N-element vector, x whose elements x_i can each take on values representing "protonated" or "deprotonated." Let $\mathbf{0}$ be a particular value of this vector chosen as a reference state, for example, the state with all sites in their neutral state, as in the definition of pK_{intr} (see below). Consider the chemical equilibrium between some state x and the reference state:

$$M(0) + \nu(x)\mathrm{H}^+ \rightleftharpoons M(x) \tag{36}$$

where $\nu(x)$ is the number of protons that would be released on going from state x to the reference state. The free energy change for going to the right in this reaction at some fixed pH is,

$$\Delta G(x, \mathrm{pH}) = -RT \ln \frac{[M(x)]}{[M(0)]} \tag{37}$$
$$= +\mu_{\mathrm{M}}^{\circ}(x) - \mu_{\mathrm{M}}^{\circ}(0) - \nu(x)(\mu_{\mathrm{H}+}^{\circ} - 2.303RT\mathrm{pH}) ,$$

where the μ° are standard chemical potentials.

Suppose $P_i(x_i)$ is the change in the protein's chemical potential for changing site i from its reference state to state x_i, while all other sites remain in the reference state. The expression for chemical potential will then contain a sum of these P_i, but more terms are needed for the site–site interactions. Since the interactions are presumed to be governed by linear equations (the Poisson or linearized Poisson-Boltzmann equations) these terms will be pairwise additive. Therefore,

$$\mu_M^{\circ}(x) = \mu_M^{\circ}(0) + \sum_i^N P_i(x_i) + \frac{1}{2} \sum_{ij, i \neq j}^N W_{ij}(x_i, x_j) , \tag{38}$$

where W_{ij} is the electrostatic interaction between sites i and j, relative to the reference state. Inserting this into 37 and writing $\nu(x)$ as a sum over sites, $\sum_i \nu_i(x_i)$, gives,

$$\Delta G(x, \mathrm{pH}) = \sum_i [P_i(x_i) - \nu_i(x_i)(\mu_{\mathrm{H}+}^{\circ} - 2.303RT\mathrm{pH})] + \frac{1}{2} \sum_{ij, i \neq j}^N W_{ij}(x_i, x_j) \tag{39}$$

It is convenient to define a quantity, the intrinsic pK or pK_{intr} of a site which is the pK_a that the site would have if all other sites were held in their neutral states. Assuming that the reference state is this neutral state, one finds, $P_i(x_i) = \nu_i(x_i)(\mu_{\mathrm{H}+}^{\circ} - 2.303RT pK_{\mathrm{intr},i})$, so that,

$$\Delta G(x, \mathrm{pH}) = \sum_i \nu_i(x_i) 2.303RT(\mathrm{pH} - pK_{\mathrm{intr},i}) + \frac{1}{2} \sum_{ij, i \neq j}^N W_{ij}(x_i, x_j) . \tag{40}$$

The intrinsic pK_a values of each site, $pK_{\mathrm{intr},i}$, can be calculated by the methods of Sec. 4.3. The $W_{ij}(x_i, x_j)$ can be calculated by considering the additional electrostatic work to change site i's charges from the reference state to x_i if site j is in state x_j instead of its reference state.

Computation of experimentally meaningful quantites typically requires a partition function or average over all possible protonation states. For example, the protonation fraction of site i at a particular pH is

$$\theta_i(\text{pH}) = \frac{\sum_x^{2^N} \nu_i(x_i) \exp[-\Delta G(x, \text{pH})/RT]}{\sum_x^{2^N} \exp[-\Delta G(x, \text{pH})/RT]} . \tag{41}$$

Since the sums grow exponentially in the number of sites, the explicit use of such formula is only feasible if the number of sites is not too large.

There are two main ways in which the multiple-site problem can be addressed. One approach makes a mean-field (or Tanford-Roxby [130]) approximation, in which sites interacts not with the particular protonation states of other sites, but with their (pH-dependent) average protonation; in effect, sidechains with a pK_a value near the solution pH will be represented with charge distributions that are intermediate between the protonated and de-protonated forms [126, 130]. The other is Monte Carlo sampling over protonation states to estimate thermal averages [126, 135].

4.5 Constant pH Simulations

Combining the above methods with molecular dynamics should, in principle, give a methodology that accounts for conformational flexibility in the pK_a prediction problem, and allows for simulations of pH effects on macromolecule conformation.

Some workers use a continuous titration coordinate, so that partially protonated sites are represented by charge distributions that interpolate between the protonated and de-protonated endpoints [136–139]. This is similar in spirit to the mean field approach to the multi-site titration problem. Recent work by Lee et al. [140] describes a novel approach for avoiding mean field approximations in a continuous protonation state model, drawing on the ideas of λ-dynamics [141]. In this constant pH model, a potential is constructed along a coordinate interpolating between the protonated and deprotonated states, and equations of motion are used to propagate the λ coordinate. Convergence to an intermediate charge state is avoided by introducing an energetic barrier centered at $\lambda = 1/2$, which forces λ toward values representing full protonation or deprotonation. The barrier height is a tunable parameter, trading off between protonation state transition rate and fraction of simulation time spent in intermediate protonation states. Convergence problems are apparent in application to proteins, but this is a nearly universal result: trying to average over both conformational states and protonation states is a daunting task for all but the simplest situations. Despite the convergence issues, the model produces predicted pKa values that are in good agreement with experimental values. The average absolute error is 1.6 for hen egg white lysozyme, 1.2 for turkey ovomucoid, and 0.9 for bovine pancreatic trypsin inhibitor (BPTI).

A variety of constant pH methods have been proposed that avoid the use of intermediate charge states. All of these employ MD simulations to sample conformations, and periodically update the force field parameters (principally the charges) based on energetic analyses of the current configuration, using a Monte Carlo method to choose the new parameters. The methods differ in the frequency with which protonation state updates are made, and in the methods of estimating the energies of protonation or deprotonation events [142–144]. Due to the computational expense of PB calculations, MC steps are relatively infrequent (every 1 to 5 ps) in these PB based methods.

Greater efficiency can be obtained by using the generalized Born (GB) model for both the dynamics and the protonation state sampling [145]. This permits much more frequent Monte Carlo protonation steps, which appears to improve convergence. These frequent changes of the protonation state do not appear to adversely affect the stability of the simulation. Unlike the PB based methods, which sample over all relevant protonation states at each MC step, the current implementation changes the state of at most one titratable group on each MC step. The method was applied to lysozyme, where 1 ns trajectories at a range of pH values are less well converged but nevertheless yield pKa values with only 0.82 RMS error with respect to experimental data.

5 Conclusions

Implicit solvent models significantly reduce the degrees of freedom in simulations by approximate treatment of solvent behavior. However, there are several types of implicit solvent models and it is often difficult to determine which model offers the efficiency and accuracy needed for a particular application.

PB theory is usually the standard for assessing implicit solvent methodologies [78, 89, 146]. Most other implicit solvent methods, such as GB, attempt to approximate the solution to the Poisson or Poisson-Boltzmann equation with varying descriptions of dielectric coefficients and ionic accessibilities. Several studies have shown that, with appropriate parameterization, PB methods can provide polar solvation energies and forces for proteins and small molecules that compare very well with results from explicit solvent simulations and experiment [17, 147, 148]. Unfortunately, PB methods are often slow. Although there have been numerous attempts to remedy this problem, the historical problems with PB efficiency have opened the door for GB models and other simpler approximations. As described above, GB methods can be very efficient at evaluating solvation energies and forces for large biomolecular systems.

However, GB methods also have their weaknesses. Recently, Onufriev and co-workers demonstrated that GB methods can, in principle, offer levels of accuracy on par to PB solvers [78]. However, despite recent advances in GB methods, it is not clear that the same levels of accuracy are achieved in the routine application of GB models [89]. While a number of groups have developed methods to address some of these accuracy issues, it appears that there is ample room for improvement in GB as well as PB methods.

As mentioned above, implicit solvent methods offer the prospect of improved simulation efficiency at the expense of *approximate* treatment of solvation. The choice of implicit solvent model for a simulation must ultimately be made by careful consideration of the benefits and caveats outlined above. Simulators should use caution and determine, perhaps a priori or by comparison of results for short model simulations, whether the acceleration offered by GB or PB methods provides the appropriate level of accuracy for the biological system of interest.

Acknowledgments

NAB thanks the NIH and Alfred P. Sloan Foundation for financial support. DAC and DB were supported by NIH grant GM-57513.

References

[1] N. A. Baker and J. A. McCammon. Electrostatic interactions. In P. Bourne and H. Weissig, editors, *Structural Bioinformatics*, pages 427–440. John Wiley and Sons, Inc., New York, 2003.

[2] M. Gilson. Introduction to continuum electrostatics. In D. A. Beard, editor, *Biophysics Textbook Online*, volume Computational Biology. Biophysical Society, Bethesda, MD, 2000.

[3] B. Roux. Implicit solvent models. In O. M. Becker, Jr. MacKerell, A. D., B. Roux, and M. Watanabe, editors, *Computational Biochemistry and Biophysics*, pages 133–152. Marcel Dekker, New York, 2001.

[4] M. E. Davis and J. A. McCammon. Electrostatics in biomolecular structure and dynamics. *Chem. Rev.*, 94:7684–7692, 1990.

[5] B. Honig and A. Nicholls. Classical electrostatics in biology and chemistry. *Science*, 268(5214):1144–9, 1995.

[6] N. A. Baker. Poisson-boltzmann methods for biomolecular electrostatics. *Meth. Enzymol.*, 383:94–118, 2004.

[7] Christian Holm, Patrick Kekicheff, and Rudolf Podgornik, editors. *Electrostatic effects in soft matter and biophysics*, volume 46 of *NATO Science Series*. Kluwer Academic Publishers, Boston, 2001.

[8] L. Onsager. Electric moments of molecules in liquids. *J. Amer. Chem. Soc.*, 58:1486–1493, 1936.

[9] S. Miertus, E. Scrocco, and J. Tomasi. Electrostatic interaction of a solute with a continuum. A direct utilization of ab initio molecular potentials for the prevision of solvent effects. *Chem. Phys.*, 55:117–129, 1981.

[10] J. Tomasi and M. Persico. Molecular interactions in solution: an overview of methods based on continuous distributions of solvent. *Chem. Rev.*, 94:2027–2094, 1994.

[11] J. Tomasi. Thirty years of continuum solvation chemistry: a review, and prospects for the near future. *Theor. Chem. Acc.*, 112:184–203, 2004.

[12] C. Lim, D. Bashford, and M. Karplus. Absolute pKa calculations with continuum dielectric methods. *J. Phys. Chem.*, 95:5610–5620, 1991.

[13] W.H. Richardson, C. Peng, D. Bashford, L. Noodleman, and D.A. Case. Incorporating solvation effects into density functional theory: Calculation of absolute acidities. *Int. J. Quantum Chem.*, 61:207–217, 1997.

[14] J.J. Klicic, R.A. Friesner, S.Y. Liu, and W.C. Guida. Accurate prediction of acidity constants in aqueous solution via density functional theory and self-consistent reaction field methods. *J. Phys. Chem. A*, 106:1327–1335, 2002.

[15] D.M. Chipman. Computation of pKa from Dielectric Continuum Theory. *J. Phys. Chem. A*, 106:7413–7422, 2002.

[16] M.S. Busch and E.W. Knapp. Accurate pKa Determination for a Heterogeneous Group of Organic Molecules. *Chemphyschem*, 5:1513–1522, 2004.

[17] D. Sitkoff, K.A. Sharp, and B. Honig. Accurate calculation of hydration free energies using macroscopic solvent models. *J. Phys. Chem.*, 98:1978–1988, 1994.

[18] M. Nina, D. Beglov, and B. Roux. Atomic radii for continuum electrostatics calculations based on molecular dynamics free energy simulations. *J. Phys. Chem. B*, 101:5239–5248, 1997.

[19] E. Gallicchio, L.Y. Zhang, and R.M. Levy. The SGB/NP Hydration Free Energy Model Based on the Surface Generalized Born Solvent Reaction Field and Novel Nonpolar Hydration Free Energy Estimators. *J. Comput. Chem.*, 23:517–529, 2002.

[20] E. Gallicchio and R.M. Levy. AGBNP: An Analytic Implicit Solvent Model Suitable for Molecular Dynamics Simulations and High-Resolution Modeling. *J. Comput. Chem.*, 25:479–499, 2004.

[21] M. Born. Volumen und hydratationswärme der ionen. *Z. Phys.*, 1:45–48, 1920.

[22] K. Linderstrøm-Lang. On the ionisation of proteins. *Comptes-rend Lab. Carlsberg*, 15(7):1–29, 1924.

[23] C. Tanford and J. G. Kirkwood. Theory of protein titration curves. I. General equations for impenetrable spheres. *J. Am. Chem. Soc.*, 79:5333–5339, 1957.

[24] M. A. Flanagan, G. K. Ackers, J. B. Matthew, G. I. H. Hanania, and F. R. N. Gurd. Electrostatic contributions to the energetics of dimer-tetramer assembly in human hemoglobin: pH dependence and effect of specifically bound chloride ions. *Biochemistry*, 20:7439–7449, 1981.

[25] J. Warwicker and H. C. Watson. Calculation of the electric potential in the active site cleft due to alpha-helix dipoles. *J. Mol. Biol.*, 157(4):671–9, 1982.

[26] E. Demchuk, D. Bashford, G. Gippert, and D.A. Case. Thermodynamics of a reverse turn motif. Solvent effects and side-chain packing. *J. Mol. Biol.*, 270:305–317, 1997.

[27] J. Srinivasan, T.E. Cheatham, III, P. Kollman, and D.A. Case. Continuum Solvent Studies of the Stability of DNA, RNA, and Phosphoramidate–DNA Helices. *J. Am. Chem. Soc.*, 120:9401–9409, 1998.

[28] P.A. Kollman, I. Massova, C. Reyes, B. Kuhn, S. Huo, L. Chong, M. Lee, T. Lee, Y. Duan, W. Wang, O. Donini, P. Cieplak, J. Srinivasan, D.A. Case, and T.E. Cheatham, III. Calculating Structures and Free Energies of Complex Molecules: Combining Molecular Mechanics and Continuum Models. *Accts. Chem. Res.*, 33:889–897, 2000.

[29] J. O. Bockris and A. K. N. Reddy. *Modern Electrochemistry: Ionics.* Plenum Press, New York, 1998.

[30] John David Jackson. *Classical Electrodynamics.* John Wiley and Sons, New York, 2nd edition, 1975.

[31] C. J. Bötcher. *Theory of Electric Polarization.* Elsevier Science, New York, 1973.

[32] B. Lee and F. M. Richards. The interpretation of protein structures: estimation of static accessibility. *J. Mol. Biol.*, 55(3):379–400, 1971.

[33] M. L. Connolly. The molecular surface package. *J. Mol. Graph.*, 11(2):139–41, 1993.

[34] W. Im, D. Beglov, and B. Roux. Continuum solvation model: electrostatic forces from numerical solutions to the Poisson-Boltzmann equation. *Comp. Phys. Commun.*, 111(1–3):59–75, 1998.

[35] J. Andrew Grant, Barry T. Pickup, and Anthony Nicholls. A smooth permittivity function for Poisson-Boltzmann solvation methods. *J. Comput. Chem.*, 22(6):608–640, 2001.

[36] M. Nina, W. Im, and B. Roux. Optimized atomic radii for protein continuum electrostatics solvation forces. *Biophys. Chem.*, 78(1-2):89–96, 1999.

[37] F. Dong, M. Vijaykumar, and H. X. Zhou. Comparison of calculation and experiment implicates significant electrostatic contributions to the binding stability of barnase and barstar. *Biophys. J.*, 85(1):49–60, 2003.

[38] J. Wagoner and N. A. Baker. Solvation forces on biomolecular structures: A comparison of explicit solvent and Poisson-Boltzmann models. *J. Comput. Chem.*, 25(13):1623–9, 2004.

[39] K. A. Sharp and B. Honig. Calculating total electrostatic energies with the nonlinear Poisson-Boltzmann equation. *J. Phys. Chem.*, 94(19):7684–7692, 1990.

[40] F. Fogolari and J. M. Briggs. On the variational approach to Poisson-Boltzmann free energies. *Chemical Physics Letters*, 281(1–3):135–139, 1997.

[41] A. M. Micu, B. Bagheri, A. V. Ilin, L. R. Scott, and B. M. Pettitt. Numerical considerations in the computation of the electrostatic free energy of interaction within the Poisson-Boltzmann theory. *J. Comput. Phys.*, 136(2):263–271, 1997.

[42] M. K. Gilson, M. E. Davis, B. A. Luty, and J. A. McCammon. Computation of electrostatic forces on solvated molecules using the Poisson-Boltzmann equation. *J. Phys. Chem.*, 97(14):3591–3600, 1993.

[43] D. Beglov and B. Roux. Solvation of complex molecules in a polar liquid: an integral equation theory. *J. Chem. Phys.*, 104(21):8678–8689, 1996.

[44] B. Egwolf and P. Tavan. Continuum description of solvent dielectrics in molecular-dynamics simulations of proteins. *J. Chem. Phys.*, 118(5):2039–56, 2003.

[45] B. Egwolf and P. Tavan. Continuum description of ionic and dielectric shielding for molecular-dynamics simulations of proteins in solution. *J. Chem. Phys.*, 120(4):2056–2068, 2004.

[46] M. Holst and F. Saied. Multigrid solution of the Poisson-Boltzmann equation. *J. Comput. Chem.*, 14(1):105–13, 1993.

[47] M. J. Holst and F. Saied. Numerical solution of the nonlinear Poisson-Boltzmann equation: developing more robust and efficient methods. *J. Comput. Chem.*, 16(3):337–64, 1995.

[48] Dietrich Braess. *Finite Elements. Theory, Fast Solvers, and Applications in Solid Mechanics.* Cambridge University Press, Cambridge, 1997.

[49] O. Axelsson and V. A. Barker. *Finite Element Solution of Boundary Value Problems. Theory and Computation.* Academic Press, Inc., San Diego, 1984.

[50] M. Holst. Adaptive numerical treatment of elliptic systems on manifolds. *Advances in Computational Mathematics*, 15(1-4):139–191, 2001.

[51] M. Holst, N. Baker, and F. Wang. Adaptive multilevel finite element solution of the Poisson-Boltzmann equation. I. Algorithms and examples. *J. Comput. Chem.*, 21(15):1319–1342, 2000.

[52] N. Baker, M. Holst, and F. Wang. Adaptive multilevel finite element solution of the Poisson-Boltzmann equation II. Refinement at solvent-accessible surfaces in biomolecular systems. *J. Comput. Chem.*, 21(15):1343–1352, 2000.

[53] C. M. Cortis and R. A. Friesner. An automatic three-dimensional finite element mesh generation system for the Poisson-Boltzmann equation. *J. Comput. Chem.*, 18:1570–1590, 1997.

[54] C. M. Cortis and R. A. Friesner. Numerical solution of the Poisson-Boltzmann equation using tetrahedral finite-element meshes. *J. Comput. Chem.*, 18:1591–1608, 1997.

[55] P E Dyshlovenko. Adaptive numerical method for Poisson-Boltzmann equation and its application. *Comp. Phys. Commun.*, 147:335–338, 2002.

[56] R. E. Bank and M. Holst. A new paradigm for parallel adaptive meshing algorithms. *SIAM Journal on Scientific Computing*, 22(4):1411–1443, 2000.

[57] N. A. Baker, D. Sept, M. J. Holst, and J. A. McCammon. The adaptive multilevel finite element solution of the Poisson-Boltzmann equation on massively parallel computers. *IBM Journal of Research and Development*, 45(3–4):427–438, 2001.

[58] R. J. Zauhar and R. S. Morgan. The rigorous computation of the molecular electric potential. *J. Comput. Chem.*, 9(2):171–187, 1988.

[59] A. H. Juffer, E. F. F. Botta, B. A. M. van Keulen, A. van der Ploeg, and H. J. C. Berendsen. The electric potential of a macromolecule in a solvent: a fundamental approach. *J. Comput. Phys.*, 97:144–171, 1991.

[60] S. A. Allison and V. T. Tran. Modeling the electrophoresis of rigid polyions: application to lysozyme. *Biophys. J.*, 68(6):2261–70, 1995.

[61] A. J. Bordner and G. A. Huber. Boundary element solution of the linear Poisson-Boltzmann equation and a multipole method for the rapid calculation of forces on macromolecules in solution. *J. Comput. Chem.*, 24:353–367, 2003.

[62] A. H. Boschitsch and M. O. Fenley. Hybrid boundary element and finite difference method for solving the nonlinear Poisson-Boltzmann equation. *J. Comput. Chem.*, 25(7):935–955, 2004.

[63] Z. Zhou, P. Payne, M. Vasquez, N. Kuhn, and M. Levitt. Finite-difference solution of the Poisson-Boltzmann equation: complete elimination of self-energy. *J. Comput. Chem.*, 17:1344–1351, 1996.

[64] Y. N. Vorobjev and H. A. Scheraga. A fast adaptive multigrid boundary element method for macromolecular electrostatic computations in a solvent. *J. Comput. Chem.*, 18(4):569–583, 1997.

[65] A. Nicholls and B. Honig. A rapid finite difference algorithm, utilizing successive over-relaxation to solve the Poisson-Boltzmann equation. *J. Comput. Chem.*, 12(4):435–445, 1991.

[66] M. E. Davis and J. A. McCammon. Solving the finite difference linearized Poisson-Poltzmann equation: a comparison of relaxation and conjugate gradient methods. *J. Comput. Chem.*, 10:386–391, 1989.

[67] N. A. Baker, D. Sept, S. Joseph, M. J. Holst, and J. A. McCammon. Electrostatics of nanosystems: Application to microtubules and the ribosome. *Proc Natl. Acad. Sci. USA*, 98(18):10037–10041, 2001.

[68] M. K. Gilson and B. H. Honig. Calculation of electrostatic potentials in an enzyme active site. *Nature*, 330(6143):84–6, 1987.

[69] K. A. Sharp and B. Honig. Electrostatic interactions in macromolecules – theory and applications. *Annu. Rev. Biophys. Biophys. Chem.*, 19:301–332, 1990.

[70] G. T. Balls and P. Colella. A finite difference domain decomposition method using local corrections for the solution of Poisson's equation. *J. Comput. Phys.*, 180(1):25–53, 2002.

[71] W. Rocchia, S. Sridharan, A. Nicholls, E. Alexov, A. Chiabrera, and B. Honig. Rapid grid-based construction of the molecular surface and the use of induced surface charge to calculate reaction field energies: applications to the molecular systems and geometric objects. *J. Comput. Chem.*, 23(1):128–37, 2002.

[72] A. Nicholls, K. A. Sharp, and B. Honig. Protein folding and association: insights from the interfacial and thermodynamic properties of hydrocarbons. *Proteins*, 11(4):281–96, 1991.

[73] D. Bashford. An object-oriented programming suite for electrostatic effects in biological molecules. In Y. Ishikawa, R. R. Oldehoeft, J. V. W. Reynders, and M. Tholburn, editors, *Scientific Computing in Object-Oriented Parallel Environments*, volume 1343 of *Lecture Notes in Computer Science*, pages 233–240. Springer, Berlin, 1997.

[74] J. D. Madura, J. M. Briggs, R. C. Wade, M. E. Davis, B. A. Luty, A. Ilin, J. Antosiewicz, M. K. Gilson, B. Bagheri, L. R. Scott, and J. A. McCammon. Electrostatics and diffusion of molecules in solution - simulations with the university of houston brownian dynamics program. *Comp. Phys. Commun.*, 91(1-3):57–95, 1995.

[75] Jr. MacKerell, A. D., B. Brooks, III Brooks, C. L., L. Nilsson, B. Roux, Y. Won, and M. Karplus. Charmm: the energy function and its parameterization with an overview of the program. In P. v. R. Schleyer, editor, *The Encyclopedia of Computational Chemistry*, volume 1, pages 271–277. John Wiley and Sons, Chichester, 1998.

[76] R. Luo, L. David, and M. K. Gilson. Accelerated Poisson-Boltzmann calculations for static and dynamic systems. *J. Comput. Chem.*, 23(13):1244–53, 2002.

[77] W.C. Still, A. Tempczyk, R.C. Hawley, and T. Hendrickson. Semianalytical treatement of solvation for molecular mechanics and dynamics. *J. Am. Chem. Soc.*, 112:6127–6129, 1990.

[78] A. Onufriev, D.A. Case, and D. Bashford. Effective Born radii in the generalized Born approximation: The importance of being perfect. *J. Comput. Chem.*, 23:1297–1304, 2002.

[79] J.D. Jackson. *Classical Electrodynamics*. Wiley and Sons, New York, 1975.

[80] M. Scarsi, J. Apostolakis, and A. Caflisch. Continuum Electrostatic Energies of Macromolecules in Aqueous Solutions. *J. Phys. Chem. A*, 101:8098–8106, 1997.

[81] M.S. Lee, F.R. Salsbury, Jr., and C.L. Brooks, III. Novel generalized Born methods. *J. Chem. Phys.*, 116:10606–10614, 2002.

[82] M.S. Lee, M. Feig, F.R. Salsbury, and C.L. Brooks. New Analytic Approximation to the Standard Molecular Volume Definition and Its Application to Generalized Born Calculations. *J. Comput. Chem.*, 24:1348–1356, 2003.

[83] M. Schaefer and C. Froemmel. A precise analytical method for calculating the electrostatic energy of macromolecules in aqueous solution. *J. Mol. Biol.*, 216:1045–1066, 1990.

[84] G.D. Hawkins, C.J. Cramer, and D.G. Truhlar. Pairwise solute descreening of solute charges from a dielectric medium. *Chem. Phys. Lett.*, 246:122–129, 1995.

[85] G.D. Hawkins, C.J. Cramer, and D.G. Truhlar. Parametrized models of aqueous free energies of solvation based on pairwise descreening of solute atomic charges from a dielectric medium. *J. Phys. Chem.*, 100:19824–19839, 1996.

[86] J. G. Kirkwood. Theory of solutions of molecules containing widely separated charges with special applications to zwitterions. *J. Chem. Phys.*, 2:351–361, 1934.

[87] D. Bashford and D.A. Case. Generalized Born Models of Macromolecular Solvation Effects. *Annu. Rev. Phys. Chem.*, 51:129–152, 2000.

[88] M. Feig and C.L. Brooks, III. Recent advances in the development and application of implicit solvent models in biomolecule simulations. *Curr. Opin. Struct. Biol.*, 14:217–224, 2004.

[89] M. Feig, A. Onufriev, M.S. Lee, W. Im, D.A. Case, and C.L. Brooks. Performance Comparison of Generalized Born and Poisson Methods in the Calculation of Electrostatic Solvation Energies for Protein Structures. *J. Comput. Chem.*, 25:265–284, 2004.

[90] M. Schaefer and M. Karplus. A comprehensive analytical treatment of continuum electrostatics. *J. Phys. Chem.*, 100:1578–1599, 1996.

[91] A. Onufriev, D. Bashford, and D.A. Case. Modification of the Generalized Born Model Suitable for Macromolecules. *J. Phys. Chem. B*, 104:3712–3720, 2000.

[92] A. Onufriev, D. Bashford, and D.A. Case. Exploring Protein Native States and Large-Scale Conformational Changes With a Modified Generalized Born Model. *Proteins*, 55:383–394, 2004.

[93] M.L. Connolly. Solvent-accessible surfaces of proteins and nucleic acids. *Science*, 221:709–713, 1983.

[94] A. Ghosh, C.S. Rapp, and R.A. Friesner. Generalized Born Model Based on a Surface Integral Formulation. *J. Phys. Chem. B*, 102:10983–10990, 1998.

[95] A. H. Elcock. Prediction of functionally important residues based solely on the computed energetics of protein structure. *J. Mol. Biol.*, 312(4):885–896, 2001.

[96] M. J. Ondrechen, J. G. Clifton, and D. Ringe. Thematics: a simple computational predictor of enzyme function from structure. *Proc. Natl. Acad. Sci. USA*, 98(22):12473–8, 2001.

[97] Z. Y. Zhu and S. Karlin. Clusters of charged residues in protein three-dimensional structures. *Proc. Natl. Acad. Sci. USA*, 93(16):8350–5, 1996.

[98] A. M. Richard. Quantitative comparison of molecular electrostatic potentials for structure-activity studies. *J. Comput. Chem.*, 12(8):959–69, 1991.

[99] R. Norel, F. Sheinerman, D. Petrey, and B. Honig. Electrostatic contributions to protein-protein interactions: Fast energetic filters for docking and their physical basis. *Prot. Sci.*, 10(11):2147–2161, 2001.

[100] S. M. de Freitas, L. V. de Mello, M. C. da Silva, G. Vriend, G. Neshich, and M. M. Ventura. Analysis of the black-eyed pea trypsin and chymotrypsin inhibitor-alpha-chymotrypsin complex. *FEBS Lett*, 409(2):121–7, 1997.

[101] L. Lo Conte, C. Chothia, and J. Janin. The atomic structure of protein-protein recognition sites. *J Mol Biol*, 285(5):2177–98, 1999.

[102] J. Janin and C. Chothia. The structure of protein-protein recognition sites. *J. Biol. Chem.*, 265(27):16027–30, 1990.

[103] V. A. Roberts, H. C. Freeman, A. J. Olson, J. A. Tainer, and E. D. Getzoff. Electrostatic orientation of the electron-transfer complex between plastocyanin and cytochrome c. *J. Biol. Chem.*, 266(20):13431–41, 1991.

[104] R. C. Wade, R. R. Gabdoulline, and F. De Rienzo. Protein interaction property similarity analysis. *International Journal of Quantum Chemistry*, 83(3–4):122–127, 2001.

[105] J. Novotny and K. Sharp. Electrostatic fields in antibodies and antibody/antigen complexes. *Prog. Biophys. Mol. Biol.*, 58(3):203–24, 1992.

[106] A. J. McCoy, V. Chandana Epa, and P. M. Colman. Electrostatic complementarity at protein/protein interfaces. *J. Mol. Biol.*, 268(2):570–84, 1997.

[107] A. Arbuzova, L. B. Wang, J. Y. Wang, G. Hangyas-Mihalyne, D. Murray, B. Honig, and S. McLaughlin. Membrane binding of peptides containing both basic and aromatic residues. experimental studies with peptides corresponding to the scaffolding region of caveolin and the effector region of marcks. *Biochemistry*, 39(33):10330–10339, 2000.

[108] Jung-Hsin Lin, Nathan Andrew Baker, and J. Andrew McCammon. Bridging the implicit and explicit solvent approaches for membrane electrostatics. *Biophys. J.*, 83(3):1374–1379, 2002.

[109] D. Murray and B. Honig. Electrostatic control of the membrane targeting of C2 domains. *Molecular Cell*, 9(1):145–154, 2002.

[110] R. R. Gabdoulline and R. C. Wade. Simulation of the diffusional association of barnase and barstar. *Biophys J*, 72(5):1917–29, 1997.

[111] S. H. Northrup, S. A. Allison, and J. A. McCammon. Brownian dynamics simulation of diffusion-influenced biomolecular reactions. *J. Chem. Phys.*, 80:1517–1524, 1984.

[112] J. L. Smart, T. J. Marrone, and J. A. McCammon. Conformational sampling with Poisson-Boltzmann forces and a stochastic dynamics monte carlo method: Application to alanine dipeptide. *J. Comput. Chem.*, 18(14):1750–1759, 1997.

[113] N. V. Prabhu, P. Zhu, and K. A. Sharp. Implementation and testing of stable, fast implicit solvation in molecular dynamics using the smooth-permittivity finite difference Poisson-Boltzmann method. *J. Comput. Chem.*, 25(16):2049–2064, 2004.

[114] Q. Lu and R. Luo. A Poisson-Boltzmann dynamics method with nonperiodic boundary condition. *J. Chem. Phys.*, 119(21):11035–11047, 2003.

[115] B.Z. Lu, W.Z. Chen, C.X. Wang, and X. Xu. Protein Molecular Dynamics With Electrostatic Force Entirely Determined by a Single Poisson-Boltzmann Calculation. *Proteins*, 48:497–504, 2002.

[116] F. Fogolari, A. Brigo, and H. Molinari. Protocol for MM/PBSA molecular dynamics simulations of proteins. *Biophys. J.*, 85(1):159–166, 2003.

[117] P. A. Kollman, I. Massova, C. Reyes, B. Kuhn, S. Huo, L. Chong, M. Lee, T. Lee, Y. Duan, W. Wang, O. Donini, P. Cieplak, J. Srinivasan, D. A. Case, and III Cheatham, T. E. Calculating structures and free energies of complex molecules: combining molecular mechanics and continuum models. *Accounts of Chemical Research*, 33(12):889–97, 2000.

[118] J. M. J. Swanson, R. H. Henchman, and J. A. McCammon. Revisiting free energy calculations: A theoretical connection to MM/PBSA and direct calculation of the association free energy. *Biophys. J.*, 86(1):67–74, 2004.

[119] J. E. Nielsen and J. A. McCammon. On the evaluation and optimization of protein x-ray structures for pK_a calculations. *Prot. Sci.*, 12(2):313–26, 2003.

[120] R. E. Georgescu, E. G. Alexov, and M. R. Gunner. Combining conformational flexibility and continuum electrostatics for calculating pK_as in proteins. *Biophys. J.*, 83(4):1731–1748, 2002.

[121] J. Warwicker. Improved pK_a calculations through flexibility based sampling of a water-dominated interaction scheme. *Prot. Sci.*, 13(10):2793–805, 2004.

[122] J. Wyman and S.J. Gill. *Binding and linkage*. University Science Books, Mill Valley, CA, 1990.

[123] R.A. Alberty. *Thermodynamics of Biochemical Reactions*. John Wiley, New York, 2003.

[124] J.B. Matthew, F.R.N. Gurd, B. Garcia-Moreno E., M.A. Flanagan, K.L. March, and S.J. Shire. pH-dependent processes in proteins. *CRC Crit. Rev. Biochem.*, 18:91–197, 1985.

[125] P. Beroza and D.A. Case. Calculations of proton-binding thermodynamics in proteins. *Meth. Enzymol.*, 295:170–189, 1998.

[126] D. Bashford. Macroscopic electrostatic models for protonation states in proteins. *Frontiers Biosci.*, 9:1082–1099, 2004.

[127] B. García-Moreno and C.A. Fitch. Structural interpretation of pH and salt-dependent processes in proteins with computaional methods. *Meth. Enzymol.*, 380:20–51, 2004.

[128] C. Tanford and J.G. Kirkwood. Theory of titration curves. I. General equations for impenetrable spheres. *J. Am. Chem. Soc.*, 79:5333–5339, 1957.

[129] C. Tanford. Theory of protein titration curves. II. Calculations for simple models at low ionic strength. *J. Am. Chem. Soc.*, 79:5340–5347, 1957.

[130] C. Tanford and R. Roxby. Interpretation of protein titration curves. *Biochemistry*, 11:2192–2198, 1972.

[131] S.J. Shire, G.I.H. Hanania, and F.R.N. Gurd. Electrostatic effects in myoglobin. Hydrogen ion equilibria in sperm whale ferrimyoglobin. *Biochemistry*, 13:2967–2974, 1974.

[132] J.B. Matthew and F.R.N. Gurd. Calculation of electrostatic interactions in proteins. *Meth. Enzymol.*, 130:413–436, 1986.

[133] A. Onufriev, D.A. Case, and G.M. Ullmann. A novel view of pH titration in biomolecules. *Biochemistry*, 40:3413–3419, 2001.

[134] D. Poland. Free energy of proton binding in proteins. *Biopolymers*, 69:60–71, 2003.

[135] P. Beroza, D.R. Fredkin, M.Y. Okamura, and G. Feher. Protonation of interacting residues in a protein by Monte Carlo method: Application to lysozyme and the photosynthetic reaction center of *Rhodobacter sphaeroides*. *Proc. Natl. Acad. Sci. USA*, 88:5804–5808, 1991.

[136] A.M. Baptista, P.J. Martel, and S.B. Petersen. Simulation of protein conformational freedom as a function of pH: Constant-pH molecular dynamics using implicit titration. *Proteins*, 27:523–544, 1997.

[137] U. Börjesson and P.H. Hünenberger. Explicit-solvent molecular dynamics simulation at constant pH: Methodology and application to small amines. *J. Chem. Phys.*, 114:9706–9719, 2001.

[138] A.M. Baptista. Comment on "Explicit-solvent molecular dynamics simulation at constant pH: Methodology and application to small amines". *J. Chem. Phys.*, 116:7766–7768, 2002.

[139] U. Börjesson and P.H. Hünenberger. pH-dependent stability of a decalysine α-helix studied by explicit-solvent molecular dynamics simulations at constant pH. *J. Phys. Chem. B*, 108:13551–13559, 2004.

[140] M.S. Lee, F.R. Salsbury, and C.L. Brooks. Constant-pH molecular dynamics using continuous titration coordinates. *Proteins*, 56:738–752, 2004.

[141] X. Kong and C.L. Brooks, III. λ-dynamics: A new approach to free energy calculations. *J. Chem. Phys.*, 105:2414–2423, 1996.

[142] A.M. Baptista, V.H. Teixeira, and C.M. Soares. Constant-pH molecular dynamics using stochastic titration. *J. Chem. Phys.*, 117:4184–4200, 2002.

[143] M. Dlugosz and J.M. Antosiewicz. Constant-pH molecular dynamics simulations: a test case for succinic acid. *Chem. Phys.*, 302:161–170, 2004.

[144] M. Dlugosz, J.M. Antosiewicz, and A.D. Robertson. Constant-pH molecular dynamics study of protonation-structure relationship in a hexapeptide derived from ovomucoid third domain. *Phys. Rev. E*, 69:021915, 2004.

[145] J. Mongan, D.A. Case, and J.A. McCammon. Constant pH molecular dynamics in generalized Born implicit solvent. *J. Comput. Chem.*, 25:2038–2048, 2004.

[146] R. Zhou, G. Krilov, and B. J. Berne. Comment on "can a continuum solvent model reproduce the free energy landscape of a -hairpin folding in water?" the poisson-boltzmann equation. *J. Phys. Chem. B*, 108(22):7528–30, 2004.

[147] J. Wagoner and N.A. Baker. Solvation Forces on Biomolecular Structures: A Comparison of Explicit Solvent and Poisson-Boltzmann Models. *J. Comput. Chem.*, 25:1623–1629, 2004.

[148] M. Nina, W. Im, and B. Roux. Optimized atomic radii for protein continuum electrostatics solvations forces. *Biophys. Chem.*, 78:89–96, 1999.

New Distributed Multipole Metdhods
for Accurate Electrostatics
in Large-Scale Biomolecular Simulations

Celeste Sagui[1], Christopher Roland[1], Lee G. Pedersen[2,3], and Thomas A. Darden[3]

[1] Center for High Performance Simulations and Department of Physics, North Carolina State University, Raleigh, NC 27695, USA
[2] Laboratory of Structural Biology, National Institute of Environmental Health Sciences, Research Triangle Park, NC 27709, USA
[3] Chemistry Department, UNC - Chapel Hill, Chapel Hill, NC 27599, USA

Abstract. It has long been known that accurate electrostatics is a key issue for improving current force fields for large-scale biomolecular simulations. Typically, this calls for an improved and more accurate description of the molecular electrostatic potential, which eliminates the artifacts associated with current point charge-based descriptions. In turn, this involves the partitioning of the extended molecular charge distribution, so that charges and multipole moments can be assigned to different atoms. As an alternate to current approaches, we have investigated a charge partitioning scheme that is based on the maximally-localized Wannier functions. This has the advantage of partitioning the charge, and placing it around the molecule in a chemically meaningful manner. Moreover, higher order multipoles may all be calculated without any undue numerical difficulties. In order to deal with the extra computational costs, we have developed an efficient Cartesian formalism for the treatment of higher order multipoles. The Ewald 'direct sum' is evaluated through a McMurchie-Davidson formalism. The 'reciprocal sum' has been implemented in three different ways: using an Ewald scheme, a Particle Mesh Ewald (PME) method and a multigrid-based approach. Even though the use of the McMurchie-Davidson formalism considerably reduces the cost of the calculation with respect to the standard matrix implementation of multipole interactions, the calculation in direct space remains expensive. When most of the calculation is moved to reciprocal space via the PME method, the cost of a calculation where all multipolar interactions (up to hexadecapole-hexadecapole) are included is only about 8.5 times more expensive than a regular AMBER 7 implementation with only charge-charge interactions. The multigrid implementation is slower but shows very promising results for parallelization.

1 Introduction

Empirical force fields for large-scale biomolecular simulations divide the atomic interactions into those that are *short-ranged bonded*, and *long-ranged nonbonded* interactions. The latter include the electrostatic, the exchange-repulsion, and the dispersion interactions, and are currently responsible for the greatest loss of accuracy in classical force fields. In particular, the correct treatment of the electrostatic interactions is one of the greatest challenges for classical force fields, not only because of the loss of accuracy inherent in the partial-charge representation, but also because they are computationally the most expensive part of the calculation [29].

In order to significantly improve the current treatment of the electrostatics for biomolecular simulations, two key issues must be addressed. First, a more accurate description – one that eliminates the artifacts associated with classical point charges and faithfully reproduces the molecular electrostatic potential (MEP) – must be found. Second, since such a description of necessity implies the introduction of new parameters, fast and computationally efficient methods for simulating the electrostatic contributions must be developed. In this article, we discuss a new method [33] of calculating the distributed multipoles of a molecular charge density based on the maximally-localized Wannier functions [4, 19]. This approach has several important advantages, not least of which is the first principles calculation of multipole moments without any additional *ad hoc* assumptions. Then we discuss a new method [32] for the efficient calculation of these costly interactions based on multipolar extensions to the Particle Mesh Ewald Summation (PME) [11, 12] and multigrid [30] methods.

The calculation of atomic charges is nontrivial because any molecular charge distribution is of necessity *extended*. The method to partition the charge among the different atoms must be both accurate (*i.e.*, able to reproduce the MEP outside the van der Waals surface) and robust (stable against conformational variations). The most popular methods for extracting charges from molecular wavefunctions are based on a fitting of the atomic charges to the MEP computed with *ab initio* methods. The charge fitting procedure consists of minimizing the root mean-squared deviation between the Coulombic potential generated by the atomic charges and the MEP. These nonbonded potentials are then expressed as a sum of *spherically isotropic* atom-atom potentials. Such a description is known to represent an important source of error for the current force fields [36]. For instance, the average error in the energy due to such a partial charge representation is of the order of 1-3 kcal/mol, while the individual errors are known in some cases to be as large as ~ 10 kcals/mol. This is enough to bias the potential energy so that the molecule gets trapped in false minima. Outside the van der Waals surface, the fit to the MEP (and hence the electrostatic energy) can be improved by either adding more charge sites [9], or by including a higher order multipole description. Near the atomic nuclei, however, the electrostatic interaction energy must be corrected for the "penetration" effects at close range [41], and the optimal values of the charges or the higher-order point multipoles may be poorly determined [2, 5, 13]. In spite of this, it has been shown that true convergence to the quantum MEP outside the van der Waals surface of a molecule requires the inclusion of up to hexadecapoles [16, 22–25, 36]. However, it should be noted that depending on the atom and its molecular context, a good representation may not require such a fourth-order description, and a combination of monopoles through quadrupoles may well be enough. *Individual* multipole moments are often less reliable as a description, since they depend significantly on conformational variation, as well as on basis-set effects for the ab initio calculations. Although their sum is expected to be reliable and stable for a representation of the electron density outside the van der Waals surface, it is often the case that such stable representation can only be achieved by including polarizability [28].

Among more sophisticated methods for assigning site multipoles are the distributed multipole analysis (DMA) introduced by several authors [6, 26, 34, 40], and Bader's topological theory of "Atoms in Molecules" (AIM) [1, 24]. The DMA method assigns multipole moments to several sites within the molecule. Such a description has been shown to converge much faster than any given single-center expansion [10, 14, 15, 34, 35]. In the original DMA formalism, a Gaussian basis set is employed, such that the electronic density is also expressed as a sum of Gaussians, and the multipoles are then distributed solely on the basis of the relative positions of the Gaussian centers with respect to the nuclei. The AIM method uses the concept of a *gradient vector path*, which follows the gradient of the electronic charge density around the atoms in the molecule in order to uniquely partition charge density into different atom-centered domains. This means that the multipoles are defined over finite volumes, and formally would avoid any convergence problems. However, to deal with the problem of infinite charge extension, the method uses the van der Waals surface for the outer atoms, and the results tend to depend on the arbitrary position of this surface.

In this work, we explore a new way of partitioning the molecular charge based on Wannier functions, familiar from solid state physics. We believe that such an approach has several important advantages. Wannier functions (WF) are obtained from a *localization* procedure, which provides for a very physical and chemically intuitive way of partitioning the electronic charge. The resulting WFs are *distributed* in space, which should allow for a more faithful representation of the charge density in terms of the different multipole moments. In contrast to several other competing methods, there is no need to make any *ad hoc* assumptions about associating the moments with any specific atom or any other location, although such choices can be made if desired. Furthermore, the Wannier Centers (WCs) are intimately related to the *polarization* [17, 19, 39] of a molecule, which is a very appealing connection. Finally, we note that the different multipole moments are efficiently calculated within the WF formalism, without any undue numerical difficulties, which is a very important consideration.

Although a higher order multipole expansion considerably improves the representation of the electronic cloud, its implementation is *not* common due to the high computational costs incurred in a simulation. For instance, a straightforward implementation which includes up to hexadecapoles (i.e. 25 independent multipolar components) may be expected to be $O(25^2 = 625)$ times more expensive than a treatment based on monopoles only [20]. Recently, Sagui et al.[32] presented an efficient simulation scheme for such a description using the Cartesian tensor formalism. It was found, that by switching most of the calculation to the reciprocal space in the PME method, a highly accurate calculation of the interactions up to hexadecapoles costs only a factor of 8.5 times more a regular AMBER simulation using monopoles (charges) only. An earlier version of this work [38] included an implementation of classical Ewald [31] and PME based treatments of fixed and induced point dipoles into the SANDER molecular dynamics module of AMBER 6 and 7 [7, 21], together with a Car-Parrinello scheme for the computation of the induced polarization. The multigrid method [30] is slower than the PME method on serial computers, but looks promising for parallel implementation. It may also provide for a natural way to interface with continuous, Gaussian-based electrostatics.

2 Calculations

The *unabridged* multipole moments $\{\langle\mu\rangle\}$ of a molecular charge distribution $\rho(\mathbf{r})$ are defined in terms of the expectation values of the position operator $r_{\alpha_1} r_{\alpha_2} r_{\alpha_3} \ldots r_{\alpha_l}$ as:

$$\langle \mu_{\alpha_1, \alpha_2, \alpha_3 \ldots \alpha_l} \rangle = \frac{1}{l!} \int_V d\mathbf{r} \, r_{\alpha_1} r_{\alpha_2} r_{\alpha_3} \ldots r_{\alpha_l} \, \tilde{\rho}(\mathbf{r}) . \tag{1}$$

where the integral is taken over the volume V of the distribution. Here r_{α_i} denotes any of the three components of the vector $\mathbf{r}$, i.e. $r_{\alpha_i} = $ x, y or z ($i = 1 \ldots l$). In order of increasing l, the multipole moments defined above are the monopole ($l = 0$), dipole ($l = 1$), quadrupole ($l = 2$), octupole ($l = 3$), and hexadecapole ($l = 4$) moments, and are denoted by the symbols $\mathbf{q}$, $\mathbf{p}$, $\mathbf{Q}$, $\mathbf{O}$, and $\mathbf{H}$. These terms correspond to a scalar, a vector, and second-rank, third-rank, and fourth-rank tensor quantities, respectively. The monopole represents the total charge of the system, while the larger moments define varying degrees of the charge separation of the distribution. To re-express the higher-order multipole tensors in terms of a set of *traceless* definitions, please refer to [2]. Given these moments, we introduce the multipolar operator $\hat{L}_i$ by:

$$\hat{L}_i = (q_i + \mathbf{p}_i \cdot \nabla_i + \mathbf{Q}_i : \nabla_i \nabla_i + \mathbf{O}_i \vdots \nabla_i \nabla_i \nabla_i + \mathbf{H}_i :: \nabla_i \nabla_i \nabla_i \nabla_i) \tag{2}$$

where the subscript i on ∇ denotes differentiation with respect to coordinate $\mathbf{r}_i$. The different "dot" products stand for the usual tensor contraction, *i.e.*, $\mathbf{H} :: \nabla\nabla\nabla\nabla = \sum_{\alpha,\beta,\gamma,\delta} H_{\alpha,\beta,\gamma,\delta} \, (d/dx_\alpha)(d/dx_\beta)(d/dx_\gamma)(d/dx_\delta)$. The electrostatic potential at $\mathbf{r}_i$ due to a set of point multipoles at $\mathbf{r}_j$ can therefore be written as $\phi(\mathbf{r}_i) = L_j |\mathbf{r}_i - \mathbf{r}_j|^{-1}$. An alternative formulation using the integral formulation of the electrostatic potential has been given by Bertaut [3].

Suppose there are N point charges and multipoles at positions $\mathbf{r}_1, \mathbf{r}_2, \ldots, \mathbf{r}_N$ within the neutral unit cell ($q_1 + q_2 + \cdots + q_N = 0$). The edges of the unit cell are denoted by vectors $\mathbf{a}_\alpha$, $\alpha = 1, 2, 3$, which need not be orthogonal. Each and every component of a multipole set $\{q_i, \mathbf{p}_i, \mathbf{Q}_i, \mathbf{O}_i, \mathbf{H}_i\}$ at position $\mathbf{r}_i$ interacts with each and every component of another multipole set $\{q_j, \mathbf{p}_j, \mathbf{Q}_j, \mathbf{O}_j, \mathbf{H}_j\}$ at positions $\mathbf{r}_j$, $j \neq i$, as well as with their periodic images at positions $\mathbf{r}_j + n_1 \mathbf{a}_1 + n_2 \mathbf{a}_2 + n_3 \mathbf{a}_3$ for all integer triples (n_1, n_2, n_3). It also interacts with its own periodic images at $\mathbf{r}_i + n_1 \mathbf{a}_1 + n_2 \mathbf{a}_2 + n_3 \mathbf{a}_3$ for all such triples with n_1, n_2, n_3 not all zero. The electrostatic energy of the unit cell is then written as:

$$U(\mathbf{r}_1, \ldots, \mathbf{r}_N) = \frac{1}{2} \sum_{\mathbf{n}}{}' \sum_i \sum_j \hat{L}_i \hat{L}_j \left(\frac{1}{|\mathbf{r}_i - \mathbf{r}_j + \mathbf{n}|} \right) \tag{3}$$

where the outer sum is over the vectors $\mathbf{n} = n_1\mathbf{a}_1 + n_2\mathbf{a}_2 + n_3\mathbf{a}_3$, the prime indicating that terms with $i = j$ and $\mathbf{n} = 0$ are omitted. Finally, the electrostatic field and force on atom i at position $\mathbf{r}_i$ are computed as the negative gradient of the electrostatic potential $\phi(\mathbf{r}_i)$ and electrostatic energy $U(\mathbf{r}_i)$, respectively: $\mathbf{E}(\mathbf{r}_i) = -\nabla_i\phi(\mathbf{r}_i)$ and $\mathbf{F}(\mathbf{r}_i) = -\nabla_i U(\mathbf{r}_i) = -\hat{L}_i\nabla_i\phi(\mathbf{r}_i)$

To calculate the different electrostatic moments quantum mechanically, say from a standard DFT calculation, it is necessary to partition the electronic cloud surrounding each of the nuclei. Given the extended nature of the wavefunction, any such scheme is of necessity *ad hoc*, and several of the more common approaches to this problem have already been mentioned in the introduction. Here, we explore a scheme based on a Wannier function decomposition, which emerges naturally from the following considerations. For periodic and supercell-type systems, it turns out that the expectation values of the position operator $\mathbf{r}$ (and other similar operators) as given in Eq.(2.1) are actually ill-defined [17, 27, 37, 39]. This is a direct consequence of expanding the electronic wavefunctions in terms of periodic Bloch orbitals $\psi_{n,\mathbf{k}}(\mathbf{r})$, which are labeled by their band n and crystal momentum wavenumber $\mathbf{k}$. To efficiently deal with the needed expectation values, it is convenient to change the representation of the electronic problem from one given in terms of these extended Bloch orbitals, to one given in terms of the more localized WFs W_n. For a cell $\mathbf{R}$, these are obtained from the Bloch functions by means of a unitary transformation:

$$W_n(\mathbf{r} - \mathbf{R}) = \frac{V}{(2\pi)^3}\int_{BZ} d\mathbf{k}\, \psi_{n,\mathbf{k}}(\mathbf{r})\, e^{-i\mathbf{k}\cdot\mathbf{R}}, \tag{4}$$

so that

$$\psi_{n,\mathbf{k}}(\mathbf{r}) = \sum_{\mathbf{R}} e^{i\mathbf{k}\cdot\mathbf{R}}\, W_n(\mathbf{r} - \mathbf{R})\,, \tag{5}$$

where the integral is taken over the volume V of the Brillouin zone (BZ). Wannier functions are currently enjoying something of a renaissance because of their deep connection to the theory of polarization, the theory of insulators, and the development of linear-scaling methods.

One of the major problems associated with WFs is their non-uniqueness. This is a direct result of the fact that Bloch orbitals are only defined up to an arbitrary phase factor for the case of a single band, and up to a unitary transformation among all the occupied orbitals for the multiband case. To resolve this issue, Marzari and Vanderbilt [19] introduced the concept of maximally localized WFs. These are obtained by requiring that the *spread*

$$S = \sum_n^N (\langle W_n|r^2|W_n\rangle - \langle W_n|\mathbf{r}|W_n\rangle^2)\,, \tag{6}$$

among the N occupied molecular orbitals be minimized.

To calculate the higher-order multipoles within the context of the single-point formulas, we have generalized the elegant method of Souza, Wilkens and Martin [37] for calculating the gauge-invariant cumulants. Once the numerical values of the cumulants have been obtained, the appropriate expectation value for the *unabridged* moments, now defined for each of the WFs, is determined by an order-by-order inversion of the cumulants using their proper definition. For details, see [2].

Higher-order multipole expansions have not been used in the past due to the high computational costs. For instance, a straightforward implementation which includes hexadecapoles (i.e. 25 independent multipolar components) may be expected to be $O(25^2 = 625)$ times more expensive than a treatment based on monopoles only [20]. Recently, Sagui et al.[32] presented an efficient simulation scheme for such a description using the Cartesian tensor formalism. The technical details of such implementation are extensive, and the interested reader is referred to [4]. Essentially, the long-range electrostatic interactions are divided in two sums according to the usual Ewald scheme: the *direct* sum, which evaluates the rapidly-varying, particle-particle interactions, considered up to a given cutoff in real space; and the *"reciprocal"* sum, which evaluates the smoothly varying, long-range part of the interaction. To speed up the evaluation of the direct part, a McMurchie-Davidson formalism [18] – originally developed for the evaluation of Gaussian-based integrals for quantum chemistry calculations– along with later extensions due to Challacombe et al. [8] for the Cartesian multipole interaction tensors, was implemented. The reciprocal part of the problem was treated in three different ways: using the standard Ewald scheme, a Particle-Mesh Ewald (PME) [11, 12] based formalism, and a multigrid-based approach [30]. A very important property of the PME method is its favorable scalability, $O(N \ln(N))$, when compared to the Ewald method ($O(N^2)$). When it comes to the implementation of the multipoles, the PME method has an additional advantage, the factorability of the spline functions (used to interpolate the point multipoles onto the grid [12]) along each dimension. The dimension of each spline function therefore increases only by a factor of l (the multipole order) with respect to the function for charges alone. As a consequence there is a significant efficiency gain when filling the charge grid array. In particular, memory access and loading becomes inexpensive, increasing only by a factor of 2 when going from charges to hexadecapoles, for a fixed cutoff and spline order. As a consequence, calculations in reciprocal space become inexpensive compared to those in the direct sum, even with the use of the McMurchie-Davidson formalism. Strategically, therefore, it becomes convenient to move the bulk of the calculation into reciprocal space. Finally, the multigrid method [30] is slower than the PME method on serial computers, but looks promising for parallel implementation.

3 Results and Discussion

To illustrate the utility of the WF formalism for calculating the electrostatic multipoles, we have investigated a select number of molecules using standard DFT methods [33]. Once the Bloch orbitals were obtained, they were transformed into the maximally localized WFs. These WFs were then analyzed for their centers, the multipole moments, and the charge overlap between the different WF densities. The multipole moments were corrected for finite-size effects and extrapolated to the "infinite-box" limit (isolated molecule) and then the MEP as generated by these moments was computed and compared to its *ab initio* counterpart. The classical MEP, as generated by the electrostatic moments, made use of our multigrid implementation of the multipolar method [32].

Due to space constraints, in this section we present results for only a single-water and a carbon dioxide molecule, concentrating primarily on the electrostatic moments and the MEP. We consider the spin-degenerate valence electrons only. The water molecule is then characterized by four doubly occupied WFs. Essentially, the WCs are arranged tetrahedrally with respect to each other. Two of the WFs are associated with the covalent O-H bonds while the other two are associated with the occupied lone-pair orbitals of the O atom . The WCs associated with the covalent bonds are located a distance of 0.526 Åaway from the O atom, while the lone-pair WCs are about 0.298Åaway. The two vectors that join the lone-pair WCs and the O atom form an angle of $126.8°$. The carbon dioxide molecule is characterized by eight different WFs arranged symmetrically about the C-atom: three are associated with each of the C-O bonds for a total of six WFs, while a further two WFs are associated with O atom lone pairs. The WCs associated with C-O bonds, whose WFs consist of a mixture of an O atom lone pair plus the bonding orbitals, display inversion symmetry with respect to the central C-atom. From the point of view of partitioning the charge, it is important to note that there is considerable overlap between the WF densities of a given molecule. Hence, unlike AIM – which partitions the charge into isolated, non-overlapping spatial regions – the charge at any given point is now decomposed in terms of its contribution from each of the occupied WFs. For the isolated water molecule, the normalized overlap between the two O-lone pair WF densities is considerable, $\approx 25\%$, and it is somewhat less between the WFs associated with the O-H bonds. For the linear CO_2 molecule, the overlap between WFs around one of the C-O bonds is considerable, while the overlap between WF densities on different sides of the C-atom is almost negligible, reflecting their localized character.

We now turn to the MEPs as generated via the WF formalism. For each of the WFs, we have calculated the extrapolated unabridged multipole moments about their own WCs. Thus, every filled WF has a charge of $-2($ units of electronic charge e) associated with it. Since the multipolar moments are computed with respect to the center of charge, there are no dipolar terms. Given the individual site multipolar moments, the MEPs are compared to the corresponding *ab initio*-based potentials. To illustrate this comparison, Figs. 1 and 2 show contour plots of these MEPs, as the maximum multipolar order included in the calculation is increased. In these figures, the top panels show the MEP on the plane of the molecule, while the lower panels show the MEP on planes perpendicular to the plane of the molecule through the center of mass. Any multipolar expansion will diverge as it approaches the origin of the charge distribution where there is a singularity. For this reason, the multipolar expansion is valid only outside the van der Waals surface, and in our figures the van der Waals region has been "cut out" from the electrostatic potential plots. Fig. 1 shows results for the water molecule. The correspondence between the WF and the DFT MEPs appears to be quite good. Not surprisingly, the MEP generated with only monopoles already appears to be adequate, which is a reflection of the chemically ap-

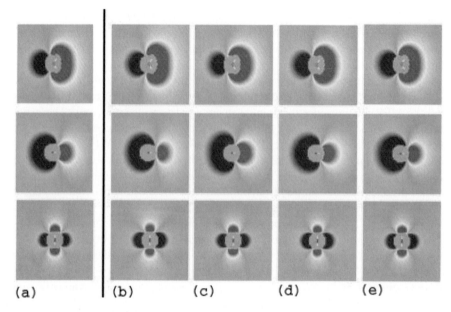

Figure 1. Views of the MEP for H_2O on three perpendicular planes that meet at the center of mass of the molecule. DFT results are to the left of the solid line and WF generated MEPs are to the right of this line. (**a**) *ab initio* MEP; (**b**) MEP generated with classical monopoles only; (**c**) MEP with monopoles through quadrupoles; (**d**) MEP with monopoles through octupoles; and (**e**) MEP with monopoles through hexadecapoles. The van der Waals radius is cut out from each of the MEPs

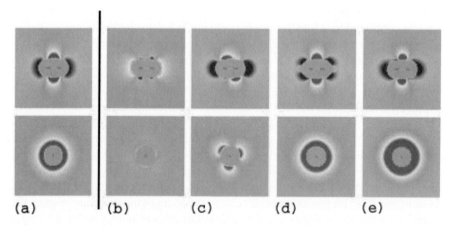

Figure 2. Views of cuts through the MEPs for CO_2: (**a**) *ab initio* MEP; (**b**) classical monopoles; (**c**) classical monopoles through quadrupoles; (**d**) monopoles through octupoles; and (**e**) monopoles through hexadecapoles. The van der Waals radius is cut out from each of the MEPs

propriate way in which the WCs are distributed about the *polar* molecule. Somewhat less good are the results for the *quadrupolar* CO_2 molecule. Clearly, from Fig. 2 it is obvious that one needs to go to at least to octupoles in order to properly reproduce the correct symmetry of the MEP; a 'partial charge' description based on monopoles and quadrupoles simply would not be correct.

We have calculated the error as a function of the distance from the atomic centers. Here, the error is defined to be the absolute value of a site-by-site difference between the *ab-initio* and WF-generated MEPs. These absolute differences are then binned up in shells according to the distance to the nearest atomic nuclei. To generate these figures, we have considered the electrostatic potential for an isolated molecule. Fig. 3 shows the error in MEP for H_2O (left panel) and CO_2 (right panel). Descriptions based on higher-order multipoles are considerably more accurate than descriptions based on the lower orders, although there are some minor variations of this as the van der Waals radius is approached. In these figures it is possible to appreciate quantitatively some of the results visually apparent in the MEP contours. Results for a polar molecule like water are very good, even for the case of using monopoles only, where the error becomes less than 1 kcal/mol at ~ 2.6 Å. When multipoles up to hexadecapoles are included (**mqoh**), the error becomes less than 0.1 kcal/mol at ~ 2.6 Å. For all the molecules we considered [33], the error becomes very small beyond the 2-3 Årange, and it worsens near the van der Waals radius, rising rapidly as it crosses this boundary. This is a consequence of the point-like nature of a multipolar description.

Historically, the reason why higher order multipolar interactions have not been implemented into macromolecular codes is their high cost. In a standard, *fixed-cutoff* implementation of these interactions, the computation time grows with the square of the multipolar components and therefore becomes expensive when compared to charge-charge interactions. The belief that –since higher-order multipole interactions decay much faster than charge interactions– the use of a cutoff alleviates the problem proves incorrect. In fact, *most of the cost of the interactions originates in the* **direct** *part*, even under relatively short cutoffs. Additionally, too short of a cutoff produces serious force and energy errors. On the other hand, moving the majority of the computation of the interactions into reciprocal space not only preserves accuracy but also has a moderate cost, which justifies the reduction of the cutoff for the direct space. This is true as far as the PME and multigrid methods are concerned; the traditional Ewald method still scales as $\sim O(625\ N^2)$ (for hexadecapoles, [20]) and therefore is not competitive.

The force and energy relative RMS errors with respect to costly 'exact' values are presented in Tables 1 and 2, below.

The parameters are chosen such that the *relative* force error ϵ stays the same for all multipole levels (in this case, $\epsilon \simeq 5 \times 10^{-4}$). This requires that as the multipolar order increases, the spline order and the Ewald parameter β be increased, while the mesh space h_x and the cutoff radius R_c (and the cutoff radius R_G for the Gaussians in the multigrid method) be decreased. Increasing β and decreasing R_c effectively moves most of the calculation into reciprocal (or 'multigrid') space. Table 1 shows the parameters used for the PME multipole implementation and the corresponding

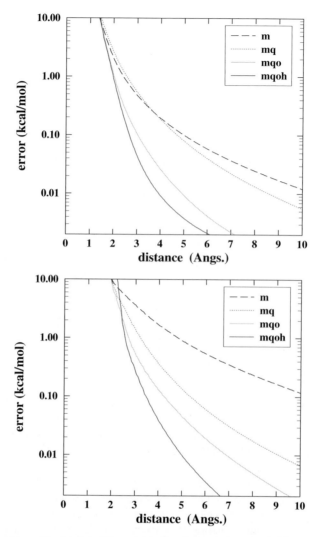

Figure 3. Semi-logarithmic plot of the error in the electrostatic potential for a H_2O molecule (*left*) and a CO_2 molecule (*right*). The different curves correspond to monopole **m**; monopole plus quadrupole **mq**; monopole plus quadrupole plus octupole **mqo**; and monopole plus quadrupole plus octupole plus hexadecapole **mqoh**

results. The 'other' time column in the table includes the contributions from the adjusted and self terms as well as a variety of tensor operations needed for the calculations. This time is relatively insignificant compared to that from the direct and reciprocal contributions. The last two rows in Table 1 give timings for one-order and two-order reduction in relative force error for the full multipolar calculation.

For reference, an AMBER 7 run with sander auto settings (8Åcutoff for the direct part; Ewald coefficient $\beta = 0.34864$; spline order 4; mesh size 0.992Å) yields a comparable relative RMS force error of 3.8×10^{-4} with direct sum time 0.23s; reciprocal 0.13s; and total Ewald sum time 0.36s. Our new implementation with all multipolar interactions up to hexadecapole-hexadecapole costs therefore only a factor of $\simeq 8.5$ more than the standard AMBER implementation with charge-charge interactions.

Table 1. This table shows results for the PME implementation of the permanent electrostatic multipole moments. The timing was performed on Intel Xeon, 3.06 GHz, 512 KB cache, 2GB memory (compiler is g77), 1 processor. The averaged timings are per step. The system consists of 4,096 water molecules (12,288 atomic sites) in a 49.6 Åcubic box. The table shows the order of the spline used for interpolation, the mesh size h_x, the Ewald parameter β and the cutoff radius R_c for the direct interactions. Timings shown are for the direct interactions (they also include the van der Waals interactions out to 8Åcutoff with continuous correction), the reciprocal interactions, the 'other' interactions (which include the adjusted and self terms and a variety of tensor operations needed for the calculations), and the time for the total force. The parameters were chosen so that the relative force error is 5×10^{-4}. The last two rows give timings for one-order and two-order reduction in relative force error (i.e. , 5×10^{-5} and 5×10^{-6} RMS force errors)

multipole level	spline order	h_x (Å)	β (Å^{-1})	R_c (Å)	time (s) direct	time (s) reciprocal	time (s) other	time (s) force
0	5	0.775	0.50	5.63	0.21	0.18	0.03	0.42
1	5	0.775	0.50	5.60	0.33	0.21	0.04	0.58
2	6	0.688	0.55	5.10	0.57	0.32	0.07	0.96
3	8	0.620	0.70	4.25	0.93	0.60	0.17	1.70
4	8	0.459	0.85	3.60	1.54	1.12	0.39	3.05
4	10	0.344	0.90	3.85	1.81	2.23	0.38	4.42
4	12	0.344	0.95	4.06	2.00	2.66	0.37	5.03

For the sake of comparison, we have also implemented a Coulomb code with an optimized McMurchie-Davidson acceleration. We find that a 'regular' cutoff of 8Åincreases the computation time by a factor of ~ 6 when compared to the total force time in the first hexadecapole row in Table 1. In addition, the relative RMS force error increases about two orders of magnitude with respect to that reported in Table 2. It is fair to argue that the cutoff should only be used for quadrupoles and higher order multipoles. We have therefore repeated the calculation where the charges and dipoles of the model have been set to zero (that is, all the interactions are quadrupole-quadrupole or higher order and should converge absolutely). In our tests we found that the relative RMS force error is greater than 1% at a 7Åcutoff and about 8×10^{-3} at an 8Åcutoff. A run with a 7Åcutoff costs approximately three times more than the optimized PME which uses a 3.6Åcutoff (and computes the additional charge and dipole interactions).

Table 2. This table shows results for the MULTIGRID implementation of the permanent electrostatic multipole moments. The timing was performed on Intel Xeon, 3.06 GHz, 512 KB cache, 2GB memory (compiler is g77), 1 processor. The averaged timings are per step. The system consists of 4,096 water molecules (12,288 atomic sites) in a 49.6 Å cubic box. The table shows the mesh size h_x, the Ewald parameter β, the cutoff radius R_c for the direct interactions, the cutoff radius R_G for the Gaussians in the Poisson's equation, and the averaged timing for the reciprocal interactions. The parameters were chosen so that the relative force error is 5×10^{-4}. The timings for "other" calculations is the same as in Table I and not shown here

multipole level	h_x (Å)	β (Å^{-1})	R_c (Å)	R_G (Å)	time (s) direct	time (s) reciprocal
0	0.620	0.60	5.20	3.50	0.20	2.30
1	0.620	0.61	5.20	3.63	0.29	2.66
2	0.516	0.70	4.80	3.45	0.52	4.64
3	0.443	0.75	4.25	3.10	0.94	8.42
4	0.388	0.79	4.25	3.05	2.21	15.71

Table 2 shows the parameters used for the multigrid multipole implementation and the corresponding results. The 'other' time (that includes the contributions from the adjusted and self terms as well as a variety of tensor operations needed for the calculations) is the same as the 'other' time column in Table 1 and therefore is not included here. The multigrid is implemented with the specific goal of parallelization. The parallel implementation of the 'reciprocal' part is a straightforward generalization of our previous method [30]. From the timings shown in Table 2 (obtained in one processor) one would expect that a relatively efficient parallel scheme would make this method quite competitive. Since the direct part is computed with the Ewald coefficient β, the width of the Gaussians that define the charge density in the Poisson's equation is determined by $\sqrt{2}\beta$. Large β's (narrow Gaussians) need smaller Gaussian cutoff radius R_G for the representation of the Gaussian on the grid. However, accuracy requires that many points of the Gaussian be sampled (since gradients of the distribution are needed), which in turn means that narrow Gaussians need smaller mesh sizes, if they are to subtend a sufficient number of grid points. Therefore, for a fixed accuracy, a maximum efficiency will be achieved as a compromise between reducing the Gaussian cutoff radius R_G and increasing the mesh size h_x (in such a way as to preserve accuracy). In principle, there is a coefficient β that will optimize the results of this compromise, but of course this β also weighs the relative contributions of U_{dir} and U_{rec}. The ideal set of h_x, β, R_C and R_G that optimize the performance of the multigrid method will be determined once parallelization of the entire code is complete.

We also performed a 1 nanosecond simulation of 216 waters using this "hybrid" water model at the hexadecapole level under NVE conditions. The model displayed good energy conservation. In order to carry out molecular dynamics, we introduced a method to calculate the torques produced by the different multipolar components at both atomic and extra-atomic sites and to convert these torques into atomic forces.

4 Conclusion

Accurate electrostatics remains an important issue for the improvement of current force fields for large-scale biomolecular simulations. It has long been known that a description based on a multipolar analysis can dramatically improve the molecular electrostatic potentials, and thereby the quality of the simulation. This, in turn, requires a partitioning of the extended molecular charge density into different spatial regions so that partial charges and multipoles may be calculated and assigned to either atoms or other points. As an alternate approach to current methods, we have examined here a charge partitioning scheme based on a Wannier-function analysis. This approach, which makes use of the recently developed maximally localized Wannier functions, has the advantage of placing the calculated partial charges and multipoles in chemically meaningful locations about the molecule. The calculation of the different multipole moments is performed without any undue numerical difficulties or *ad hoc* assumptions. In particular, the "exact" reference state for our errors is an *exact ab initio* calculation, and not an integration over a finite volume where part of the charge density has been discarded. We have compared the MEPs generated both by DFT methods and classical multipolar calculations based on the WF approach for selected isolated molecules and water dimers, and found very good agreement. While a description based on monopoles and quadrupoles may be sufficient –especially for polar molecules– improvements for more "difficult" non-polar molecules can be achieved when higher order multipoles are included.

Until recently, the major hurdle in the implementation of classical multipoles in biomolecular simulations has been the associated computational cost. However, a recent implementation of higher order Cartesian multipoles for classical biomolecular molecular dynamics simulations using both PME and multigrid methods has dramatically overcome this difficulty [32], as illustrated in Tables 1 and 2. While the results obtained in this work are highly encouraging, there are several important outstanding issues that need to be addressed in order to further improve and quantify the WF description. Among others, these include penetration effects, polarization and dynamical effects. At present we are tackling some of these issues. Ultimately, we believe that a WF-based description (or similar localized wavefunction description) with a polarization scheme, when combined with recent advances in simulating multipolar descriptions [32], will lead to molecular dynamics simulations of biomolecules that are considerably more accurate and reliable.

Acknowledgments

Support for this work has been provided by NSF grants ITR-0121361 and CAREER 0348039.

References

[1] Bader R.: Atoms in Molecules: A Quantum Theory. Clarendon Press, Oxford (1990)

[2] Bayly C., Cieplak P., Cornell W., Kollman P.: A well-behaved electrostatic potential based method using charge restraints for deriving atomic charges - the resp model. J. Phys. Chem., **97**, 10269–10280 (1993)

[3] Bertaut E.F.: The equivalent charge concept and its application to the electrostatic energy of charges and multipoles. J. Phys., **39**, 1331 (1978)

[4] Boys S.F.: Construction of Some Molecular Orbitals to Be Approximately Invariant for Changes from One Molecule to Another. Rev. Mod. Phys., **32**, 296 (1960)

[5] Chipot C., Angyan J., Millot C.: Statistical analysis of distributed multipoles derived from molecular electrostatic potentials. Molecular Physics, **94**, 881–895 (1998)

[6] Claverie P.: Elaboration of approximate formulas for the interaction between large molecules: Application to Organic Chemistry. In: Pullman B. (ed.) Intermolecular Interactions: From Diatomics to Biopolymers, vol.1, 69, Wiley Interscience, New York (1978)

[7] Case D.A., Pearlman D.A., Caldwell J.W., Cheatham III T.E., Wang J., Ross W.S., Simmerling C., Darden T., Merz K., Stanton R., Cheng A., Vincent J., Crowley M., Tsui V., Gohlke H., Radmer R., Duan Y., Pitera J., Massova I., Seibel G., Singh U., Weiner P., Kollman P.: AMBER 7. University of California, San Francisco (2002)

[8] Challacombe M., Schwegler E., Almlöf J.: Recurrence relations for calculation of the cartesian multipole tensor. Chem. Phys. Letters, **241**, 67–72 (1995)

[9] Dixon R., Kollman P.: Advancing beyond the atom-centered model in additive and non-additive molecular mechanics. J. Comp. Chem., **18**, 1632–1646 (1997)

[10] Dykstra C.: Electrostatic interaction potentials in molecular force fields. Chem. Rev., **93**, 2339–2353 (1993)

[11] Darden T.A., York D.M., Pedersen L.G.: Particle mesh Ewald: An N log(N) method for Ewald sums in large systems. J. Chem. Phys., **98**, 10089–10092 (1993)

[12] Essmann U., Perera L., Berkowitz M.L., Darden T., Lee H., Pedersen L.G.: A smooth particle mesh Ewald method. J. Chem. Phys., **103**, 8577–8593 (1995)

[13] Francl M.M., Chirlian L.A.: The pluses and minuses of mapping atomic charges to electrostatic potentials. In: Lipkowitz K., Boyd D.B. (eds.) Reviews in Computational Chemistry, volume 14, VCH Publishers, New York, NY (1999)

[14] Hättig C., Jansen G., Hess B., Ángyán J.: Intermolecular interaction energies by topologically partitioned electric properties: II. Dispersion energies in one-centre and multicentre multipole expansions. Molec. Phys., **91**, 145–160 (1997)

[15] Jansen G., Hättig C., Hess B., Ángyán J.: Intermolecular interaction energies by topologically partitioned electric properties: 1. electrostatic and induction energy in one-centre and multicentre multipole expansions. Molec. Phys., **88**, 69–92 (1996)

[16] Kosov D.S., Popelier P.L.A.: Atomic partitioning of molecular electrostatic potentials. J. Phys. Chem. A, **104**, 7339–7345 (2000)

[17] King-Smith R., Vanderbilt D.: Theory of polarization of crystalline solids. Phys. Rev. B, **47**, 1651 (1993)

[18] McMurchie L., Davidson E.: One- and two-electron integrals over cartesian gaussian functions. J. Comput. Phys, **26**, 218–231 (1978)

[19] Marzari N., Vanderbilt D.: Maximally localized generalized Wannier functions for composite energy bands. Phys. Rev. B, **56**, 12847–12865 (1997)

[20] Computationally, due to diverse technical reasons, we also found that the use of traceless tensors, which have 25 independent components up to hexadecapoles, is more expensive than the use of unabridged moments, with 35 components. Of course, both representations produce the same electrostatic potential.

[21] Pearlman D.A., Case D.A., Caldwell J.W., Ross W.S., Cheatham III T.E., De-Bolt S., Ferguson D., Seibel G., Kollman P.: Amber, a package of computer programs for applying molecular mechanics, normal mode analysis, molecular dynamics and free energy calculations to simulate the structural and energetic properties of molecules. Comp. Phys. Comm., **91**, 1–41 (1995)

[22] Popelier P.L.A., Joubert L., Kosov D.S.: Convergence of the electrostatic interaction based on topological atoms. J. Phys. Chem. A, **105**, 8254–8261 (2001)

[23] Popelier P.L.A., Kosov D.S.: Atom■atom partitioning of intramolecular and intermolecular Coulomb energy. J. Chem. Phys., **114**, 6539–6547 (2001)

[24] Popelier P.: Atoms in Molecules: An Introduction. Prentice Hall, Harlow, England (2000)

[25] Price S.: Toward more accurate model intermolecular potentials for organic molecules. In: Lipkowitz K., Boyd D.B. (eds.) Reviews in Computational Chemistry, volume 14, VCH Publishers, New York, NY (1999)

[26] Pack G.R., Wang H.Y., Rein R.: A quantitative demonstration of the domain of multipole representations of molecular potentials. Chem. Phys. Lett., **17**, 381 (1972)

[27] Resta R.: Macroscopic polarization in crystalline dielectrics: the geometric phase approach. Rev. Mod. Phys., **66**, 899 (1994)

[28] Ren P., Ponder J.W.: A consistent treatment of inter- and intramolecular polarization in molecular mechanics calculations. J. Comput. Chem., **23**, 1497–1506 (2002)

[29] Sagui C., Darden T.A.: Molecular dynamics simulations of biomolecules: Long-range electrostatic effects. Annu. Rev. Biophys. Biomol. Struct., **28**, 155–179 (1999)

[30] Sagui C., Darden T.A.: Multigrid methods for classical molecular dynamics simulations of biomolecules. J. Chem. Phys., **114**, 6578–6591 (2001)

[31] Smith W.: Point multipoles in the ewald summation. CCP5 Information Quarterly, **4**, 13–25 (1982)

[32] Sagui C., Pedersen L., Darden T.A.: Towards an accurate representation of electrostatics in classical force fields: Efficient implementation of multipolar interactions in biomolecular simulations. J. Chem. Phys., **120**, 73–87 (2004)

[33] Sagui C., Pomorski P., Darden T.A., Roland C.: Ab initio calculation of electrostatic multipoles with Wannier functions for large-scale biomolecular simulations. J. Chem. Phys., **120**, 4530–4544 (2004)

[34] Stone A.J.: Distributed multipole analysis, or how to describe a molecular charge distribution. Chem. Phys. Letters, **83**, 233–239 (1981)

[35] Stone A.J.: Distributed polarizabilities. Molec. Phys., **56**, 1065–1082 (1985)

[36] Stone A.J.: The Theory of Intermolecular Forces. Clarendon Press, Oxford (1996)

[37] Souza I., Wilkens T., Martin R.: Polarization and localization in insulators: Generating function approach. Phys. Rev. B, **62**, 1666–1683 (2000)

[38] Toukmaji A., Sagui C., Board J.A., Darden T.: Efficient PME-based approach to fixed and induced dipolar interactions. J. Chem. Phys., **113**, 10913–10927 (2000)

[39] Vanderbilt D., King-Smith R.: Electric polarization as a bulk quantity and its relation to surface charge. Phys. Rev. B, **48**, 1993 (1993)

[40] Vigné-Maeder F., Clavérie P.: The exact multicenter multipolar part of a molecular charge distribution and its simplified representations. J. Chem. Phys., **88**, 4934 (1988)

[41] Wheatley R., Mitchell J.: Gaussian multipoles in practice: Electrostatic energies for intermolecular potentials. J. Comp. Chem., **15**, 1187–1198 (1994)

Quantum-Chemical Models for Macromolecular Simulation

Towards Fast and Reliable Quantum Chemical Modelling of Macromolecules

Yaoquan Tu[1] and Aatto Laaksonen[2]

[1] Theoretical Chemistry, Royal Institute of Technology, SE-10691 Stockholm, Sweden
[2] Division of Physical Chemistry, Arrhenius Laboratory, Stockholm University, 106 91 Stockholm, Sweden

Abstract. Two possible routes to fast and reliable quantum chemical modelling and simulations of large macromolecules are outlined; a wave-function-based and an electron-density-based. In the former method, the NDDO (Neglect of Diatomic Differential Overlap) approximation, a widely used basis for many semi-empirical molecular orbital approaches is examined and a new model, overcoming the inherent deficiencies in the NDDO model is proposed. In the new model, first order correction term is added to the electron-electron Coulomb interactions, thereby improving the balance between the core-electron and the electron-electron interactions. Using the new model, non-empirical calculations are performed showing that the total energies from this model are consistently slightly higher than those from corresponding *ab initio* calculation but closer to the *ab initio* results than when the standard NDDO is used. Before introducing the latter scheme, current *ab initio* tight-binding methods, developed from density functional theory are overviewed. Thereafter some improvements, especially those leading to faster and more accurate integral evaluations developed in our group, are presented. Aeries of calculations on sample molecules show that the results from our *ab initio* tight-binding method have nearly the same accuracy as those obtained from the corresponding frozen-core density functional calculations but at much faster speed. Both methods, the extended NDDO and the *ab initio* tight-binding, show a great potential as future schemes to model and simulate macromolecular systems at *ab initio* level of accuracy but at the speed comparable to today's semi-empirical calculations.

Key words: Quantum Chemistry, MD simulation, NDDO, *ab initio* tight-binding, DFT

1 Introduction

Current molecular modelling is about employing *interacting* structural models in computers and routinely visualizing the results using computer graphics. Many conceptual ideas from pre-computer modelling of atoms and molecules from early 1900's are, somewhat surprisingly, still "the-state-of-the-art" and frequently used in modern techniques of macromolecular modelling and simulations. It may have been the phenomenally rapid development of computer technology during the last decades hampering the development of molecular simulation methods and models, or it may have been that these old models were highly robust already when presented. Most likely it is the combination of both.

In atomistic computer modelling and simulations of molecules, a total energy of the whole system is repeatedly calculated by summing up the contributions from a number of separate, very simple potential energy functions with the instantaneous atom-atom distances and selected angles between covalently bonded atoms as parameters. Conceptually these terms are divided into so-called "bonded" and "non-bonded", or alternatively, "intra"- and "intermolecular" interactions. Together with the molecular geometries this collection of empirically parameterized potential functions, known as the *force field*, are the primary input to standard modelling software. The various contributions in force fields are assumed to be additive, thereby giving an intuitively simple picture of the interactions within molecules and with the surrounding molecules. In simple terms, force fields provide a mathematical tool to guide each atom in a system towards a new direction under the influence of all the other atoms in its surrounding. Although continuously improved, the force fields have obvious inherent limitations. Not so much because they are such simple, man-made constructions designed to do their job at normal conditions close to equilibrium. Not because their quality rely on a very limited number of adjustable parameters. But rather because they lack the essential means to model chemical processes. They simply lack the electrons. The electrons make the Chemistry go round!

The interactions between atoms and molecules should be ideally described based on quantum mechanics. Within close distances between the molecules, there is a complex interplay of forces of quantum mechanical origin. Although the various physical mechanisms behind the interactions can be described, at least approximately, the inherent mathematical complexity and the large number of degrees of freedom effectively prohibits a rigorous treatment of condensed matter systems of a large number of interacting particles. However, with the computers becoming faster and faster, we can either improve the lower-end quantum mechanical methods to become more accurate, or we can introduce reasonable approximations to high-end methods and develop efficient schemes to speed up the computations to treat larger and more complex systems. In this work we will give one example of both these strategies to improve the quantum chemical modelling of molecular systems with the primary aim being to create accurate and efficient methods to model macromolecular systems in their natural environment.

Molecular mechanical force fields will continue to be useful, not only in conventional modelling and simulation methods but also when hybridized to first principles simulations[1]. Combining quantum chemical calculations with empirical force fields is often a good compromise to study the molecular processes and reactions in order to learn how surrounding medium, concentration changes, pressure and temperature affect various molecular properties or reactions[2]. The method of molecular dynamics can also be introduced into quantum mechanics as was done in an elegant way by Car and Parrinello[3]. However, being several orders of magnitude more CPU time consuming compared to corresponding classical molecular mechanical methods, quantum chemical calculations provide currently a bottleneck in such computer simulation schemes. This severely restricts applicability of these simulation methods to molecular systems with a rather limited number of atoms. For example the Car-Parrinello methods are regularly used today for systems up to 100+ atoms over times of a few tens of picoseconds. Besides, application of Car-Parrinello method on much larger systems appears not to be possible due to plane waves being used as basis functions.

It can be noticed that the demand for developing both efficient and reliable quantum chemistry methods does not only come from inside the quantum chemistry community but also from the simulation community. As an example of the latter we wish to present two quantum chemistry methods we expect to be useful in computer simulations of macromolecular systems:

- In the first one the objective is to improve the traditional Neglect of Diatomic Differential Overlap (NDDO) model[4] to obtain accurate and reliable semi-empirical molecular orbital (MO) methods in future. We present an improvement beyond the NDDO model thereby allowing it to be used in future parameterizations of new semi-empirical schemes which allow for example calculations of relative energies of macromolecules or computer simulations where the forces are calculated based on MO theory.
- The second method is based on the Density Functional Theory (DFT)[5, 6]. Our purpose is to develop a type of *ab initio* tight-binding method which is as reliable as the conventional DFT methods but nearly as fast as the currently used semi-empirical MO methods.

2 Extended NDDO Approximation (PART I)

Semi-empirical MO methods can normally be divided into two main categories. In the first one, there are those without explicit electron-electron interactions. Instead, the interactions are implicitly included in an effective Hamiltonian. This type includes for example the Hückel MO (HMO)[7], and the Extended HMO (EHMO)[8] methods. The major advantage of these methods is of course their great simplicity, thus they can be used in the study of large molecular systems with very limited computer resources. However, the neglect of the explicit treatment of the electron-electron interactions makes these methods unable to study many important systems, for example open-shell molecules. In the second type there are those with the electron-electron interactions explicitly treated. Therefore, these methods are more general and can be used to study various kinds of molecular properties. In order to reduce the calculation demand, these methods can be simplified with many approximations and furnished with (semi-)empirical parameters. The so-called Zero Differential Overlap (ZDO) of the integrals is one important overall approximation. Within the ZDO approximation, all the three- and four-center two-electron repulsion integrals (TERIs), as well as some two-center TERIs are neglected. In the semi-empirical methods using the ZDO approximation, the remaining integrals are usually calculated approximately or simply replaced by some empirical parameters. Among the models adopting the ZDO approximation are Complete Neglect of the Differential Overlap (CNDO)[4, 9], Intermediate Neglect of Differential Overlap (INDO)[10], and Neglect of Diatomic Differential Overlap (NDDO)[4]. Since the NDDO model neglects a least number of electron repulsion integrals, those semi-empirical methods which are based on the NDDO model should in principle give more accurate results. Among the semi-empirical methods that are developed from the NDDO model are MNDO[11], AM1[12], and PM3[13]. These are most widely used in the studies of molecular properties.

The main attractive feature in semi-empirical MO methods is their fast calculation speed, offering the possibility to use quantum chemistry calculations to study fairly large systems. For example, for studies of systems containing more than a thousand atoms, semi-empirical calculations are still the only practical choice. Semi-empirical MO calculations can also be used as the precursor to *ab initio* calculations. For calculations of some molecular properties, such as the ionization potential, dipole moment, and the energy for a chemical reaction, as examples, the absolute errors in AM1[12] and PM3[14] methods are usually fairly small. The biggest problem in these semi-empirical methods is that the errors in the results from calculations are non-systematic and unpredictable. This makes it difficult to obtain reliable results for many molecular properties from the semi-empirical calculations. For example the current semi-empirical calculations fail sometimes badly in studies of internal rotational barriers and conformations of large molecules, as well as hydrogen-bonding systems. The conventional semi-empirical MO methods have been improved by adding some correction terms to the core-electron interaction matrices[15–17]. After this the new methods have given improved results for the internal rotational barrier of a molecule and the bonding energies for the hydrogen-bonding systems[15–17]. However, they are still unable to give reliable results in conformational studies of large molecules[18].

Spanget-Larsen[19, 20] has pointed out that an imbalanced treatment of the core-electron and electron-electron interactions is the main problem in the semi-empirical methods based on the ZDO approximation and the problem is difficult to solve by further optimization of the parameters. An elegant improvement in the orbital equations was also proposed by him to keep the balance between the core-electron and electron-electron interactions[19, 20]. The semi-empirical method thus developed has shown to be successful in the studies of molecular spectroscopic properties[20]. However, it was not shown by Spanget-Larsen[19, 20] how to keep a balanced treatment of core-electron and electron-electron interactions in the total energy. Therefore, it is not clear how this particular method can be used in the studies of some molecular properties related to the total energies, such as the conformations.

To improve the standard NDDO model, it is necessary to first have a deeper understanding of the basic model when only the valence atomic orbitals are used as basis set. In the first part of this work, the NDDO model is re-examined based on non-empirical frozen-core calculations to find out:

1. if the NDDO model can be justified when orthonormal valence atomic orbitals are used as basis set?

2. how well does the Roby model perform? This model is considered as the practical implementation of the NDDO model where the TERIs are calculated over non-orthonormal atomic orbitals.

Based on detailed studies of the above two models, an improvement going beyond the NDDO approximation will be presented by reformulating the basic equations.

2.1 Theoretical Models

The theoretical basis for many conventional approximate MO methods is the Hartree-Fock-Roothaan (HFR) equations. The semi-empirical MO methods developed from the NDDO model can also be considered as the approximate ways to solve the HFR equations. For simplicity, we discuss below only the closed-shell system.

When the LCAO (linear combination of atomic orbitals) approximation is used to express the MOs, the ith MO ψ_i can be written as:

$$\psi_i = \sum_\mu C_{\mu i} \phi_\mu ,$$ (1)

where ϕ_μ denotes an atomic orbital. ψ_i satisfies the orthonormal condition:

$$\langle \psi_i \mid \psi_j \rangle = \delta_{ij} .$$

The total electronic energy of the system is given (in atomic units) as:

$$E_{el} = \sum_{\mu v} P_{\mu v} h_{\mu v} + \frac{1}{2} \sum_{\mu v} \sum_{\lambda \sigma} P_{\mu v} P_{\lambda \sigma} \left[(\mu v | \lambda \sigma) - \frac{1}{2} (\mu \sigma | \lambda v) \right] ,$$ (2)

where $h_{\mu v}$ represents the one-electron Hamiltonian matrix and $P_{\mu v}$ the density matrix, respectively. Here $P_{\mu v}$ is:

$$P_{\mu v} = 2 \sum_i^{N_{occ}} C_{\mu i} C_{vi} ,$$ (3)

with the summation only over the occupied orbitals. The best MO coefficients $C_{\mu i}$ can be obtained by solving the following secular equation self-consistently:

$$\mathbf{FC} = \mathbf{SC}\varepsilon,$$ (4)

where $\mathbf{S}$ is the overlap matrix and $\mathbf{F}$ the Fock matrix. For a closed-shell system, the Fock matrix can be written as:

$$F_{\mu v} = h_{\mu v} + \sum_{\lambda \sigma} P_{\lambda \sigma} \left[(\mu v | \lambda \sigma) - \frac{1}{2} (\mu \sigma | \lambda v) \right].$$ (5)

If the orthonormal atomic orbitals, such as those obtained from the Löwdin orthogonalization[21, 22], are used as basis set, then the total electronic energy can be written as:

$$E_{el} = \sum_{\mu\nu} {}^{\lambda}P_{\mu\nu} {}^{\lambda}h_{\mu\nu}$$
$$+ \frac{1}{2} \sum_{\mu\nu} \sum_{\lambda\sigma} {}^{\lambda}P_{\mu\nu} {}^{\lambda}P_{\lambda\sigma} \left[\left({}^{\lambda}\mu {}^{\lambda}\nu | {}^{\lambda}\lambda {}^{\lambda}\sigma \right) - \frac{1}{2} \left({}^{\lambda}\mu {}^{\lambda}\sigma | {}^{\lambda}\lambda {}^{\lambda}\nu \right) \right] \qquad (6)$$

where ${}^{\lambda}\mu$, ${}^{\lambda}\nu$, ${}^{\lambda}\lambda$ and ${}^{\lambda}\sigma$ represent the orthonormal atomic orbitals. ${}^{\lambda}h_{\mu\nu}$ is the one-electron Hamiltonian matrix and ${}^{\lambda}P_{\mu\nu}$ the density matrix under the orthonormal basis set. ${}^{\lambda}P_{\mu\nu}$ can be expressed as:

$$^{\lambda}P_{\mu\nu} = 2 \sum_{i}^{N_{occ}} {}^{\lambda}C_{\mu i} {}^{\lambda}C_{\nu i} . \qquad (7)$$

Correspondingly, the secular equation and Fock matrix in the orthonormal basis set are

$$^{\lambda}\mathbf{F}^{\lambda}\mathbf{C} = {}^{\lambda}\mathbf{C}\varepsilon \qquad (8)$$

and

$$^{\lambda}F_{\mu\nu} = {}^{\lambda}h_{\mu\nu} + \sum_{\lambda\sigma} {}^{\lambda}P_{\lambda\sigma} \left[\left({}^{\lambda}\mu {}^{\lambda}\nu | {}^{\lambda}\lambda {}^{\lambda}\sigma \right) - \frac{1}{2} \left({}^{\lambda}\mu {}^{\lambda}\sigma | {}^{\lambda}\lambda {}^{\lambda}\nu \right) \right], \qquad (9)$$

respectively. By comparing (8) with (4), we can find the following relations between ${}^{\lambda}F$ and F, ${}^{\lambda}C$ and C, respectively,

$$^{\lambda}\mathbf{F} = \mathbf{S}^{-\frac{1}{2}} \, \mathbf{F} \, \mathbf{S}^{-\frac{1}{2}} \qquad (10)$$

and

$$^{\lambda}\mathbf{C} = \mathbf{S}^{\frac{1}{2}} \, \mathbf{C} . \qquad (11)$$

The NDDO Model Over the Orthogonal Basis Set

Under the LCAO-MO approximation, many TERIs need to be calculated. For a larger molecule, the evaluation of the TERIs is a severe bottleneck in the whole calculation. In semi-empirical MO methods, most of the TERIs are neglected and the remaining integrals are usually calculated approximately. The NDDO model is one of the important approximate models in neglecting the TERIs. The conventional NDDO model uses implicitly orthonormal atomic orbitals as basis set and applies the approximation:

$$\left({}^{\lambda}\mu_A {}^{\lambda}\nu_B | {}^{\lambda}\lambda_C {}^{\lambda}\sigma_D \right) = \left({}^{\lambda}\mu_A {}^{\lambda}\nu_A | {}^{\lambda}\lambda_C {}^{\lambda}\sigma_C \right) \delta_{AB}\delta_{CD} , \qquad (12)$$

where A, B, C, and D denote atom. Therefore, in the NDDO model, the integral $\left({}^{\lambda}\mu {}^{\lambda}\nu | {}^{\lambda}\lambda {}^{\lambda}\sigma \right)$ is calculated only if ${}^{\lambda}\mu$ and ${}^{\lambda}\nu$ are on the same atom A, and ${}^{\lambda}\lambda$ and ${}^{\lambda}\sigma$ are on the same atom C.

The total electronic energy in NDDO can be written as:

$$E_{el} = \sum_{\mu\nu} {}^{\lambda}P_{\mu\nu}\,{}^{\lambda}h_{\mu\nu}$$

$$+\frac{1}{2}\sum_{\mu_A\nu_A}\sum_{\lambda_B\sigma_B} {}^{\lambda}P_{\mu_A\nu_A}\,{}^{\lambda}P_{\lambda_B\sigma_B}\left({}^{\lambda}\mu_A{}^{\lambda}\nu_A\big|{}^{\lambda}\lambda_B{}^{\lambda}\sigma_B\right)$$

$$-\frac{1}{4}\sum_{\mu_A\sigma_A}\sum_{\lambda_B\nu_B} {}^{\lambda}P_{\mu_A\nu_B}\,{}^{\lambda}P_{\lambda_B\sigma_A}\left({}^{\lambda}\mu_A{}^{\lambda}\sigma_A\big|{}^{\lambda}\lambda_B{}^{\lambda}\nu_B\right) \tag{13}$$

The corresponding Fock matrix is:

$$^{\lambda}F_{\mu_A\nu_A} = {}^{\lambda}h_{\mu_A\nu_A} + \sum_{\lambda_B\sigma_B} {}^{\lambda}P_{\lambda_B\sigma_B}\left({}^{\lambda}\mu_A{}^{\lambda}\nu_A\big|{}^{\lambda}\lambda_B{}^{\lambda}\sigma_B\right)$$

$$-\frac{1}{2}\sum_{\lambda_A\sigma_A} {}^{\lambda}P_{\lambda_A\sigma_A}\left({}^{\lambda}\mu_A{}^{\lambda}\sigma_A\big|{}^{\lambda}\lambda_A{}^{\lambda}\nu_A\right) \tag{14}$$

and

$$^{\lambda}F_{\mu_A\nu_B} = {}^{\lambda}h_{\mu_A\nu_B} - \frac{1}{2}\sum_{\lambda_B\sigma_A} {}^{\lambda}P_{\lambda_B\sigma_A}\left({}^{\lambda}\mu_A{}^{\lambda}\sigma_A\big|{}^{\lambda}\lambda_B{}^{\lambda}\nu_B\right), \tag{15}$$

$$(A \neq B)$$

The Roby Model

Roby model[23, 24] can be considered as the practical implementation of the NDDO model over the orthonormal basis set. From the above NDDO model, we can see that many of the TERIs are neglected formally. However, for a polyatomic system, the remaining integrals, such as $\left({}^{\lambda}\mu_A{}^{\lambda}\nu_A\big|{}^{\lambda}\lambda_C{}^{\lambda}\sigma_C\right)$, are still linear combinations of one-, two-, three- and four-center TERIs. Therefore, if the integrals in the NDDO model were calculated rigorously, we would not obtain any gain from the model. Currently, in many implementations of the NDDO model, the following approximation is further made:

$$\left({}^{\lambda}\mu_A{}^{\lambda}\nu_A\big|{}^{\lambda}\lambda_C{}^{\lambda}\sigma_C\right) \approx \left(\mu_A\nu_A\big|\lambda_C\sigma_C\right) \tag{16}$$

That is, in the Roby model, the TERIs are calculated over the normal (non-orthonormal) atomic orbitals. The total electronic energy and Fock matrix in the model can thus be written, respectively, as:

$$E_{el} = \sum_{\mu\nu} {}^{\lambda}P_{\mu\nu}\,{}^{\lambda}h_{\mu\nu}$$

$$+\frac{1}{2}\sum_{\mu_A\nu_A}\sum_{\lambda_B\sigma_B} {}^{\lambda}P_{\mu_A\nu_A}\,{}^{\lambda}P_{\lambda_B\sigma_B}\left(\mu_A\nu_A\big|\lambda_B\sigma_B\right)$$

$$-\frac{1}{4}\sum_{\mu_A\sigma_A}\sum_{\lambda_B\nu_B} {}^{\lambda}P_{\mu_A\nu_B}\,{}^{\lambda}P_{\lambda_B\sigma_A}\left(\mu_A\sigma_A\big|\lambda_B\nu_B\right) \tag{17}$$

$$^{\lambda}F_{\mu_A\nu_A} = {}^{\lambda}h_{\mu_A\nu_A} + \sum_{\lambda_B\sigma_B} {}^{\lambda}P_{\lambda_B\sigma_B}\left(\mu_A\nu_A\big|\lambda_B\sigma_B\right) \tag{18}$$

$$-\frac{1}{2}\sum_{\lambda_A\sigma_A} {}^{\lambda}P_{\lambda_A\sigma_A}\left(\mu_A\sigma_A\big|\lambda_A\nu_A\right)$$

and

$$^{\lambda}F_{\mu_A v_B} = {}^{\lambda}h_{\mu_A v_B} - \frac{1}{2} \sum_{\lambda_B \sigma_A} {}^{\lambda}P_{\lambda_B \sigma_A} \left(\mu_A \sigma_A | \lambda_B v_B \right) \tag{19}$$

$$(A \neq B)$$

An Improvement Beyond the NDDO Model

In either the NDDO model or the Roby model[23, 24], the errors are caused by the approximate treatment of the electron-electron interactions. If we take the electron-electron interactions in the above two models as the zero:th order approximation to those in the *ab initio* MO theory, we can improve the above models by adding first order and higher order corrections. In the improvement made here, we add the first order correction to the electron-electron Coulomb interaction in the Roby model. The derivation of the improved model can be found in [25]. After the correction, the total electronic energy of the system can be expressed as:

$$E_{el} = E_{el}^{\text{Roby}} + \sum_{\mu v} \sum_{\lambda_C} {}^{\lambda}P_{\mu v} {}^{\lambda}P_{\lambda_C \lambda_C} \left({}^{\lambda}\mu^{\lambda}v | \phi_{Sc} \phi_{Sc} \right)$$
$$- \sum_{\mu_A v_A} \sum_{\lambda_C} {}^{\lambda}P_{\mu_A v_A} {}^{\lambda}P_{\lambda_C \lambda_C} \left(\mu_A v_A | \phi_{Sc} \phi_{Sc} \right), \tag{20}$$

where E_{el}^{Roby} is given by (17), ϕ_{Sc} is the S type atomic orbital on atom C. The corresponding Fock matrix we used are:

$$^{\lambda}F_{\mu_A v_A} = {}^{\lambda}F_{\mu_A v_A}^{\text{Roby}} + \sum_{\lambda_C} {}^{\lambda}P_{\lambda_C \lambda_C} \left({}^{\lambda}\mu_A {}^{\lambda}v_A | \phi_{Sc} \phi_{Sc} \right)$$
$$- \sum_{\lambda_C} {}^{\lambda}P_{\lambda_C \lambda_C} \left(\mu_A v_A | \phi_{Sc} \phi_{Sc} \right) \tag{21}$$

and

$$^{\lambda}F_{\mu_A v_B} = {}^{\lambda}F_{\mu_A v_B}^{\text{Roby}} + \sum_{\lambda_C} {}^{\lambda}P_{\lambda_C \lambda_C} \left({}^{\lambda}\mu_A {}^{\lambda}v_B | \phi_{Sc} \phi_{Sc} \right), \tag{22}$$

$$(A \neq B)$$

where F^{Roby} is given by (18) and (19).

2.2 Results and Discussion

In this section, we will discuss the performances of the above outlined three models; the NDDO model, the Roby model[23, 24], and our improvement (extended NDDO), based on the non-empirical calculations of some sample molecules. In the implementations of the models, the effective-core potential (ECP) of Stevens et al.[26] are used together with their minimal basis set. The non-empirical calculations can put the models to severe test, avoiding possible effects that empirical parameters may cause. For comparison, the corresponding *ab initio* frozen-core MO calculations with the same ECP and basis set are also carried out for the same set of molecules. In all the calculations, the molecules are fixed in the *ab initio* optimized geometries. The *ab initio* frozen-core calculations are carried out by using the Gaussian98 package[27].

Can the NDDO Model be Justified when the Orthonormal Valence Atomic Orbitals are Used?

In Table 1, we listed the results from the NDDO model and the corresponding *ab initio* frozen-core calculations. From the Table, we can see that the total molecular energies obtained from the NDDO model with the orthonormal valence atomic orbitals through the Löwdin orthogonalization are much higher than those from the corresponding *ab initio* frozen-core calculations. This indicates that even though the orthonormal valence atomic orbitals are used as basis set, the Diatomic Differential Overlap (DDO) type of integrals that are neglected in the NDDO model are not small enough to be omitted. This observation is totally in contradiction to that from the all-electron NDDO model[28]. The possible reasons for such difference is explained elsewhere[25].

Table 1. Total energies (in a.u.) calculated based on standard NDDO and the corresponding *ab initio* calculations

Molecule	NDDO	Ab initio	Error
H_2	-1.11756	-1.11751	-0.00005
HF	-24.13809	-24.38734	0.24925
C_2	-10.15466	-10.51693	0.36227
CO	-20.63927	-20.96118	0.32191
N_2	-18.72804	-19.14124	0.41320
F_2	-47.26035	-47.49321	0.23286
H_2O	-16.28820	-16.77084	0.48264
NH_3	-10.72334	-11.34169	0.61835
CH_4	-7.07612	-7.75203	0.67591
CH_3OH	-22.32937	-23.33448	1.00511
CH_3NH_2	-16.74129	-17.91242	1.17113

In all-electron calculations using the NDDO model, the radial part of the atomic orbitals from the same atom are orthogonal each other by using the Schmidt orthogonalization, and the orbitals from different atoms are orthogonal through the Löwdin orthogonalization[21, 22]. Therefore, some atomic orbitals have radial nodes. This makes the DDO type of integrals small enough so that they can be neglected. However, when only the valence atomic orbitals are used as basis set, the orbitals do not have a radial node. Therefore the DDO types of integrals are not small enough to be omitted. Obviously, neglect of such (DDO) type of integrals cause some significant errors in the molecular properties, especially in the total energy. The larger the molecule, the larger the number of neglected DDO type of integrals, and thus the larger are the errors. Therefore, we conclude that the NDDO model cannot be justified when the orthonormal valence atomic orbitals are used.

How Well Does the Roby Model Perform?

Table 2 lists the total energies for the same set of sample molecules calculated from the Roby model[23, 24]. From the Table we can find that the total molecular energies from the Roby model are closer to those from the *ab initio* frozen-core calculations. In the Roby model, the remaining TERIs in the NDDO model are in fact calculated with the normal (non-orthogonal) atomic orbitals, which partly compensates the errors in the NDDO model caused by the neglect of the DDO type integrals. This probably could be used to justify why many approximate MO methods use non-orthogonal valence atomic orbitals in the calculation of the remaining TERIs in their implementations[16, 17, 29]. However, it seems that this kind of compensation is fairly arbitrary and thus the errors in the total energies from the Roby model remain non-systematic and unpredictable. This can also be seen from Table 2.

Table 2. Total energies (in a.u.) calculated based on Roby model and the corresponding *ab initio* calculations

Molecule	Roby model	Ab initio	Error
H_2	-1.12002	-1.11751	-0.00251
HF	-24.58560	-24.38734	-0.19826
C_2	-10.41269	-10.51693	0.10424
CO	-21.14675	-20.96118	-0.18557
N_2	-19.12310	-19.14124	0.01814
F_2	-47.57380	-47.49321	-0.08059
H_2O	-16.95644	-16.77084	-0.18560
NH_3	-11.45568	-11.34169	-0.11399
CH_4	-7.74359	-7.75203	0.00844
CH_3OH	-23.80864	-23.33448	-0.47416
CH_3NH_2	-18.31354	-17.91242	-0.40112

For most of the molecules, the total energies calculated from the Roby model are lower than those from the corresponding *ab initio* frozen-core calculations. However, for some molecules, such as C_2, N_2, and CH_4, the total energies from the Roby model are higher than the corresponding *ab initio* results. We believe that this problem that the errors are non-systematic is caused by the model itself and is therefore difficult to solve by any semi-empirical parameterization scheme.

What is the Quality of the Improvement Beyond the NDDO Model?

In Table 3, we list the total energies for the sample molecules calculated from our improved model. In our improvement, we added a first order correction term to

Table 3. Total energies (in a.u.) calculated based on our improvement of the NDDO model and the corresponding *ab initio* calculations

Molecule	This work	*Ab initio*	Error
H_2	-1.11737	-1.11751	0.00014
HF	-24.33254	-24.38734	0.05480
C_2	-10.48385	-10.51693	0.03308
CO	-20.82892	-20.96118	0.13226
N_2	-19.03930	-19.14124	0.10194
F_2	-47.41386	-47.49321	0.07935
H_2O	-16.67632	-16.77084	0.09452
NH_3	-11.23293	-11.34169	0.10876
CH_4	-7.67744	-7.75203	0.07459
CH_3OH	-23.13057	-23.33448	0.20391
CH_3NH_2	-17.68410	-17.91242	0.22832

the electron-electron Coulomb interactions. Therefore, it is expected that the results from the improved model would be better than those from the NDDO model (see Table 1) and the Roby model (see Table 2). From Table 3 we can see that this is indeed the case. Compared with the NDDO model and the Roby model, the improved model gives total molecular energies closer to those from the corresponding *ab initio* frozen-core calculations. In addition, we can also find from Table 3 that by adding the first order correction to the electron-electron interactions, the balance between the core-electron and electron-electron interactions is also improved, leading to the errors in total molecular energies more systematic. It is possible to introduce a first order correction term also in the electron-electron exchange interaction. However, we believe that the current improvement is good enough for using it in the semi-empirical schemes. Such implementations, due to the introduction of the first order correction in the electron-electron interactions, give more balanced treatment of the core-electron and electron-electron interactions and therefore is expected to overcome the non-systematic characteristics in the errors currently residing in many semi-empirical MO methods.

3 An Efficient *ab initio* Tight-Binding Method (PART II)

Tight-binding (TB) methods[30] are widely used theoretical approaches in materials science. Empirical TB methods are generally accepted as originated from the important work by Slater and Koster[31] and thus have almost as long history as conventional semi-empirical MO methods. As far as the methods themselves are concerned, many empirical TB approaches appear even simpler than the conventional semi-empirical MO methods because they lack the explicit treatment of electron-electron interactions. Indeed, many empirical TB methods[32, 33] are very similar to the semi-empirical extended Hückel MO methods[8]. Empirical TB methods were previously used mainly in the studies of properties of materials, such as band structures etc. In recent years, empirical TB methods have been extended to the studies of the electronic structures of molecules and their reactions[32]. Empirical TB methods are usually very simple. After careful parameterization, they can give remarkable accuracy in the calculation results. However, in order to obtain high accuracy, different parameters are usually required even for the same element in different environment. This makes the transferability of the parameters rather limited and thus limits the general applicability of empirical TB methods.

Obviously, to keep the simplicity of empirical TB methods while developing non-empirical schemes for general use in the studies of molecular properties is more valuable. Foulkes et.al.[34] have pointed out the relationship between DFT and TB methods. Consequently it is possible to develop non-empirical TB methods based on DFT. The TB methods thus developed are theoretically more robust and can be used where conventional DFT calculations are adequate. Sankey et.al.[35] first proposed an *ab initio* TB scheme based on the Harris-Foulkes functional[34, 36]. The work is further developed by Lin and Harris[37], Horsfield[38], and Sankey et al.[39]. A simplified non-empirical TB method is also proposed by Elstner et.al.[40]. The results from several calculations show that the developed *ab initio* TB methods are reliable and reasonably accurate in studies of molecular properties. Due to their fast calculation speed, non-empirical TB methods have a great potential in studies of large molecular systems as well as in computer simulations.

In this second part, we will first give an overview of the theoretical basis of the non-empirical TB methods and thereafter present a general and efficient *ab initio* TB method developed in our group.

3.1 Theoretical Background

DFT and Harris-Foulkes Functional

As mentioned above, DFT[5, 6, 41] is the theoretical basis for most non-empirical TB methods. In modern quantum chemistry, the widely used DFT formulation was proposed by Kohn and Sham[6]. In the Kohn-Sham formulation, the total electron density of a system is assumed as (in atomic units):

$$\rho\left(\mathbf{r}\right) = \sum_{i}^{N_{occ}} n_i \left|\psi_i\left(\mathbf{r}\right)\right|^2 , \tag{23}$$

where ψ_i is the i:th occupied one-electron orbital (known as the Kohn-Sham orbital) and n_i the occupation number of the orbital. The total energy functional of an electronic system is given by:

$$E_{KS}\left[\rho\right] = \sum_{i}^{N_{occ}} n_i \left\langle \psi_i \left| -\frac{1}{2}\nabla^2 \right| \psi_i \right\rangle + \int V_{ext}\left(\mathbf{r}\right)\rho\left(\mathbf{r}\right) d\mathbf{r}$$
$$+ \frac{1}{2} \int \int \frac{\rho\left(\mathbf{r}_1\right)\rho\left(\mathbf{r}_2\right)}{r_{12}} d\mathbf{r}_1 d\mathbf{r}_2 + E_{xc}\left[\rho\right] + V_{I-I} , \tag{24}$$

where V_{ext} is the potential due to the external field. E_{xc} is the exchange-correlation energy functional. V_{I-I} is the ion-ion interaction term. By minimizing the total energy functional, we obtain the Kohn-Sham equations as:

$$\left[-\frac{1}{2}\nabla_1^2 + V_{ext}(\mathbf{r}_1) + \int \frac{\rho(\mathbf{r}_2)}{r_{12}}d\mathbf{r}_2 + V_{xc}(\mathbf{r}_1)\right]|\psi_i(\mathbf{r}_1)\rangle = \varepsilon_i|\psi_i(\mathbf{r}_1)\rangle , \quad (25)$$

with

$$V_{xc}(\mathbf{r}) = \frac{\delta E_{xc}[\rho(\mathbf{r})]}{\delta \rho(\mathbf{r})}, \quad (26)$$

where V_{xc} is the so-called exchange-correlation potential.

The Kohn-Sham equations are similar to the Hartree-Fock equations. Therefore, the way to solve the Kohn-Sham equations is similar to that of the Hartree-Fock equations. However, the integrals related to the exchange-correlation potential and energy functional are nonlinear with respect to $\rho(\mathbf{r})$, so they have to be solved by numerical integration schemes. Since DFT involves contributions from the electron correlation interaction, it can give more accurate results than Hartree-Fock theory. Compared to the other quantum chemistry methods involving the electron correlation, such as MP2, and CI etc, DFT calculations are simpler and faster. Currently, DFT calculations are among the most popular techniques used in studies of the material and molecular properties.

If we expand the electron-electron interactions in the total energy functional in DFT (24) with respect to a reference density $\tilde{\rho}$, keeping only the first order correction and neglecting the second order and higher order corrections, we can obtain the so-called Harris-Foulkes (HF) functional as:

$$\begin{aligned}
E_{HF}[\rho] = {}& \sum_i^{N_{occ}} n_i \langle \psi_i | -\tfrac{1}{2}\nabla^2 | \psi_i \rangle + \int V_{ext}(\mathbf{r})\rho(\mathbf{r})\,d\mathbf{r} \\
& + \int\int \frac{\rho(\mathbf{r}_1)\tilde{\rho}(\mathbf{r}_2)}{r_{12}}d\mathbf{r}_1 d\mathbf{r}_2 - \frac{1}{2}\int\int \frac{\tilde{\rho}(\mathbf{r}_1)\tilde{\rho}(\mathbf{r}_2)}{r_{12}}d\mathbf{r}_1 d\mathbf{r}_2 \\
& + \int \rho(\mathbf{r})V_{xc}(\tilde{\rho}(\mathbf{r}))d\mathbf{r} + E_{xc}[\tilde{\rho}] - \int \tilde{\rho}(\mathbf{r})V_{xc}(\tilde{\rho}(\mathbf{r}))d\mathbf{r} + V_{I-I}.
\end{aligned} \quad (27)$$

By variationally minimizing E_{HF}, we obtain the corresponding one-electron orbital equation:

$$\left[-\frac{1}{2}\nabla_1^2 + V_{ext}(\mathbf{r}_1) + \int \frac{\tilde{\rho}(\mathbf{r}_2)}{r_{12}}d\mathbf{r}_2 + V_{xc}(\tilde{\rho}(\mathbf{r}_1))\right]|\psi_i(\mathbf{r}_1)\rangle = \varepsilon_i|\psi_i(\mathbf{r}_1)\rangle . \quad (28)$$

Therefore, the Harris-Foulkes functional can also be written as:

$$\begin{aligned}
E_{HF}[\rho] = {}& \sum_i^{N_{occ}} \varepsilon_i - \frac{1}{2}\int\int \frac{\tilde{\rho}(\mathbf{r}_1)\tilde{\rho}(\mathbf{r}_2)}{r_{12}}d\mathbf{r}_1 d\mathbf{r}_2 \\
& + E_{xc}[\tilde{\rho}] - \int \tilde{\rho}(\mathbf{r})V_{xc}(\tilde{\rho}(\mathbf{r}))d\mathbf{r} + V_{I-I} .
\end{aligned} \quad (29)$$

Harris-Foulkes functional is stable with respect to the reference density $\tilde{\rho}$. That is, for a small change of $\tilde{\rho}$, the error in the total energy is only in the second order. From (28) and (29) we can see that the Harris-Foulkes functional is determined by

$\tilde{\rho}$. For a given $\tilde{\rho}$, we can find $\{\psi_i\}$ from (28) and thus calculate the total electron density and the energy of the system. Because the major part in the second order correction is the Coulomb interaction between the electrons and is always positive-definite, the total energy from the Harris-Foulkes functional is below that from the corresponding Kohn-Sham functional. Clearly, the choice of $\tilde{\rho}$ is essential for the calculation accuracy and speed. We need to find a $\tilde{\rho}$ that can make the neglected second order correction as small as possible and the calculation as simple as possible. We will in the following section discuss ways to choose $\tilde{\rho}$.

3.2 *Ab initio* TB Methods

Ab initio TB methods start from the Harris-Foulkes (HF) functional. However, rather than solving the HF functional rigorously, *ab initio* TB methods introduce a few simplifications. This is because solving the Harris-Foulkes functional rigorously would not give too much gain compared to solving the Kohn-Sham functional simply because the exchange correlation part in the orbital equations and energy functional would need to be calculated accurately by numerical integration schemes. On the other hand the *ab initio* TB methods avoid introducing any empirical parameters to simplify the Harris-Foulkes functional, making the *ab initio* TB calculations reliable and applicable for more general cases.

Although many *ab initio* TB implementations are based on the original Harris-Foulkes functional, we make a further simplification of the functional in order to make the calculations simpler and faster. In our work, the simplified Harris-Foulkes functional is expressed as:

$$
\begin{aligned}
\tilde{E}\left[\rho\right] = &\sum_i^{N_{occ}} n_i \left\langle \psi_i \left| -\tfrac{1}{2}\nabla^2 \right| \psi_i \right\rangle + \int V_{ext}\left(\mathbf{r}\right)\rho\left(\mathbf{r}\right) d\mathbf{r} \\
&+ \int\int \frac{\rho(\mathbf{r}_1)\tilde{\rho}(\mathbf{r}_2)}{r_{12}} d\mathbf{r}_1 d\mathbf{r}_2 - \tfrac{1}{2}\int\int \frac{\tilde{\rho}(\mathbf{r}_1)\tilde{\rho}(\mathbf{r}_2)}{r_{12}} d\mathbf{r}_1 d\mathbf{r}_2 \\
&+ \int \rho\left(\mathbf{r}\right) V_{xc}\left(\rho^{(0)}\left(\mathbf{r}\right)\right) d\mathbf{r} + E_{xc}\left[\rho^{(0)}\right] \\
&- \int \rho^{(0)}\left(\mathbf{r}\right) V_{xc}\left(\rho^{(0)}\left(\mathbf{r}\right)\right) d\mathbf{r} + V_{I-I} .
\end{aligned}
\tag{30}
$$

The corresponding one-electron orbital equation is given as:

$$
\left[-\tfrac{1}{2}\nabla_1^2 + V_{ext}\left(\mathbf{r}_1\right) + \int \frac{\tilde{\rho}\left(\mathbf{r}_2\right)}{r_{12}} d\mathbf{r}_2 + V_{xc}\left(\rho^{(0)}(\mathbf{r}_1)\right) \right] \left| \psi_i\left(\mathbf{r}_1\right) \right\rangle = \varepsilon_i \left| \psi_i\left(\mathbf{r}_1\right) \right\rangle .
\tag{31}
$$

Compared with the original Harris-Foulkes functional, our simplification is in the exchange-correlation part. (30) is equivalent to expanding the Coulomb and exchange-correlation interactions between the electrons in the Kohn-Sham energy functional with respect to $\tilde{\rho}$ and $\rho^{(0)}$, respectively, and neglecting all the second order and higher order corrections. Later, we can find that due to the simplification made here, the number of time-consuming integrals related to the exchange-correlation part decreases greatly. It can be proven that the error in the total energy caused by such simplification is also only in the second order and would have only a minor effect on the calculation results.

In many *ab initio* TB calculations, the frozen-core approximation is usually used and the effects of the nucleus and core electrons of an atom are modelled by a pseudopotential. This is also the case in our implementation. Therefore,

$$V_{ext} = \sum_I V_I^{(PP)} , \qquad (32)$$

where $V_I^{(PP)}$ is the pseudo-potential for atom I. In our work, the norm-conserving separable dual-space pseudo-potentials devised by Goedecker, Teter, and Hutter[42] are used. After the frozen-core approximation, the electron density of a molecular system depends on the behaviour of the valence electrons only.

A natural question here is how to choose the reference density in Harris-Foulkes functional? Harris used originally a superposition of spherically distributed electron densities of a neutral atom as reference density [36]. That is

$$\tilde{\rho}(\mathbf{r}) \approx \rho^{(0)}(\mathbf{r}) = \sum_I \rho_I^{(0)}(\mathbf{r}) , \qquad (33)$$

where $\rho_I^{(0)}(\mathbf{r})$ is the spherically distributed electron density of neutral atom I at point $\mathbf{r}$. By using the above reference density, it is not necessary to make self-consistent-field (SCF) iterations in solving the Harris-Foulkes equations. The potential curves of N_2 and He_2 thus calculated show that such a scheme can be used satisfactorily[36]. Remarkable good results can be obtained even for the ionic compounds like NaCl[43]. For molecular systems, Averill and Painter[44] suggested a better reference density that is the superposition of the spherically distributed atomic orbital densities of the occupied shells,

$$\tilde{\rho}(\mathbf{r}) \approx \sum_I \tilde{\rho}_I(\mathbf{r}) = \sum_I \sum_{l_I} n_{l_I} f_{l_I}(r_I) , \qquad (34)$$

where f_{l_I} is the spherically averaged single electron density of a shell with the angular quantum number l_I. Further $r_I = |\mathbf{r} - \mathbf{R_I}|$ is the distance between $\mathbf{r}$ and atomic site $\mathbf{R_I}$. The coefficient n_{l_I} is obtained by minimizing the second order error in the total energy, which is equivalent to:

$$\frac{\partial E_{HF}}{\partial n_{l_I}} = 0 . \qquad (35)$$

When the reference density expressed in (34) is chosen, the Harris-Foulkes equations are solved by the SCF iteration method. Lin and Harris[37] used a similar reference density in their implementation.

In this work, we choose the two reference densities as:

$$\tilde{\rho}(\mathbf{r}) = \sum_I \tilde{\rho}_I(r_I)$$
$$\rho^{(0)}(\mathbf{r}) = \sum_I \rho_I^{(0)}(r_I) \qquad (36)$$

with

$$\tilde{\rho}_I(r_I) = \rho_I^{(0)}(r_I) + \Delta n_I f_I(r_I) , \tag{37}$$

where $\rho_I^{(0)}$ is the valence electron density of neutral atom I, and $f_I(r_I)$ corresponds to the density of a single electron in the highest occupied orbital. Both $\rho_I^{(0)}$ and $f_I(r_I)$ are spherically averaged. Δn_I can be considered as the net number of electrons that atom I obtains in a molecular system and is determined by:

$$\frac{\partial \tilde{E}}{\partial \Delta n_I} = 0 . \tag{38}$$

This is equivalent to minimizing the second order error in the electron-electron Coulomb interaction. The choice of $\tilde{\rho}_I(r_I)$ here is similar to that by Horsfield[38]. In our work, $\rho_I^{(0)}$ and $f_I(r_I)$ are further expanded as linear combinations of $1S$ type gaussians, that is:

$$\rho_I^{(0)}(r_I) = \sum_j c_{jI} e^{-\alpha_{jI} r_I^2} \tag{39}$$

and

$$f_I(r_I) = \sum_j d_{jI} e^{-\alpha_{jI} r_I^2} , \tag{40}$$

where c_{jI} and d_{jI} are combination coefficients. α_{jI}, c_{jI} and d_{jI} are obtained by fitting to the corresponding spherically distributed densities.

In the practical implementations of the TB methods, the LCAO-MO approximation is usually used. When the frozen-core approximation is used, the atomic orbitals used in the LCAO-MO are only valence atomic orbitals. These atomic orbitals can be obtained by the DFT calculations on the corresponding atoms. Normally, a confining potential is also applied in solving the DFT equation of an atom. The orbitals thus obtained are of finite range, that is, the "tail" in an atomic orbital approaches zero faster than that in the normal case. There are several advantages in using the finite-range atomic orbitals in *ab initio* TB methods. The first one is that such orbitals can reflect the fact that atomic orbitals could "contract" when going from atomic to molecular environment. The second advantage is that for a large molecular system, the number of integrals to be computed decreases greatly. More importantly, the use of finite-range atomic orbitals makes the resulting matrix sparser thus the so-called linear scaling methods can be used more efficiently. The idea of using finite-range atomic orbitals in *ab initio* TB method is first proposed by Sankey et.al[35]. In their implementation, the finite-range atomic orbital is obtained by placing the atom in the center of a spherical potential well and solving the corresponding DFT equation for the atom. Later, different ways to obtain the finite-range atomic orbitals have been proposed[38, 45–48]. The widely used approach is to add a confining potential of the form $\left(\frac{r}{r_0}\right)^n$ (where r_0 is a parameter and n could be chosen as 2 in [45], 4 in [47] or 6 in [38]). For a better understanding of the confining potentials and finite-range atomic orbitals, the reader is referred to the excellent paper of Junquera et.al.[48]. Usually, such atomic orbitals are expressed in table form. In our implementation, the atomic orbitals are expressed as linear combination of gaussians, that is:

$$\phi_\mu = \sum_k d_{k\mu} \chi_k \,, \tag{41}$$

where $\{\chi_k\}$ are normalized primitive gaussians and $\{d_{k\mu}\}$ the combination coefficients. In this work ϕ_μ is the finite-range atomic orbital originally from CP2K and slightly optimized according to the dipole moments of some sample molecules.

Under the LCAO-MO approximation, the molecular orbital is given as:

$$\psi_i = \sum_\mu C_{\mu i} \phi_\mu. \tag{42}$$

If we insert (36) and (37) into (30), we obtain the total energy as:

$$
\begin{aligned}
\tilde{E}\left[\rho\right] = {} & \sum_i^{N_{OCC}} n_i \varepsilon_i \\
& -\frac{1}{2}\sum_{I,J}\int\int \frac{\left(\rho_I^{(0)}(\mathbf{r}_1)+\Delta n_I f_I(\mathbf{r}_1)\right)\cdot\left(\rho_J^{(0)}(\mathbf{r}_2)+\Delta n_J f_J(\mathbf{r}_2)\right)}{r_{12}} d\mathbf{r}_1 d\mathbf{r}_2 \\
& +E_{XC}\left[\rho^{(0)}\right] - \int V_{XC}\left(\rho^{(0)}(\mathbf{r})\right)\rho^{(0)}(\mathbf{r})d\mathbf{r} + V_{I-I}
\end{aligned}
\tag{43}
$$

where

$$
\begin{aligned}
E_{XC}\left[\rho^{(0)}\right] &= \int \rho^{(0)}(\mathbf{r})\,\varepsilon_{XC}\left(\rho^{(0)}(\mathbf{r})\right)d\mathbf{r} \\
&= \sum_I \int \rho_I^{(0)}(\mathbf{r})\,\varepsilon_{XC}\left(\rho^{(0)}(\mathbf{r})\right)d\mathbf{r}
\end{aligned}
\tag{44}
$$

and

$$\int V_{XC}\left(\rho^{(0)}(\mathbf{r})\right)\rho^{(0)}(\mathbf{r})d\mathbf{r} = \sum_I \int \rho_I^{(0)}(\mathbf{r})V_{XC}\left(\rho^{(0)}(\mathbf{r})\right)d\mathbf{r} \tag{45}$$

where ϵ_{XC} is known as the exchange-correlation energy density. The orbital energy ϵ_i can be obtained by solving the following equation self-consistently:

$$\mathbf{FC} = \mathbf{SC}\varepsilon \,, \tag{46}$$

where

$$
\begin{aligned}
F_{\mu v} = {} & \langle\phi_\mu| -\tfrac{1}{2}\nabla^2 |\phi_v\rangle + \langle\phi_\mu| V_{ext} |\phi_v\rangle \\
& + \langle\phi_\mu(\mathbf{r}_1)| \int \frac{\tilde{\rho}(\mathbf{r}_2)}{r_{12}} d\mathbf{r}_2 |\phi_v(\mathbf{r}_1)\rangle + \langle\phi_\mu(\mathbf{r})| V_{XC}\left(\rho^{(0)}(\mathbf{r})\right) |\phi_v(\mathbf{r})\rangle
\end{aligned}
$$

$$
\begin{aligned}
= {} & \langle\phi_\mu| -\tfrac{1}{2}\nabla^2 |\phi_v\rangle + \sum_I \left(\langle\phi_\mu| V_I^{(PP)} |\phi_v\rangle + \langle\phi_\mu(\mathbf{r}_1)| \int \frac{\rho_I^{(0)}(\mathbf{r}_2)}{r_{12}} d\mathbf{r}_2 |\phi_v(\mathbf{r}_1)\rangle \right) \\
& + \sum_I \Delta n_I \langle\phi_\mu(\mathbf{r}_1)| \int \frac{f_I(\mathbf{r}_2)}{r_{12}} d\mathbf{r}_2 |\phi_v(\mathbf{r}_1)\rangle + \langle\phi_\mu(\mathbf{r})| V_{XC}\left(\rho^{(0)}(\mathbf{r})\right) |\phi_v(\mathbf{r})\rangle
\end{aligned}
\tag{47}
$$

Because $V_I^{(PP)}$, $\rho_I^{(0)}$ and $f_I(\mathbf{r})$ are expressed in analytical forms, it is clear that when the atomic orbitals are expressed as linear combinations of primitive gaussians, all the integrals except for those related to the exchange-correlation potential and energy functional can be calculated analytically. The integrals related to the exchange-correlation part could be calculated by a numerical integration scheme. However, because the numerical integration is very time consuming, approximate calculation schemes are always employed in the *ab initio* TB methods to avoid direct numerical integration in each calculation. In the scheme proposed by Sankey et.al.[35], the exchange-correlation potential and energy density are expanded with respect to a constant reference density. For example, $\varepsilon_{XC}(\rho)$ is expanded as:

$$\varepsilon_{XC}(\rho) \approx \varepsilon_{XC}(\bar{\rho}) + ... \tag{48}$$

The constant density $\bar{\rho}$ is obtained by minimizing the corresponding second order matrix neglected. In the implementation of Lin and Harris[37], the exchange-correlation energy density or potential are expanded with respect to the density. For example:

$$\varepsilon_{XC}(\rho) = a_0 + a_1\rho + a_2\rho^2 + ... \tag{49}$$

Obviously, such expansion schemes make it easier to calculate the related integrals. For the Local Density Approximation (LDA), the accuracies of the expansions can in principle increase with the inclusion of more terms in the expansions. However, the above expansions are clearly not so robust for the GGA (General Gradient Approximation) exchange-correlation energy functional. Currently, in some improved *ab initio* TB methods the many-center expansion schemes of the integrals are employed[38, 39]. For example,

$$\left\langle \phi_{I\alpha} \left| V_{xc}\left(\rho^{(0)}\right) \right| \phi_{I\beta} \right\rangle \approx \left\langle \phi_{I\alpha} \left| V_{xc}\left(\rho_I^{(0)}\right) \right| \phi_{I\beta} \right\rangle$$
$$+ \sum_{J(\neq I)} \left\langle \phi_{I\alpha} \left| V_{xc}\left(\rho_I^{(0)} + \rho_J^{(0)}\right) - V_{xc}\left(\rho_I^{(0)}\right) \right| \phi_{I\beta} \right\rangle \tag{50}$$

and

$$\left\langle \phi_{I\alpha} \left| V_{xc}\left(\rho^{(0)}\right) \right| \phi_{J\beta} \right\rangle \approx \left\langle \phi_{I\alpha} \left| V_{xc}\left(\rho_I^{(0)} + \rho_J^{(0)}\right) \right| \phi_{J\beta} \right\rangle$$
$$+ \sum_{\substack{K(\neq I,J)}} \left\langle \phi_{I\alpha} \left| V_{xc}\left(\rho_I^{(0)} + \rho_J^{(0)} + \rho_K^{(0)}\right) - V_{xc}\left(\rho_I^{(0)} + \rho_J^{(0)}\right) \right| \phi_{J\beta} \right\rangle \tag{51}$$
$$(I \neq J)$$

where α and β denote atomic orbitals and I, J, and K atoms, respectively.

Usually, the many-center expansion has a higher accuracy and is valid not only for the LDA functional but also for the GGA functional. Horsfield[38] has pointed out that the two-term approximation to the on-site integral (50) is not accurate enough and occasionally may cause some ghost states. To improve the accuracy, Horsfield add a correction that is calculated by a numerical integral scheme with much less number of grids. In the improved *ab initio* TB method of Sankey et.al.[39], the on-site integral is only expanded to the second term too and no numerical integral scheme is incorporated. However, the second term in the expansion is scaled down to reduce the errors caused by the fact that the on-site integral is only expanded as linear combinations of one- and two-center integrals.

Indeed, we find that for the on-site integral $\left\langle \phi_{I\alpha} \left| V_{xc}\left(\rho^{(0)}\right) \right| \phi_{I\beta} \right\rangle$ that the expansion (50) can sometimes cause significant errors. In order to save the calculation time and improve the integral accuracy, we expand the on-site integral to higher orders and no numerical integral scheme is incorporated. For example, in our work the integrals related to the exchange-correlation potential are approximated as

$$
\begin{aligned}
\left\langle \phi_{I\alpha} \left| V_{xc}\left(\rho^{(0)}\right) \right| \phi_{I\beta} \right\rangle &\approx \left\langle \phi_{I\alpha} \left| V_{xc}\left(\rho_I^{(0)}\right) \right| \phi_{I\beta} \right\rangle \\
&+ \sum_{J(\neq I)} \left\langle \phi_{I\alpha} \left| V_{xc}\left(\rho_I^{(0)} + \rho_J^{(0)}\right) - V_{xc}\left(\rho_I^{(0)}\right) \right| \phi_{I\beta} \right\rangle \\
&+ \frac{1}{2} \sum_{\substack{J(\neq I) \\ K(\neq I, J)}} \left\langle \phi_{I\alpha} \left| \begin{array}{c} V_{xc}\left(\rho_I^{(0)} + \rho_J^{(0)} + \rho_K^{(0)}\right) + V_{xc}\left(\rho_I^{(0)}\right) \\ - V_{xc}\left(\rho_I^{(0)} + \rho_J^{(0)}\right) - V_{xc}\left(\rho_I^{(0)} + \rho_K^{(0)}\right) \end{array} \right| \phi_{I\beta} \right\rangle,
\end{aligned}
\tag{52}
$$

and

$$
\begin{aligned}
\left\langle \phi_{I\alpha} \left| V_{xc}\left(\rho^{(0)}\right) \right| \phi_{J\beta} \right\rangle &\approx \left\langle \phi_{I\alpha} \left| V_{xc}\left(\rho_I^{(0)} + \rho_J^{(0)}\right) \right| \phi_{J\beta} \right\rangle \\
&+ \sum_{K(\neq I, J)} \left\langle \phi_{I\alpha} \left| V_{xc}\left(\rho_I^{(0)} + \rho_J^{(0)} + \rho_K^{(0)}\right) - V_{xc}\left(\rho_I^{(0)} + \rho_J^{(0)}\right) \right| \phi_{J\beta} \right\rangle
\end{aligned}
\tag{53}
$$

Actual calculations show that the above equations work well for many systems. The absolute error caused by (52) is usually less than 0.001 a.u.. Obviously, the form of the approximation to the on-site integrals can be extended to integrals related to the exchange-correlation functional. Therefore, we have:

$$
\begin{aligned}
\int \rho_I^{(0)}(\mathbf{r}) \, \varepsilon_{XC}\left(\rho^{(0)}(\mathbf{r})\right) d\mathbf{r} &= \int \rho_I^{(0)}(\mathbf{r}) \, \varepsilon_{XC}\left(\rho_I^{(0)}(\mathbf{r})\right) d\mathbf{r} \\
&+ \sum_{J(\neq I)} \int \rho_I^{(0)}(\mathbf{r}) \left[\varepsilon_{XC}\left(\rho_I^{(0)}(\mathbf{r}) + \rho_J^{(0)}(\mathbf{r})\right) - \varepsilon_{XC}\left(\rho_I^{(0)}(\mathbf{r})\right) \right] d\mathbf{r} \\
&+ \frac{1}{2} \sum_{\substack{J(\neq I) \\ K(\neq J, I)}} \int \rho_I^{(0)}(\mathbf{r}) \left[\begin{array}{c} \varepsilon_{XC}\left(\rho_I^{(0)}(\mathbf{r}) + \rho_J^{(0)}(\mathbf{r}) + \rho_K^{(0)}(\mathbf{r})\right) + \varepsilon_{XC}\left(\rho_I^{(0)}(\mathbf{r})\right) \\ - \varepsilon_{XC}\left(\rho_I^{(0)}(\mathbf{r}) + \rho_J^{(0)}(\mathbf{r})\right) - \varepsilon_{XC}\left(\rho_I^{(0)}(\mathbf{r}) + \rho_K^{(0)}(\mathbf{r})\right) \end{array} \right] d\mathbf{r}
\end{aligned}
\tag{54}
$$

There are many one-, two-, and three-center integrals in the many-center expansion schemes. Usually, these integrals are obtained first by finding the integrals in local coordinate systems through the interpolation of the look-up tables and then by rotating the integrals from the local coordinate systems to those in the molecular coordinate system.

The extensive application of the look-up tables is one of the important features in *ab initio* TB methods. By applying the look-up table technique, we can greatly save the time in the evaluation of integrals. Except in those for the one-center integrals, in look-up tables are integrals calculated on a pre-defined mesh of inter-atomic distances in the local coordinate systems. The efficiency of the methods depends on the number of integrals in the look-up tables and the way the integrals interpolated. In order to reduce the number of integrals in the look-up tables and save the computing time, integrals of similar summation type and same rotational property calculated from the same pre-defined inter-atomic distance can be combined and stored as one set of tables. For example, the expansion of $\langle \phi_\mu (\mathbf{r}_1)| \int \frac{\rho^{(0)}(\mathbf{r}_2)}{r_{12}} d\mathbf{r}_2 |\phi_\upsilon (\mathbf{r}_1)\rangle$ involves two-, and three-center integrals in exactly the same way as that of $\langle \phi_\mu (\mathbf{r})| V_{XC} (\rho^{(0)} (\mathbf{r})) |\phi_\upsilon (\mathbf{r})\rangle$, and the corresponding integrals in the two expansions have the same rotational properties so that they can be combined to produce one set of tables. Normally, the look-up tables are of three types, corresponding to one-, two-, and three-center integrals. Because the number of three-center integrals is much greater than those of one- and two-center integrals, it is very important in choosing a mesh of inter-atomic distances for three-center integrals. Sankey[35] and Horsfield[38] devised their own meshes for the three-center integrals and the Chebychev polynomials are used for the interpolation. We also devised a simple logarithmic mesh. The geometry used in the tables is shown in Fig. 1. The point (r_i, s_j, t_k) in the mesh is determined by the following equations:

$$r_i = r_{\max} - (r_{\max} - r_{\min}) \frac{\ln (N_r + 1 - i)}{\ln (N_r)} \tag{55}$$

$$s_j = s_{\max} - (s_{\max} - s_{\min}) \frac{\ln (N_s + 1 - j)}{\ln (N_s)} \tag{56}$$

and

$$t_k = t_{\max} - (t_{\max} - t_{\min}) \frac{\ln (N_t + 1 - k)}{\ln (N_t)} \tag{57}$$

Usually, when $r_i = r_{\max}$, or $s_j = s_{\max}$, or $t_k = t_{\max}$, most of the integrals approach to some values that can be neglected. The exceptions are those of the form $\langle \phi_\mu (\mathbf{r}_1)| \int \frac{f_J(\mathbf{r}_2)}{r_{12}} d\mathbf{r}_2 |\phi_\upsilon (\mathbf{r}_1)\rangle$. For these integrals, a similar treatment to that outlined in Ref([49]) is used.

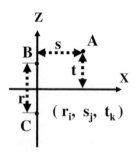

Figure 1. The geometry used in the look-up tables in this work

In practical calculations, the corresponding integrals in the local coordinate systems are first obtained by polynomial interpolation of the relevant ones read from the look-up tables and if necessary, the asymptotic values are added to some integrals. These integrals are then transformed to the molecular coordinate system to form the required molecular integrals.

3.3 Some Preliminary Results from the *ab initio* TB Calculations

In this section, we present some preliminary results from our work. In order to evaluate the performance of the *ab initio* TB method thus developed, we also made comparison the results with those from the corresponding frozen-core DFT calculations and experiment. The results discussed cover the equilibrium geometries, dipole moments, and the energies for hydrogenation reactions for some molecules containing C, H, N and O. In our *ab initio* TB calculations, we used BLYP exchange-correlation functional that consists of Becke's GGA exchange functional[50] and the GGA correlation functional of Lee, Yang, and Parr[51]. In the frozen-core DFT calculations, the same (BLYP) exchange-correlation functional and basis set are used with the effects of the nuclei and core electrons modeled by a compact effective-core potential of Stevens et.al.[26]. The frozen-core DFT calculations are carried out by GAUSSIAN98[27] and the experimental values are from Reference ([52, 53]).

First, we discuss the equilibrium geometries. In Tables 4 and 5 are listed the equilibrium geometries (bond lengths in Å and angles in degrees) for some molecules from this work, frozen-core DFT calculations and experiment. We can expect that the Harris-Foulkes functional can predict reasonably well the equilibrium geome-

Table 4. Equilibrium geometries for selected molecules. See the text for details

Molecule	Symmetry	Parameter	DFT(frozen core)	This Work	Experiment
C_2H_4	D2h	R(CC)	1.409	1.422	1.339
		R(CH)	1.169	1.158	1.085
		$\angle$(HCH)	115.9	116.4	117.8
C_2H_6	D3d	R(CC)	1.583	1.572	1.531
		R(CH)	1.171	1.162	1.096
		$\angle$(HCH)	107.1	107.1	107.8
NH_3	C3v	R(NH)	1.104	1.118	1.012
		$\angle$(HNH)	105.7	100.7	106.7
HCN	C_∞	R(CN)	1.258	1.232	1.153
		R(CH)	1.119	1.146	1.065
CH_2NH	Cs	R(CN)	1.362	1.338	1.273
		$R(CH_{syn})$	1.190	1.170	1.103
		$R(CH_{anti})$	1.181	1.167	1.081
		R(NH)	1.141	1.133	1.023
		$\angle(H_{syn}CN)$	128.2	127.4	123.4
		$\angle(H_{anti}CN)$	118.3	118.9	119.7
		$\angle$(HNC)	109.0	109.0	110.5
CH_3NH_2	Cs	R(CN)	1.498	1.533	1.471
		$R(CH_{tr})$	1.194	1.171	1.099
		$R(CH_g)$	1.177	1.165	1.099
		$R(NH_a)$	1.100	1.118	1.010
		$\angle(NCH_{tr})$	117.7	116.7	113.9
		$\angle(H_gCH_g)$	106.4	107.0	108.0
		$\angle$(HNH)	108.7	101.1	107.1

Table 5. Equilibrium geometries for selected molecules. See the text for details

Molecule	Symmetry	Parameter	DFT(frozen core)	This Work	Experimental
N_2	D∞h	R(NN)	1.257	1.195	1.098
N_2H_2	C2h	R(NN)	1.342	1.310	1.252
		R(NH)	1.190	1.153	1.028
		∠(HNN)	113.5	113.4	106.9
N_2H_4	C2	R(NN)	1.563	1.535	1.449
		R(NH$_{int}$)	1.132	1.127	1.021
		R(NH$_{ext}$)	1.132	1.127	1.021
		∠(NNH$_{int}$)	103.4	103.3	106.0
		∠(NNH$_{ext}$)	103.4	103.3	112.0
		Θ(H$_{int}$NNH$_{ext}$)	75.2	77.0	91.0
H_2O	C2v	R(OH)	1.102	1.093	0.958
		∠(HOH)	99.3	99.6	104.5
H_2O_2	C2	R(OO)	1.617	1.512	1.452
		R(OH)	1.134	1.109	0.965
		∠(OOH)	97.2	102.2	100.0
		Θ(HOOH)	139.0	116.7	119.1
CH_3OH	Cs	R(CO)	1.530	1.508	1.421
		R(CH$_{tr}$)	1.176	1.166	1.094
		R(CH$_g$)	1.184	1.173	1.094
		R(OH)	1.108	1.096	0.963
		∠(OCH$_{tr}$)	107.1	106.8	107.2
		∠(H$_g$CH$_g$)	107.5	107.4	108.5
		∠(COH)	104.4	105.5	108.0

tries because the first order correction to the total energy functional is kept in the functional. The calculated results show that this is the case. For almost all the molecules studied, the *ab initio* TB equilibrium geometries are quite close to those from the frozen-core DFT calculations. Both types of theoretical calculations give similar trends in the calculated parameters. For example, compared with the experiment values, both the *ab initio* TB and DFT calculations give longer bond lengths and for most molecules smaller bond angles. These results are in agreement with those from the other *ab initio* TB calculations [38]. The major difference is found for the calculated bond angles ∠HNH in NH_3 and CH_3NH_2. The *ab initio* TB gives too small angles while the frozen-core DFT results are close to the experiment. Another example is the dihedral angle HOOH in H_2O_2. The frozen-core DFT calculation shows too large value while the *ab initio* TB result is much closer to the experiment.

In Table 6, we listed the dipole moments for some molecules calculated from this work and the frozen-core DFT. In the table, we also listed the available experimental dipole moments for the molecules. The dipole moment of a molecule reflects the charge distribution that is affected by the pseudo-potential due to the nuclei and core electrons as well as the potentials created by the valence electrons. In the frozen-core DFT calculations, the Coulomb and exchange-correlation potentials from the valence

Table 6. Dipole moments (in Debye) for selected molecules

Molecule	Symmetry	DFT(frozen core)	This work	Experiment
NH_3	C3v	1.72	1.68	1.47
HCN	C∞v	1.92	1.94	2.98
CH_2NH	Cs	1.30	1.46	
CH_3NH_2	Cs	1.29	1.40	1.31
H_2O	C2v	2.08	1.87	1.85
H_2O_2	C2	1.84	1.70	1.57
CO	C∞h	0.49	0.31	0.11
H_2CO	C2v	2.11	2.14	2.33
CH_3OH	Cs	1.70	1.73	1.70

electrons are calculated from the self-consistent density while in the *ab initio* TB calculations, the two potentials are calculated from the reference electron densities. Therefore, we expect some differences between the dipole moments calculated from the DFT and those from the *ab initio* TB. From Table 6, we find that the dipole moments from the two theoretical methods are still rather close each other. Generally, the dipole moments from the theoretical calculations are in agreement with those from the experiment. The exception is that for the HCN molecule. For this molecule, the theoretical dipole moments are 1.92D (from DFT) and 1.94D (from *ab initio* TB), much smaller than that from the experiment (2.98D). We attribute this error to the minimal basis set we used in the calculations.

Finally, we compare the energies for the hydrogenation reactions. In Table 7 are listed the reaction energies calculated from the frozen-core DFT, this work (*ab initio* TB), as well as using the Hartree-Fock approach with STO-3G as basis set. The corresponding experimental values are also listed in the same table[53]. From the table we can see that the results from the three theoretical methods (DFT, *ab initio* TB,

Table 7. Selected hydrogenation reactions (in kcal/mol)

Reaction	DFT(frozen core)	This work	HF/STO-3G	Experiment
$C_2H_6+H_2 \rightarrow 2CH_4$	18	17	19	19
$C_2H_4+2H_2 \rightarrow 2CH_4$	72	66	91	57
$C_2H_2+3H_2 \rightarrow 2CH_4$	137	119	154	105
$CH_3OH+H_2 \rightarrow CH_4+H_2O$	17	18	16	30
$H_2O_2+H_2 \rightarrow 2H_2O$	32	28	31	86
$H_2CO+2H_2 \rightarrow CH_4+H_2O$	47	45	65	59
$CO+3H_2 \rightarrow CH_4+H_2O$	74	57	72	63
$N_2+3H_2 \rightarrow 2NH_3$	25	19	36	37
$N_2H_2+2H_2 \rightarrow 2NH_3$	52	48	75	68
$N_2H_4+H_2 \rightarrow 2NH_3$	31	26	28	48
$HCN+3H_2 \rightarrow CH_4+NH_3$	81	73	97	76
$CH_3NH_2+H_2 \rightarrow CH_4+NH_3$	17	21	20	26
$CH_2NH+2H_2 \rightarrow CH_4+NH_3$	57	55	78	64

and HF/STO-3G) are quite similar to one another. The theoretical results can reflect correctly the trend shown in the experimental results. Yet there are some differences between the calculated reaction energies and the experimental ones. Also the *ab initio* TB results are closer to those from the frozen-core DFT than to HF/STO-3G results.

In a summary, for all the molecular properties discussed here, the *ab initio* TB results are quite close to those from the corresponding frozen-core DFT calculations, showing the great potential of the *ab initio* TB method. As we know, the DFT calculation results can be improved by increasing the size of the basis set. Therefore using the basis set larger than minimal one would also be expected to improve the *ab initio* TB results. This is what we are working on now.

4 Conclusion

In this work we have outlined two methods we believe will have a great potential when used in quantum chemical computer simulations of large biomolecular systems. Both methods are about two orders of magnitude faster than conventional *ab initio* schemes in quantum chemistry but nearly as accurate. We have applied the methods on smaller molecular systems here in order to be able to compare the results with corresponding *ab initio* results as well as available experimental values. We are currently developing general-purpose software where the described methods will be used as force engines in molecular dynamics simulations. The software will be parallelized to run on parallel computers, for examples PC clusters.

References

[1] A. Laaksonen and Y. Tu. *Methods of incorporating quantum mechanical calculations into molecular dynamics simulations*. Molecular Dynamics: from classical to quantum methods (P. B. Balbuena and J.M. Seminario Eds.). Elsevier, Amsterdam, 1999.

[2] L.A. Eriksson and A. Laaksonen. *Hybrid density functional theory molecular dynamics simulations of energetic and magnetic properties of radicals and radical matrix interactions*. Recent Research Developments in Physical Chemistry. Transworld Research Network, 1998.

[3] R. Car and M. Parrinello. Unified approach for molecular dynamics and density functional theory. *Phys. Rev. Lett.*, 55:2471–2474, 1985.

[4] J. A. Pople, D. P. Santry, and G. A. Segal. Approximate self-consistent molecular orbital theory. *J. Chem. Phys.*, 43:S129–135, 1965.

[5] P. Hohenberg and W. Kohn. Inhomogeneous electron gas. *Phys. Rev.*, 136:B864–871, 1964.

[6] W. Kohn and L.J. Sham. Self-consistent equations including exchange and correlation effects. *Phys. Rev.*, 140:A1133–A1138, 1965.

[7] E. Huckel. Quatum theoretical contributions to the benzene problem. I. the electron configuration to benzene and related compounds. *Z. Physik*, 70:204–286, 1931.

[8] R. Hoffmann. An Extended Hückel theory. hydrocarbons. *J. Chem. Phys.*, 39:1397, 1963.

[9] J.A. Pople and G.A. Segal. Approximate self-consistent molecular orbital theory. II. calculations with complete neglect of differential overlap. *J. Chem.Phys.*, 43:S136–151, 1965.

[10] Pople J.A., D.L. Beveridge, and P.A. Dobosh. Approximate self-consistent molecular orbital theory v. intermediate neglect of differential overlap. *J. Chem.Phys.*, 47:2026, 1967.

[11] Dewar M.J.S. and Thiel W. Ground states of molecules. 38. the MNDO method. approximations and parameters. *J. Am. Chem. Soc.*, 99:4899–4907, 1977.

[12] M.J.S. Dewar, E.G. Zoebisch, E.F. Healy, and J.J.P. Stewart. AM1: A new general purpose quantum mechanical molecular model. *J. Am. Chem. Soc.*, 107:3902, 1985.

[13] J.J.P. Stewart. Optimization of parameters for semiempirical methods. I. method. *J. Comput. Chem.*, 10:209–220, 1989.

[14] J.J.P. Stewart. Optimization of parameters for semiempirical methods. II. applications. . *Comput. Chem.*, 10:221–264, 1989.

[15] M. Kolb and W. Thiel. Beyond the MNDO model: Methodical considerations and numerical results. *J. Comput. Chem.*, 14:775–789, 1993.

[16] G.M. Zhidomirov, N.U. Zhanpeisov, I.L. Zilberberg, and I.V. Yudanov. On some ways of modifying semi-empirical quantum methods. *Int. J. Quantum Chem.*, 58:175–184, 1996.

[17] W. Weber and W. Thiel. Orthogonalization corrections for semiempirical methods. *Theor. Chem. Acc.*, 103:495–506, 2000.

[18] K. Möhle, H. Hofmann, and W. Thiel. Description of peptide and protein secondary structures employing semiempirical methods. *J. Comp. Chem.*, 22:509–520, 2001.

[19] J. Spanget-Larsen. On bridging the gap between Extended Huckel and NDO type LCAO-MO theories. *Theoret. Chim. Acta*, 55:165–172, 1980.

[20] J. Spanget-Larsen. The alternant hydrocarbon pairing theorem and all-valence electrons theory. an approximate LCOAO theory for the electron absorption and mcd spectra of conjugated organic compounds, part 2. *Theoret. Chem. Acc.*, 98:137–153, 1997.

[21] P.O. Löwdin. On the non-orthogonality problem connected with the use of atomic wave functions in the theory of molecules and crystals. *J. Chem. Phys.*, 18:367–370, 1950.

[22] P.O. Löwdin. On the nonorthogonality problem. *Adv. Quantum Chem.*, 5:185–199, 1970.

[23] K.R. Roby. On the justifiability of neglect of differential overlap molecular orbital methods. *Chem. Phys. Lett.*, 11:6–10, 1971.

[24] K.R. Roby. Fundamentals of an orthonormal basis set molecular orbital theory. *Chem. Phys. Lett.*, 12:579–582, 1972.

[25] Y. Tu, S.P. Jacobsson, and A. Laaksonen. Re-examination of the NDDO approximation and introduction of a new model beyond it. *Mol. Phys.*, 101:3009, 2003.

[26] W.J. Stevens, H. Basch, and M. Krauss. Compact effective potentials and efficient shared-exponent basis sets for the first- and second row atoms. *J. Chem. Phys.*, 81:6026–6033, 1984.

[27] M. J. Frisch, G. W. Trucks, H. B. Schlegel, G. E. Scuseria, M. A. Robb, J. R. Cheeseman, V. G. Zakrzewski, J. A. Montgomery Jr., R. E. Stratmann, J. C. Burant, S. Dapprich, J. M. Millam, A. D. Daniels, K. N. Kudin, M. C. Strain, O. Farkas, J. Tomasi, V. Barone, M. Cossi, R. Cammi, B. Mennucci, C. Pomelli, C. Adamo, S. Clifford, J. Ochterski, G. A. Petersson, P. Y. Ayala, Q. Cui, K. Morokuma, D. K. Malick, A. D. Rabuck, K. Raghavachari, J. B. Foresman, J. Cioslowski, J. V. Ortiz, A. G. Baboul, B. B. Stefanov, G. Liu, A. Liashenko, P. Piskorz, I. Komaromi, R. Gomperts, R. L. Martin, D. J. Fox, T. Keith, M. A. Al-Laham, C. Y. Peng, A. Nanayakkara, M. Challacombe, P. M. W. Gill, B. Johnson, W. Chen, M. W. Wong, J. L. Andres, C. Gonzalez, M. Head-Gordon, E. S. Replogle, and J. A. Pople. Gaussian98. Technical Report Rev. A9, Gaussian Inc., Pittsburgh PA, 1998.

[28] D.B. Cook, P.C. Hollis, and R. McWeeny. Approximate ab initio calculations on polyatomic molecules. *Mol. Phys.*, 13:553–571, 1967.

[29] B. Ahlswede and K. Jug. Consistent modifications of SINDO1: I. approximations and parameters. *J. Comput. Chem.*, 20:563–571, 1999.

[30] C.M. Goringe, D.R. Bowler, and E. Hernandez. Tight-binding modelling of materials. *Rep. Prog. Phys.*, 60:1447–1512, 1997.

[31] J.C. Slater and G.F. Koster. Simplified LCAO method for the periodic potential problem. *Phys. Rev.*, 94:1498–1524, 1954.

[32] H. Zhou, P. Selvan, K. Hirao, A. Suzuki, D. Kamei, S. Takami, M. Kubo, A. Imanura, and A. Miyamoto. Development and application of a novel quantum-chemical molecular dynamics method for degradation dynamics of organic lubricant under high pressures. *Tribol. Lett.,*, 15:155–162, 2003.

[33] Z. Zhao and J. Lu. A non-orthogonal tight-binding total energy model for molecular simulation. *Phys. Lett. A*, 319:523–529, 2003.

[34] W.M.C. Foulkes and R. Haydock. Tight-binding models and density-functional theory. *Phys. Rev. B*, 39:12520–12536, 1989.

[35] O.F. Sankey and D.J. Niklewski. Ab initio multicenter tight-binding model for molecular dynamics simulations and other applications in covalent systems. *Phys. Rev.*, B40:3979–3995, 1989.

[36] J. Harris. Simplified method for calculating the energy of weakly interacting fragments. *Phys. Rev. B*, 31:1770–1779, 1985.

[37] Z. Lin and J. Harris. A localized-basis scheme for molecular dynamics. *J. Phys.: Condens. Matter*, 4:1055–1080, 1992.

[38] A.P. Horsfield. Efficient ab initio tight binding. *Phys. Rev. B*, 56:6594–6602, 1997.

[39] J.P. Lewis, K.R. Glaesemann, G.A. Voth, J. Fritsch, A.A. Demkov, J. Ortega, and O.F. Sankey. Further developments in the local-orbital density-functional-theory tight-binding method. *Phys. Rev. B*, 64:195103.1–195103.10, 2001.

[40] M. Elstner, D. Porezag, G. Jungnickel, J. Elsner, M. Haugk, T. Frauenheim, S. Suhai, and G. Seifert. Self-consistent-charge density-functional tight-binding method for simulation of complex materials properties. *Phys. Rev. B*, 58:7260–7268, 1998.

[41] R.G. Parr and W. Yang. *Density-functional theory of atoms and molecules*. Oxford University Press, Oxford, 1989.

[42] S. Goedecker, M. Teter, and J. Hutter. Separable dual-space Gaussian pseudopotentials. *Phys. Rev. B*, 54:1703–1710, 1996.

[43] H.M. Polatoglou and M. Methfessel. Cohesive properties of solids calculated with the simplified total-energy functional of Harris. *Phys. Rev. B*, 37:10403–10406, 1988.

[44] F.W. Averill and G.S. Painter. Harris functional and related methods for calculating total energies in density functional theory. *Phys. Rev. B*, 41:10344–10353, 1990.

[45] D. Porezag, T. Frauenheim, T. Köhler, G. Seifert, and R. Kaschner. Construction of tight-binding-like potentials on the basis of density functional theory: Application to carbon. *Phys. Rev. B*, 51:12947–12957, 1995.

[46] S.D. Kenny, A. P. Horsfield, and H. Fujitani. Transferable atomic-type orbital basis sets for solids. *Phys. Rev. B*, 62:4899–4905, 2000.

[47] K. Koepernik and H. Eschrig. Full-potential nonorthogonal local-orbital minimum-basis band-structure scheme. *Phys. Rev. B*, 59:1743–1757, 1999.

[48] J. Junquera, O. Paz, D. Sánchez-Portal, and E. Artacho. Numerical atomic orbitals for linear-scaling calculations. *Phys. Rev. B.*, 64:235111.1 235111.9, 2001.

[49] A.A. Demkov, J. Ortega, O.F. Sankey, and M.P. Grumbach. Electronic structure approach for complex silicas. *Phys. Rev. B*, 52:1618–1630, 1995.

[50] A.D. Becke. Density-functional exchange-energy approximation with correct asymptotic-behavior. *Phys. Rev. A*, 38:3098–3100, 1988.

[51] C. Lee, W. Yang W, and R.G. Parr. Development of the Colle-Salvetti correlation energy formula into a functional of the electron density. *Phys Rev B*, 37:785, 1988.

[52] D.R. Lide, editor. *Handbook of Chemistry and Physics*. CRC, Boca Raton, 85 edition, 2004.

[53] W.J. Hehre, L. Radom, P.V.R. Schleyer, and J.A. Pople. *Ab initio molecular orbital theory*. John Wiley and sons, New York, 1986.

Quantum Chemistry Simulations of Glycopeptide Antibiotics

Jung-Goo Lee, Celeste Sagui and Christopher Roland

Center for High Performance Simulations and Department of Physics, North Carolina State University, Raleigh, NC 27695, USA

Abstract. The recent rise of vancomycin-resistant enterococci (VREs) has given new impetus to the study of the binding between glycopeptide antibiotics and bacterial cell-wall termini. Here, we summarize the results of recent quantum chemistry simulations of the binding between vancomycin and teicoplanin with Ac-D-Ala-D-Ala (characteristic of non-VREs) and Ac-D-Ala-D-Lac (characteristic of VREs). Binding of both antibiotics to Ac-D-Ala-D-Ala was found to be stronger by about 3-5 kcal/mol, and due primarily to the oxygen-oxygen lone-pair repulsion characteristic of the antibiotic/D-Ala-D-Lac complex. These results are in good agreement with recent experimental findings.

Key words: glycopeptide antibiotics, quantum chemistry, simulations

1 Introduction

Vancomycin is the parent compound of the family of glycopeptide antibiotics, which includes such important members as teicoplanin, ristocetin, and avoparcin [1, 2]. First isolated from soil samples in 1956 by the pharmaceutical company Eli Lilly, vancomycin is produced by the *Amycolatopis Orientalis* microorganism, and has been in clinical use since 1959 [3]. Specifically, vancomycin is used with patients allergic to β-lactams, to fight infections during chemotherapy, and as a "drug-of-last-resort" for the treatment of infections caused by Gram-positive bacteria [4]. However, because the last three decades have seen extensive use of this antibiotic, along with the use of the related glycopeptide antibiotic avoparcin in animal feeds [5], vancomycin-resistant Gram-positive bacteria have emerged: vancomycin-resistant enterococci (VRE) in 1987 [6] and vancomycin-resistant staphylococcus aureus (VRSA) in 1998 [7]. These developments represent a new, world-wide health risk, and have given new urgency to understanding the detailed structure and action of glycopeptide antibiotics.

Vancomycin and other glycopeptide antibiotics act by inhibiting cell wall biosynthesis [8]. Prokaryotic bacterial cell walls are surrounded by a mesh-like peptidoglycan layer, which provides mechanical support to the cell. This peptidoglycan layer is composed of linear polysaccharide chains, with alternative residues composed of the carbohydrates N-acetyl glucosamine and N-acetyl muramic acid. These saccharide chains run parallel to each other, with every second unit connected to a peptide moiety consisting of five amino acids L-Ala-D-Glu-L-Lys-D-Ala-D-Ala. These peptide chains crosslink, thereby forming a network which imparts a tremendous amount of strength to the cell wall [4]. Glycopeptide antibiotics act as follows. As the peptidoglycan precursor is synthesized by the cell, the glycopeptide antibiotics selectively bind to the precursor N-acyl-D-Ala-D-Ala, thereby preventing the crucial peptidoglycan crosslinking [9, 10]. The gives is a weakened cell wall, which cannot resist the osmotic pressures experienced by the cell. The ultimate result is bactericide by lysis. However, VRE have developed a sophisticated strategy for neutralizing the action of glycopeptide antibiotics. VREs, characterized by the vanA, vanB, vanD, or vanH gene cluster, evade the action of vancomycin through a modification of the antibiotic target. Specifically, the termination of the pentapeptide on the peptidoglycan precursor D-Ala-D-Ala is substituted by D-Ala-D-Lac, and to a lesser extend with D-Ala-D-Ser [10]. This changing of the last peptide does not have any effect on the quality of the peptidoglycan layer, or its ability to crosslink the glycan strands. It does, however, reduce the affinity of the antibiotic for the peptidoglycan layer by a factor of over 1000 – at least for D-Ala-D-Lac – thereby *de facto* rendering the antibiotic ineffective. Clearly, the binding betwent the antibiotic and the modified precursors is the key issue that needs understanding. In this short paper, we review the results of an extensive quantum chemistry study of the binding between prototypical glycopeptide antibiotics and bacterial cell wall analogs [11].

2 Calculations

In order to theoretically investigate the antibiotic complexes, we conducted quantum chemistry calculations based on density functional theory (DFT) and Hartree-Fock (HF) calculations [12, 13]. In DFT calculations, the total energy is expressed in terms of a functional of the electron density $\rho(r) = \sum |\psi_i(r)|^2$. A variational minimization of the total energy given by DFT leads to an effective Schrödinger equation for the Kohn-Sham orbitals $\psi_i(r)$, i=1,..N:

$$[-\nabla^2 + V_N(\mathbf{r}) + \int \frac{2\rho(\mathbf{r}')d\mathbf{r}'}{|\mathbf{r} - \mathbf{r}'|} + E_{xc}(\rho, \nabla\rho)]\psi_i(\mathbf{r}) = \epsilon_i\psi_i(\mathbf{r}) , \qquad (1)$$

for the N electrons in the system. Here, the first term represents the kinetic energy, the second the electrostatic potential due to the nuclei, and the third terms the classical electron-electron repulsion potential. E_{xc} represents the so-called exchange and correlation potential, which describes the average effects of antisymmetry requirements of the total many-body wavefunction, and is typically derived from electron gas theory. Here, we specifically made use of the popular B3LYP gradient-corrected functional for this term [14], which is known to give good results with most organic-based compounds. In the HF calculations, the exchange term is explicitly accounted for by a standard integral, and the correlation term is zero. All the calculations made use of the accurate 6-31G* basis set, and the somewhat less accurate 4-31G basis set.

Figure 1 shows a schematic of vancomycin and teicoplanin, the two clinically most important glycopeptide antibiotics. Both antibiotics are characterized by a heptapeptide core of seven amino acid residues, with the side chains of residues 2-4, 4-6, and 5-7 covalently joined. Vancomycin consists of N-methyl leucine at residue 1, asparagines at residue 3, and substituted phenylglycines at positions 2,4,5,6 and 7. Note that the peptide bond between residues 5 and 6, is in the *cis* configuration, which is characteristic of almost all glycopeptide antibiotics. The structure of teicoplanin is very similar to vancomycin. It is, however, characterized by different sugar substituents and covalently linked aromatic rings at residues 1 and 3.

All calculations were carried out with the standard Gaussian03 package [15], and here we will report on results for the binding of the aglycon forms of vancomycin ($C_{52}H_{51}N_8O_{17}Cl_2^+$) and teicoplanin ($C_{58}H_{46}N_7O_{18}Cl_2^+$) with the dipeptides Ac-D-Ala-D-Ala ($C_8H_{13}N_2O_{19}$), Ac-D-Ala-D-Lac ($C_8H_{12}NO_5^-$), Ac-D-Ac-Gly ($C_7H_{11}N_2O_4^-$) and Ac-D-Ala-D-Ser ($C_8H_{13}N_2O_5^-$), which are the analogs of the bacterial cell wall termini. The structures were optimized by means of the analytic gradient method, and the calculations employed between 1500 to 1700 basis functions, depending on the calculation. The basis sets used – especially the B3LYP/6-31G* calculations – are known to have a relatively small basis set superposition error. Furthermore, for the most part, we focus on *the relative energy differences between configurations*, which further serves to minimize this effect. Finally, as discussed, all the results show the same trend, with the 6-31G* results in good semi-quantitative agreement with the experimental results. Each of the calculations were performed on 8 processors of an IBM 690 Power IV, and took between 15 to 25 days to complete.

3 Results and Discussion

The salient results on the antibiotic binding to the cell wall termini for vancomycin and teicoplanin are shown in Figs. 2,3. Both antibiotics buckle in such a way as form a pocket which makes the important contact with the dipeptides. In order of increasing importance, the antibiotic complexes are stabilized through the dispersive van der Waals interaction (whose contribution to the energy differences is estimated to be small), the nonbonding electrostatic interaction, and, most importantly, through a series of five hydrogen bonds[16]. The latter were identified through a careful examination of the bond lengths, the orbital overlap, and electrostatic considerations. This bonding pattern holds for the case of Ac-D-Ala-D-Gly, Ac-D-Ala-D-Ala, and Ac-D-Ala-D-Ser. The D-Ala-D-Lac terminus is different insofar as it involves the substitution of a linking ester for an amide, through the exchange of a single ligand (*i.e.*, X=NH goes to X=0). This substitution acts to decrease the binding of the antibiotic complex through the removal of a single hydrogen bond, and through the repulsive interaction between two oxygen lone-pairs that have now been created (second position from the left, marked by arrows in Fig. 2). For the vancomycin/Ac-D-Ala-D-Ala complex, the *average* O–HN hydrogen bond is about 2.010 long, while the

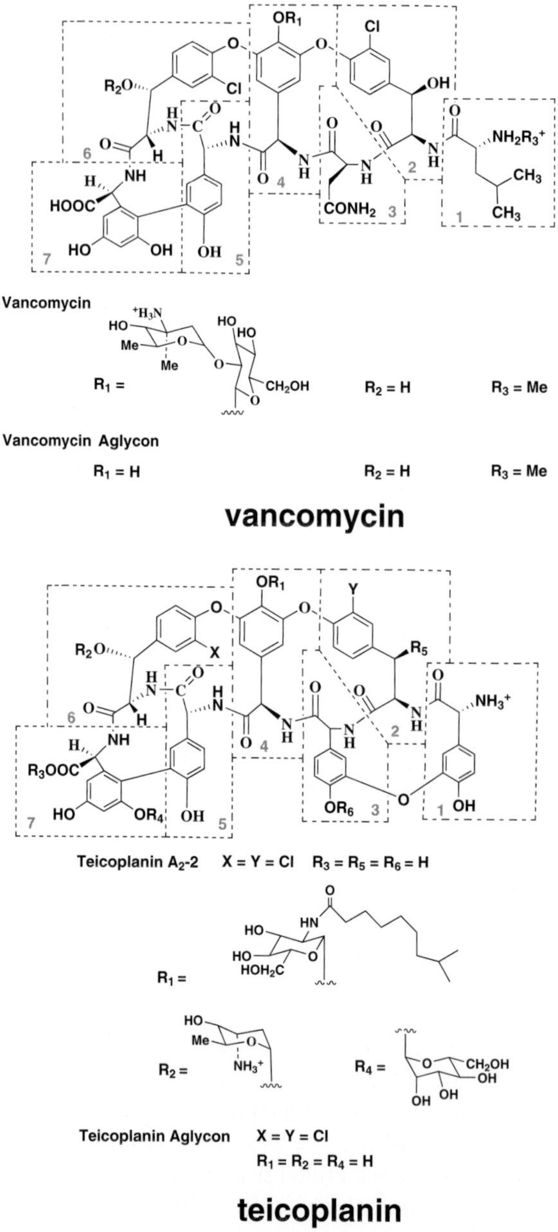

vancomycin

teicoplanin

Figure 1. Chemical structure of vancomycin and teicoplanin, the clinically most important glycopeptide antibiotics

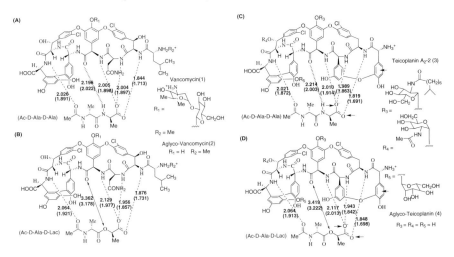

Figure 2. Schematic of (**a**) vancomycin/Ac-D-Ala-D-Ala, (**b**) vancomycin/D-Ala-D-Lac, (**c**) teicoplanin/D-Ala-D-Ala, (**d**) teicoplanin/D-Ala-D-Lac, complexes. The dotted lines mark the hydrogen bonds stabilizing the complexes, while the arrows in (b),(d) marks the O-O lone-pair repulsion chiefly responsible for the loss of binding with Ac-D-Ala-D-Lac. HF (DFT) bond distances in are also marked. In (c),(d), the small arrows mark the weak interaction between the resonating teicoplanin bonds and the D-Ala/D-Lac terminations

corresponding average O-O distance is a much further 3.185. In addition, except for the fourth hydrogen bond from the left, all the other hydrogen bonds which stabilize the complex are pushed apart by a small amount, which further serves to decrease the binding energy. These features are in agreement with previous models based on classical potentials and consistent with experimental results [17, 18].

Table I summarizes the calculated DFT binding energies for the geometry optimized structures. In broad terms, the two antibiotics show similar trends. The order of the bonding, from the strongest to weakest is D-Ala-Gly, D-Ala-D-Ala, D-Ala-D-Ser, and D-Ala-D-Lac. Roughly speaking, glycine is the smallest amino acid, and as such it can enter somewhat deeper into the antibiotic pocket. This is reflected in

Table 1. Binding energies (E) in kcal/mol for both vancomycin and teicoplanin aglycon to bacterial cell wall analogs. Binding energy differences (ΔE) are measured with respect to D-Ala-D-Lac configuration

dipeptide	vancomycin		teicoplanin	
	E	ΔE	E	ΔE
Ac-D-Ala-Gly	120.6	4.6	124.2	4.4
Ac-D-Ala-D-Ala	119.6	3.6	123.4	3.6
Ac-D-Ala-D-Ser	117.7	1.7	121.5	1.7
Ac-D-Ala-D-Lac	116.0	0	119.8	0

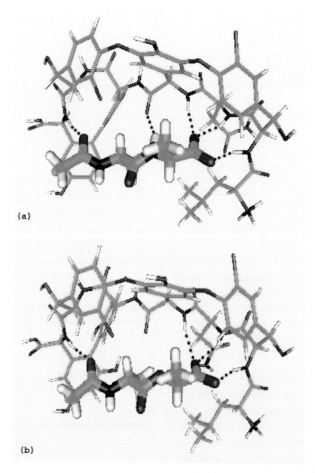

Figure 3. Optimized three-dimensional structure of the (**a**) vancomycin/D-Ala-D-Ala, (**b**) vancomycin/D-Ala-D-Lac complex. The position of the hydrogen bonds is marked by a *dotted line*, and the structure is to be compared to the exploded view shown in Fig. 2

shorter hydrogen bond distances, especially the second hydrogen bond from the left. By contrast, serine is the largest of the three amino acids, and fits slightly less well into the antibiotic pocket. The results for D-Ala-D-Ala are intermediate to the two cases. These results are in good qualitative agreement with the experimental results, with the exception of the D-Ala-Gly case [18]. For the latter, the estimated experimental binding energy is 3.7 kcal/mol as opposed to our estimate of 4.4 kcal/mol. We speculate that solvation effects, which are not accounted for here, play a role in stabilizing conformations whose binding energy is less than the ideal gas-phase configurations explored here.

Another interesting feature is that the binding of the teicoplanin-based complexes are somewhat stronger than those based on vancomycin. In part, this feature is due to a weak $\pi - \pi$ bonding associated with the stabilization of the "O-C=0" resonance at the end of the D-Ala/D-Lac terminus, and the phenyl sidegroups of teicoplanin. The small arrows on Figs. 2(c,d) mark the specific overlap of the $\pi - \pi$ bonds.

Comparing the HF and DFT binding energies, we note that the former are larger. For instance, HF binding energies with a 6-31G* basis set for vancomycin aglycon complex with Ac-D-Ala-D-Ala (Ac-D-Ala-D-Lac) are 109.8 (104.9) kcal/mol, respectively. This gives an energy difference of 4.9 kcal/mol, which is to be compared to the DFT-based energy difference of about 3.6 kcal/mol. The corresponding numbers for teicoplanin aglycon with Ac-D-Ala-D-Ala (Ac-D-Ala-D-Lac) are 119.0 (113.9) kcal/mol for an energy difference of 5.1 kcal/mol. As already noted, these numbers bracket the experimental estimate of 4.4 kcal/mol [17, 18]. At room temperature, one can therefore expect that the binding of the antibiotic to D-Ala-D-Ala is stronger by a factor of 500 to 4000 over D-Ala-D-Lac [11]. For completeness, we note that calculations with the 4-31G basis set preserve the relative order of the binding energies, but that the estimates of the numerical values are somewhat larger. For example, for teicoplanin, the difference in the binding energy between Ac-D-Ala-D-Ala and Ac-D-Ala-D-Lac is 5.8 kcal/mol.

Finally, to gain more quantitative insight into the quality of the calculations, we have calculated the RMDS of antibiotic complexes, as calculated with DFT and HF. As is clear from Fig. 4, the results from each of the two types of calculations resemble each other to a high degree, with some deviations at residue 1, which reflect its "floppy" nature. Quantitatively, the RMSD between the HF/6-31G* and DFT/6-31G* calculations with 157 atoms is 0.23 . Similarly, the comparison betwen the experimental [19], and our theoretical structures is also quite good. For example, Fig. 5 shows such a comparison between the "back-bone" atoms (hydrogens and cell wall analog atoms excluded), with a calculated RMDS of 0.69 .

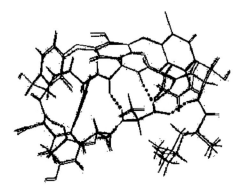

Figure 4. Wire mesh overlay, comparing the three-dimensional structure of the vancomycin aglycon/Ac-D-Ala-D-Ala complex as calculated by the HF/6-31G* (light) and DFT/6-31G* (dark) methods

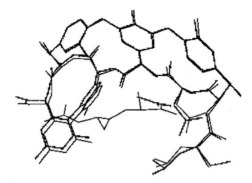

Figure 5. Wire mesh overlay comparing the experimental (light) [19] and DFT-based theoretical three-dimensional structure of the vancomycin/D-Ala-D-Ala complex. Here the hydrogens and the Ac-D-Ala-D-Ala are not shown

4 Conclusion

In summary, we report here on an extensive quantum chemistry study of the binding of the two, clinically most important glycopeptide antibiotics – vancomycin and te-icoplanin – to bacterial cell wall analogs [11]. Except for the D-Ala-D-Lac case, the antibiotic complexes are stabilized by a series of five hydrogen bonds. For the D-Ala-D-Lac case, one of the hydrogen bonds is lost and the resulting binding is weaker by an estimated 3-5 kcal/mol, with respect to the binding to D-Ala-D-Ala. This accounts for the diminished potency of the drugs towards combating VRE strains. These results are in good agreement with current experimental results.

Acknowledgements

Support for this work has been provided by NSF grants ITR-0121361 and CAREER 0348039. We also acknowledge and NSF-NCSA grant for extensive computer support.

References

[1] Williams, D.H., Bardsley, B., Angew. Chem. Int. Ed. **38**, 1172 (1999).
[2] Nicolaou, K.C., Broddy, C.N., Brase, S., Winssinger, H., Angew. Chem. Int. Ed. **38**, 2096 (1999).
[3] McCormick, M.H., Stark, W.M., Pittenger, G.E., Pittenger, R.C., McGuie, J.M., Antibiot. Annu. **1955-56**, 606.
[4] Walsh, C., *Antibiotics - Actions, Origins, Resistance*, (ASM Press, Washington DC, 2003).
[5] Robredo, B., Singh, K.V., Baquero, F., Murray, B.E., Torres, C., Antimicrob. Agents Chemother. **43**, 1137 (1999).

[6] Uttley, A.H.C., Collins, C.H., Naidoo, J., George, R.C., Lancet **1**, 57 (1988);
     Woodford, N.J., Med. Microbiol. **47**, 849 (1998).

[7] Sieradzki, K., Roberts, R.B., Haber, S.W., Tomasz, A., N. Eng. J. Med. **340**, 517
     (1999); Smith, T.L., Pearson, M.L., Wilcox, K.R., Cruz, C., Lancaster, M.V., *et
     al*, N. Eng. J. Med. **340**, 493 (1999); Novak, R., Henriques, B., Charpentier, E.,
     Normark, S., Tuomanen, E., Nature (London) **399**, 590 (1999).

[8] Perkins, H.R., Biochem. J. **111**, 195 (1969); Chatterjee, A.M., Perkins, H.R.,
     Biochem. Biophys. Res. Commun. **24**, 489 (1966); Nieto, M., Perkins, H.R.,
     Biochem. J. **123**, 789 (1971).

[9] Williams, D.H., Butcher, D.W., J. Am. Chem. Soc. **103**, 5697 (1981).

[10] Bugg, T.D., Wright, G.D., Dutka-Malen, S., Arthur, M., Courvalin, P., Walsh,
     C.T., Biochemistry **30**, 10408 (1991).

[11] Lee, J.-G., Sagui, C., Roland, C., J. Am. Chem. Soc. **126**, 8384 (2004); Lee,
     J.-G., Sagui, C., Roland, C., J. Am. Chem. Soc., submitted.

[12] Szabo, A., Ostlund, N.S., *Modern Quantum Chemistry*, (Dover, 1982).

[13] Helgaker, T., Jorgensen, P., Olsen, J., *Molecular Electronic Structure Theory*,
     (Wiley, N.Y. 2000).

[14] Lee, C., Yang, Y., Parr, R.C., Phys. Rev. B **37**, 785 (1988); Becke, A.D., J.
     Chem. Phys. **98**, 5649 (1993).

[15] Gaussian03, Rev. B.01, Frisch, M.J., *et al*, Gaussian Inc., Pittsburgh, P.A.
     (2003).

[16] For a discussion of different bonding types, see for example: Cramer, C.J. *Es-
     sentials of Computational Chemistry*, (Wiley, 2002). Stone, A.J. *The Theory of
     Intermolecular Forces*, (Claredon Press, 1996). Leeds, A.R. *Molecular Mod-
     elling*, (Laymen, 1996).

[17] McCormas, C.C., Crowley, B.M., Boger, D.L., J. Am. Chem. Soc. **125** 9314
     (2003).

[18] Li, D.H., Sreenivasan, U., Juranic, N., Macura, S., Puga, F.J.L., Frohnert, P.M.,
     Axelsen, P.H., J. Mol. Recogn. **10**, 73 (1997).

[19] Loll, P.J., Bevivino, A.E., Korty, B.D., Axelsen, P.H., J. Am. Chem. Soc. **119**,
     1516 (1997).

Panel Discussion

Comments below were given at a panel discussion on the third day of the 2004 Algorithms for Macromolecular Modelling meeting involving all participants. The panel consisted of David Case, Aatto Laaksonen, Alan Mark, Harold Scheraga, and Christof Schütte. The moderator was Robert Skeel.

Moderator: What have been the important advances in algorithm, methods, and models in the past five (or ten) years?

David Case: There have been two major algorithmic advances in the past decade. The first was the development of Ewald-based methods to handle long-range electrostatics; this general category included particle-mesh Ewald (PME) and particle-particle-particle-mesh schemes, as well as "true" Ewald summations. I view this development as a "win-win", because one got both faster and "better" simulations; "better" in the sense of removing the clearly bad features of truncation methods. There are still potential artifacts caused by imposing periodicity on what is physically a non-periodic system, but these in principle should go away simply by moving to ever-larger unit cells. The second algorithmic development involves multiple-timestep methods, that allow trajectories of high quality to be computed faster, as long as there is an available separation of forces into "rapidly varying but cheap" and "slowly varying but expensive" categories. Although the basic principles have been known for some time, there is still work to do (in my view) to take full advantage of these methods in practical biomolecular simulations. As processors and algorithms change, what is "cheap" vs. what is "expensive" also changes; furthermore, with new types of force fields (such as those that include polarizability), new developments in integration methods will also be required.

Harold Scheraga: Innovation has been in the development of global optimization procedures, with considerable progress in the last 5 years. In particular, our hierarchical approach, starting with a united-residue force field, has enabled us to search the multi-dimensional conformational space quite efficiently to locate the REGION of the global minimum of the potential function. The resulting clusters of low-energy united-residue models are then converted to an all-atom representation, and the search for the global minimum is then continued with an all-atom force field. A second development, but one that is just in its initial stages, may provide the possibility to extend the time-scale of molecular dynamics, hopefully to the millisecond range, by use of the united-residue model, thereby facilitating the search for the global minimum.

Alan Mark: I want to begin by saying that I agree with Dave Case that the two most significant advances in biomolecular simulation has been first, the systematic improvement in the force fields in particular the trend to parameterize the force field incorporating thermodynamic data as well as structural data and second, algorithmic improvements such as codes that allow the use of efficient lattice sum methods for electrostatics. It is difficult, however, to point to any specific breakthroughs over the last 10 years as the advances have been incremental. Equally, just because lattice sum methods for electrostatics can be made computationally efficient it is very wrong to suddenly claim we can treat electrostatic interactions "correctly". In many cases lattice sum methods are inappropriate or inefficient and it is very worrying that sometimes such methods are propagated more because of fashion than by science. The same is true for force fields. Major improvements have been made in the abilities of biomolecular force fields to address specific problems such as peptide folding but it is foolish to claim that either CHARMM, AMBER, Gromos or OPLS is now correct. The other area where I believe there have been exciting developments over the last 5 years is in relation to proposals such as the Jarzynski equality in free energy calculations and the use of replica exchange methods for searching conformational space in large systems. While I believe the jury is still out on whether these have in reality all the advantages that have been claimed they have stimulated new ways of looking at some long standing and very difficult problems.

Aatto Laaksonen: I completely agree with with Harold and Alan concerning particle-mesh Ewald and multiple time-step methods being among the most exciting and innovative improvements. However, I would like to add multiscale type of simulations and also quantum mechanical simulation methods. In particular I would like to mention three kinds of methods here: all of them are coming originally from materials science but have been developed further even for biological systems. One of them is tight-binding DFT based MD. I really think it is a big achievement and has a future in simulations of macromolecules. The second method is kinetic Monte Carlo, a way to add time to MC type simulation. With KMC you can go to arbitrary time-scales. The problem is that it is not a "first-principles" simulation method: you need to know transition states and transition rates in advance as input. You then get the time step by inverting the transition rate and the size of the time step varies depending on where you are going along reaction coordinates and how high are the barriers you have to cross. The third scheme is dissipative particle dynamics which can be used to model soft matter systems. Now we are in the process of adding electrostatic interactions to that method and certainly a lot will happen in the future when the method has matured.

Moderator: Are there any questions or comments from the audience?

Tamar Schlick: I think that there are a lot of interesting developments in sampling algorithms with the acute realization that better integrators for dynamics simulations would not buy us significantly more coverage time. Enhanced sampling methods are making an impact on our ability to simulate longer biomolecular processes with unprecedented information on the details of molecular motions that are key to interpreting structure/function relationships.

Alan Mark: I think we should also be realistic to say that there hasn't been revolutionary changes in the past five years in molecular simulation. There hasn't been any single thing that has been done that has caused a major advance. When we had the discussion at the last of these meetings we were talking about the same things. It's not like someone has said that "Wow I've discovered Monte Carlo"

Moderator: For which systems and questions are current all-atom force fields inadequate and how can these force fields best be improved?

David Case: I don't have a good answer. Force fields are inadequate for almost all interesting systems. The surprising thing when you start looking at quantum mechanics, anyone who talks to a chemist can come up with a small molecular problem conformational analysis problem where even your favorite quantum mechanics will be off by 2Kcal/mol in relative engy and your force field will be off by a great deal when you compare relative minima. There are a whole variety of things that are off relative to kT but somehow we put together thousands of these things to make a model for a protein and it really still is amazing that somehow what comes out looks as close to a real protein as it does. We are always hitting the limits of methodology almost as much as we are hitting the limits of sampling.

Harold Scheraga: Well I agree completely that there are a lot of force fields out there and I don't think anybody can make the claim that one is better than another. They are really all inadequate: there really is a lot of room for improvement. This showed up most dramatically in May when I was at a meeting of the Cambridge Crystallographic Database Centre which was a CASP-like meeting to predict crystal structures for just small organic molecules. The degree of success was practically zero. In other words nobody was really able to predict the native structure of a crystal as the global minimum of an energy function. However, most everybody with their force fields was able to predict a structure somewhere in the list of minima that they submitted, but not at the lowest energy. The encouraging thing is that you are finding the crystal structure somewhere in the list of low energy states. Now being at least a believer in the concept that potential functions should be transferable from one molecule to another if you have got them right, I believe that crystal calculations offer the possibility to improve potential functions. For example when we superimpose a calculated crystal structure and an observed one, it fits pretty well but there are errors in the atom positions and lattice constants. it seems that all you have to do is tweak the potential parameters and get the calculation to superimpose and you have the possibility to improve. That is the direction in which we have been going. I do have to say that all the potentials that have been used up to now start with the observed structure and tweak the potentials around the observed structure. A much more hopeful process is to do a global optimization to search the whole space not just near the observed crystal structure. We do have good global optimization techniques. In global optimization we should satisfy three requirements: (1) the computed structure should be as close as possible to the observed, (2) the observed structure should have the lowest energy, and (3) where there are experimental lattice energies available you should be able to calculate the lattice energies. With these three requirements and a global optimization procedure I think there is the possibility to make a significant improvement in our force fields that hopefully should be transferable from the crystals of amino acids and peptides to proteins.

Christof Schütte: I'm the mathematician on this panel. I'm not so interested in really large proteins, more in medium-sized peptides, and I'm not so interested in global minimization of landscapes but more in dynamics. I feel that in order to have good models for that it could be a good idea to use more data from spectroscopy. And I know of an approach in this direction of Benny Gerber in Jerusalem where he really tries to design force fields using vibrational spectroscopy of small molecules. And I really like this approach so that we could be able to have force fields that have reproduction of dynamical features for example through spectroscopy. And so I am not so interested in improving landscapes and global minimization techniques but in reproducing dynamical behaviour.

Alan Mark: When I think about force fields, first I would say that current force fields aren't very good for anything that they have not been parameterized for, which is basically saying that we operate with empirical force fields which are parameterized against a set of data and in fact they can only reproduce that data. I love a comment I heard Peter Kollman say when asked why didn't he go on to a more simplified force field to do protein folding and he said as somebody who came from quantum mechanics that the force field was already unbelievably simple. He said that the force fields that we are using have to be chosen for the level of the problem that we try to solve. For example, take, say, Gromos or CHARMM, and add a calcium ion and watch where it binds. The codes completely fail. They cannot predict such a thing. It's not in the model; it's not in the parameterization. So I think you really have to question when you say what they are adequate or inadequate for, if you want to say look at folding we (we means Wilfred van Gunsteren and others of us who have been looking at folding) believe that partition data should be related to folding. We can reparameterize the force field to make good partition data and hopefully this will improve the folding characteristics of our force field. Because we know that several of the force fields completely failed to reproduce the most simple features, such as whether a residue like tryptophan likes to be in water or hexane, and some of the force fields get the wrong sign for some of the most basic properties. And yet when you mix them all together they'll hold your proteins lovely and stable. So it is a difficult question to be able to address and I don't think you will be able to address it in a global sense.

Aatto Laaksonen: Well again I completely agree. I can certainly point out one situation where these atomic force fields are inadequate: that is in chemical reactions since electrons are needed there. Also trying to follow dynamics to seconds or minutes since on such long scales dynamics is not currently computable. It is true that forces are developed with water as solvent in mind and normal quantities and temperatures and so on, with harmonic potentials for bond stretching so they just work when you are at the minimum of those potential wells. No dissociation is possible. Also Christof mentioned that he would like to parameterize force fields to spectroscopic data and again it is just these harmonic potentials that are the bottleneck you would need anharmonic potentials to do it properly.

Harold Scheraga: I was going to emphasize the same point. I would be happier with a spectroscopic approach to force field development such as the one pioneered by the late Shneior Lifson. You do normal mode analysis but you are stuck within the harmonic approximation. I can cite many examples where force fields like CHARMM go wrong on bond angles because they are using harmonic approximations which are good only around the minimum; with that Lennard-Jones steep potential when two atoms are clashing they open up a bond angle very far from the minimum. OK the program is clever enough to bring it back but they may bring it back to the wrong angle, so you get big errors in the bond angle because they are using harmonic potentials and allowing bond angle bending. That's a problem that has to be faced in using spectroscopic data to parameterize force fields.

Moderator: Questions or comments from the audience?

Anna Choumalina: I would like to ask why you talk about the global minimum. Actually the native structure does not have to be a global minimum. It is a structure that describes how a protein folds in a cell and under the conditions in which it exists in the cell. Why not describe the conditions in the cell: how a protein gets into this conformation.For example if you boil an egg it is still the same protein but it gets into a different conformation, maybe a different local minimum. Maybe not the global minimum. But it looks like a different protein with different properties.

Harold Scheraga: When you boil an egg you unfold the protein you change it to another local minimum with a greater energy than the global minimum. And in that conformation its going to aggregate and so your boiled egg is aggregated. And you are never going to reverse it. even though if you could keep it from aggregating then the dogma is that the unfolded form would refold. That's one point, but another point is that when looking for the minimum of the global energy we have other ways to add the entropy in so we really do get the free energy. In other words you look for all the low-lying energy states. From the Hessian you can get (again in a harmonic approximation) the entropy corresponding to the different minima. And there are two sources of the entropy. First the vibrational entropy. and secondly...we've seen many cases where you have a golf hole type minimum narrow low energy compared to nearby ones (which are however much broader) and when you take the entropy into account you can get the stable state on a free energy scale. So there are ways to add in the entropy once you have dealt with the minimization problem. Finally, in looking for the thermodynamically most stable state, you have to take into account the whole system, i.e. the protein and its surroundings.

Alan Mark: I would also think it is a question of practicality. For small proteins it probably is a global minimum of an isolated protein in its folded state. For example we know that the serpines aren't in their global minimum and in fact they are in high energy states when they kinetically fold and are trapped, but under reaction they will actually go to a different minimum which is much lower and this is a serious problem for protein studies. But for small ones, it is likely to already be the global minimum and we could probably not handle the complexity of the environment in the cell. But you also have the problem that for example occurs with prions many proteins may actually have a prion-like structure as their global minimum as well but then you are dealing with aggregation and that's already a more complex phenomenon.

Jamshed Anwar: The current approach in trying to optimize potentials particularly for small molecules is to take into account structural data, typically crystallographic data and maybe lattice data. But I think we have reached the stage where we could take the next jump. I have an interest in computing coexistence points, melting points, phase transition points. And of course even for small molecules we are not quite to the point of computing these. But we are just about to the point of calculating and attempting to reproduce these coexistence points and optimize potentials with respect to the coexistence points. Maybe that is the next step.

Hayden Eastwood: At what tempearture does the force field break down in approximating the system?

Alan Mark: We don't know the temperature of our force fields. Temperature is a force field parameter in itself that we don't know...We don't even know if the force fields give the appropriate states at 300 degrees. We don't know if we simulate at 300K if it is really 300K.

Hayden Eastwood: If we parameterize for a given situation at say 700K and you are doing dynamics at that temperature is the simulation going to give you a wildly inaccurate picture? Is the energy of the system from the vibrations of the system going to give you a wildly inaccurate energy?

Alan Mark: In our simulations SPC water, the approximation was breaking down below 400K. Other people were using this all the way to 600K. I think you would be safe in the range 270K-310K. Once I am getting outside of that range I would really have doubts.

Voichita Dadarlat: Following up on Bogdan Lesyng's talk yesterday, he said that the partial charges depend on local conformation change by about 10 %. I have calculated partial charges on a block teranine dipeptide molecule I have found differences as big as 20 %. How worried should we be about MD programs that keep charges constant.

Harold Scheraga: This coupling of pK and conformation is a very serious problem and you cannot come to grips with it for a large molecule. However, we have tried to treat it for a small molecule, say 20 residues, with about 8 ionizable groups. With 8 ionizable groups you have 256 states of ionization for each conformation. We have calculated the conformations for all these states of ionization. It's not 10 or 20 % - it depends on where the charges are located. In some cases it can make a 100 % change in the charge! There is a very serious coupling. You would never think of doing this for a larger protein, but it is feasible for a small molecule.

Bogdan Lesyng: If the topology of a molecule is fixed then we do not consider the dissociation of protons and then if there are no strong intermolecular interactions then changes in the charges on atoms are relatively small say 10 %, but of course this influences results. If possible, such changes should be included. Of course if there are changes in topology then changes at sides can be much larger, say even 100 %. The problem with higher order approximations or 3-body terms is a technical challenge: we know how the terms should look but we do not know how to simplify those terms.

David Case: You might think that wasn't the question you asked, but in fact a lot of the discussion of polarizable force fields is looking at very small changes without changing the number of protons and so forth whereas for real proteins, exactly the kind of problem that Harold was referring to is a much more gigantic problem that we usually gloss over: that we fix the protonation state. So whether it is 10 % or 20 % for an alanine dipeptide is an interesting question but for function and structure of proteins the kind of problem that Harold was referring to is almost an order of magnitude bigger and very difficult and worthy of investigation.

Harold Scheraga: I recently saw a paper in which someone did a calculation on a protein at pH 7 to compare it to an NMR structure at pH 3. The charge at pH 3 is very different than at pH 7. You really have to take into account the pH dependence

Celeste Sagui: The partial charge description is a very poor description because it is mathematically underdetermined: you will always have these problems. You need something else: continuous electrostatics, higher order multipole, etc. Otherwise you will always have this conformational dependence. It has to be dealt with in a more accurate manner.

Alan Mark: You have to deal with it in an appropriate manner. It is problem dependent. You can't use multipoles in a protein folding simulation because it is too expensive. If we all use multipoles we would just eliminate 90 % of the work that we do!

David Wales: I would like to agree that it is completely clear that you should never trust an empirical potential anywhere that it wasn't fitted, and obviously that includes DFT functionals. How can the situation best be improved? Obviously there is a grave difficulty with using polarizabilities, but I wonder if some of the tricks that people have learned from studying intermolecular potentials and van der Waals molecules in hydrogen bonded complexes could be used. And to get the right answers there you can get away with distributed multipoles, and sometimes point but often distributed polarizabilities. Yes it makes it more expensive, but to get the right structure you need to put in some of these terms. For example one of the best force fields for water needs isotropic dispersion and charge transfer terms. You need to choose the cheapest alternative that is appropriate to your problem.

David Case: It's not as expensive as you might think to have distributed multipoles and distributed polarizabilities. That's not going add more an order of mag to what we do. I'm pretty convinced that long range Ewald summation can be transferred over to this setting so it's only the short range stuff that causes problems.

Moderator: For what size of systems are current sampling techniques adequate and which methods are best? A statement was made by that with the advent of replica exchange we have now solved the sampling problem?

Harold Scheraga: It depends on the question you are asking. If you want thermodynamics that's one thing. If you want structure its another. For example for thermodynamics you can use entropy sampling Monte Carlo or multicanonical sampling which is very good for thermodynamic properties. But for global optimization it does not succeed. As for the size of the sample, well we did not think that with all atom potentials we could go to 46 residues. But it looks like we can go higher. We still can't do 1-2 angstrom rmsd for a large protein but there seems to be steady improvement and we are getting 4-5 angstrom rmsd for 100-200 residues.

Alan Mark: The comment on replica exchange: a method that works beautifully on a small model will look very diffferent when you apply it to a large protein. You have to ask which method is good for a particular problem.

Moderator: Would it be worthwhile to organize purely computational contests without worrying about comparison to experiments?

David Case: The bad thing about CASP is that it promotes the notion that science is competitive. You can have a really good idea and it might not be useful in the next casp but only in the 5th one down the road.

Harold Scheraga: I do not use the word competiton when talking about CASP. I call it a CASP exercise. But the only way to tell if our methodology is working is in a blind test and I think you have to make a comparison with experiment. In fact we do not even feel that we should do a calculation until we can make a comparison with experiment. There's a lot of things you can say bad about CASP and I would agree with you about some of them. Among other things it wastes our time all summer to prepare results for this when we should be doing science. But it does provide a blind test and it motivates students to participate. If you give up the comparison with experiment I think you are shooting blindly.

Christof Schütte: I feel that we have errors at all stages of the problem: modelling, numerics, everywhere. I would like to have a competition, exercise whatever you wish to call it, where we fix the model and then we could compare the algorithms and we would learn about the different algorithms and sources of error. For my taste it is all a little too mixed up. We have errors from everything at once. If we compared algorithms only, we could talk about convergence, accuracy, efficiency of different schemes.

Alan Mark: I think tests like CASP are only good when there is self-selection. The only reason why it has to be done with a blind test is that we could not trust people to publish their results when their methods failed. People were selecting the model where it works well and saying "aren't I great" and in this situation CASP works and you need a blind test. I would also totally agree that if we don't compare against experiment we are fooling ourselves about what we do. And we have to compare against diffraction patterns, NOE intensities, real data not just somebody's model.

Aatto Laaksonen: I would like to go back to Christof Schuette's idea. I like it very much. Now there are lots of people doing coding, mainly students, and they work hard and they don't always have a background in computer science, so they invent a lot of things which are never used. A lot of this work does not pay off. It really would be nice to have a kind of competition in different areas of algorithm design. I like the idea very much.

Ben Leimkuhler: In response to what Alan and Harold said, I agree that if you want to be sure that your methods are relevant for biological simulations you have to compare against experiment, but there are a lot of mathematicians at this meeting–increasing numbers–and you may have noticed that they're not actually studying the same sorts of problems that you're studying. Mathematicians like myself are always a little ways away from the practitioners. We could use something like this to bring us up to speed.

Alan Mark: Don't you do this in the literature already? I mean compare methods?

Ben Leimkuhler: Very rarely do people compare methods carefully. Mostly people propose methods, publish papers about methods, but very few people compare methods carefully on identical problems.

Tamar Schlick: I think that CASP has in a way become a victim of its own success. The idea of this exercise - not competition - is a great one, but CASP has become so popular that jealousy and competition have entered into the picture, because winners of CASP are known to receive prestigious accolades.How to implement this successful idea of critical assessment of the field's progress with respect to algorithms and applications is the challenge. We could all do more, as Ben and Alan were discussing, in each of our papers. Many journals already require full details of the models when new algorithms are presented. Unfortunately, scientists often use model variations or differences that not not easily described in full. Still, groups of people from this room, for example, could get together to conduct some comparative studies using the exact same models and different methods and vice versa for systematic evaluations. Larger initiatives are much harder to conduct.

Chris Woods: I find it frustrating in the field that its actually very difficult to compare with other people's work since there is normally not enough information in a journal article to repeat the experiment the researchers actually performed. For example people do not publish the initial data and parameters of the simulation. So we actually have to spend a lot of time having to guess what the other researchers have actually done.It would be good to have a mechanism to enhance reproducibility of results.

Christof Schütte: It is unusual for me but I want to make an experiment. Would everyone in the room please raise your hand if you feel like a mathematician or theoretician?[about 1/5 of the room raises a hand]Who would want to have a competition to compare methods and algorithms on the basis of a given model? Raise your hand.[a similar number of people raise a hand]Now you see it is almost the same people.

Ron Elber: The reason why I did not vote for this is that you run into the possible danger that you are going to optimize your method for a given model, say water, and then you are going to run into another system, say polarizable force fields and you are going to have to change your whole method. I've seen in computer science that there is a tendency to design an algorithm for an exercise then it becomes a purely mathematical exercise.

Moderator: Final Thoughts?

Tamar Schlick: A question to the audience: To me it seems the role that we thought perhaps a decade ago that the speed of computers would revolutionize this field, that we could perform much longer simulations or solve better models. I wonder if people feel that this has been a disappointment in general or if it has allowed us to go into more areas. What's the effect of advances in computers on the field?

Alan Mark: I love an answer that Wilfred van Gunsteren made years ago to a similar question: if we were not expecting advances in computers then we should give up. Since then we would be trapped to a single problem. It's the advances in computers that allow us to keep going on to more and more interesting and more and more challenging problems. If we were stuck where we were 10 years ago we would be talking about a few thousand water molecules and free energy calculations on butane and now we can talk about spontaneous vesicle fusion and things people would not have dreamed of doing a few years ago. Without a doubt the field is advancing.

Editorial Policy

§1. Volumes in the following three categories will be published in LNCSE:

i)   Research monographs
ii)   Lecture and seminar notes
iii) Conference proceedings

Those considering a book which might be suitable for the series are strongly advised to contact the publisher or the series editors at an early stage.

§2. Categories i) and ii). These categories will be emphasized by Lecture Notes in Computational Science and Engineering. **Submissions by interdisciplinary teams of authors are encouraged.** The goal is to report new developments – quickly, informally, and in a way that will make them accessible to non-specialists. In the evaluation of submissions timeliness of the work is an important criterion. Texts should be well-rounded, well-written and reasonably self-contained. In most cases the work will contain results of others as well as those of the author(s). In each case the author(s) should provide sufficient motivation, examples, and applications. In this respect, Ph.D. theses will usually be deemed unsuitable for the Lecture Notes series. Proposals for volumes in these categories should be submitted either to one of the series editors or to Springer-Verlag, Heidelberg, and will be refereed. A provisional judgment on the acceptability of a project can be based on partial information about the work: a detailed outline describing the contents of each chapter, the estimated length, a bibliography, and one or two sample chapters – or a first draft. A final decision whether to accept will rest on an evaluation of the completed work which should include

– at least 100 pages of text;
– a table of contents;
– an informative introduction perhaps with some historical remarks which should be accessible to readers unfamiliar with the topic treated;
– a subject index.

§3. Category iii). Conference proceedings will be considered for publication provided that they are both of exceptional interest and devoted to a single topic. One (or more) expert participants will act as the scientific editor(s) of the volume. They select the papers which are suitable for inclusion and have them individually refereed as for a journal. Papers not closely related to the central topic are to be excluded. Organizers should contact Lecture Notes in Computational Science and Engineering at the planning stage.

In exceptional cases some other multi-author-volumes may be considered in this category.

§4. Format. Only works in English are considered. They should be submitted in camera-ready form according to Springer-Verlag's specifications.
Electronic material can be included if appropriate. Please contact the publisher.
Technical instructions and/or TEX macros are available via
http://www.springer.com/sgw/cda/frontpage/0,11855,5-40017-2-71391-0,00.html
The macros can also be sent on request.

General Remarks

Lecture Notes are printed by photo-offset from the master-copy delivered in camera-ready form by the authors. For this purpose Springer-Verlag provides technical instructions for the preparation of manuscripts. See also *Editorial Policy*.

Careful preparation of manuscripts will help keep production time short and ensure a satisfactory appearance of the finished book.

The following terms and conditions hold:

Categories i), ii), and iii):
Authors receive 50 free copies of their book. No royalty is paid. Commitment to publish is made by letter of intent rather than by signing a formal contract. Springer-Verlag secures the copyright for each volume.

For conference proceedings, editors receive a total of 50 free copies of their volume for distribution to the contributing authors.

All categories:
Authors are entitled to purchase further copies of their book and other Springer mathematics books for their personal use, at a discount of 33,3 % directly from Springer-Verlag.

Addresses:

Timothy J. Barth
NASA Ames Research Center
NAS Division
Moffett Field, CA 94035, USA
e-mail: barth@nas.nasa.gov

Michael Griebel
Institut für Numerische Simulation
der Universität Bonn
Wegelerstr. 6
53115 Bonn, Germany
e-mail: griebel@ins.uni-bonn.de

David E. Keyes
Department of Applied Physics
and Applied Mathematics
Columbia University
200 S. W. Mudd Building
500 W. 120th Street
New York, NY 10027, USA
e-mail: david.keyes@columbia.edu

Risto M. Nieminen
Laboratory of Physics
Helsinki University of Technology
02150 Espoo, Finland
e-mail: rni@fyslab.hut.fi

Dirk Roose
Department of Computer Science
Katholieke Universiteit Leuven
Celestijnenlaan 200A
3001 Leuven-Heverlee, Belgium
e-mail: dirk.roose@cs.kuleuven.ac.be

Tamar Schlick
Department of Chemistry
Courant Institute of Mathematical
Sciences
New York University
and Howard Hughes Medical Institute
251 Mercer Street
New York, NY 10012, USA
e-mail: schlick@nyu.edu

Mathematics Editor at Springer: Martin Peters
Springer-Verlag, Mathematics Editorial IV
Tiergartenstrasse 17
D-69121 Heidelberg, Germany
Tel.: *49 (6221) 487-8185
Fax: *49 (6221) 487-8355
e-mail: martin.peters@springer.com

Lecture Notes
in Computational Science
and Engineering

Vol. 1   D. Funaro, *Spectral Elements for Transport-Dominated Equations.* 1997. X, 211 pp. Softcover. ISBN 3-540-62649-2

Vol. 2   H. P. Langtangen, *Computational Partial Differential Equations.* Numerical Methods and Diffpack Programming. 1999. XXIII, 682 pp. Hardcover. ISBN 3-540-65274-4

Vol. 3   W. Hackbusch, G. Wittum (eds.), *Multigrid Methods V.* Proceedings of the Fifth European Multigrid Conference held in Stuttgart, Germany, October 1-4, 1996. 1998. VIII, 334 pp. Softcover. ISBN 3-540-63133-X

Vol. 4   P. Deuflhard, J. Hermans, B. Leimkuhler, A. E. Mark, S. Reich, R. D. Skeel (eds.), *Computational Molecular Dynamics: Challenges, Methods, Ideas.* Proceedings of the 2nd International Symposium on Algorithms for Macromolecular Modelling, Berlin, May 21-24, 1997. 1998. XI, 489 pp. Softcover. ISBN 3-540-63242-5

Vol. 5   D. Kröner, M. Ohlberger, C. Rohde (eds.), *An Introduction to Recent Developments in Theory and Numerics for Conservation Laws.* Proceedings of the International School on Theory and Numerics for Conservation Laws, Freiburg / Littenweiler, October 20-24, 1997. 1998. VII, 285 pp. Softcover. ISBN 3-540-65081-4

Vol. 6   S. Turek, *Efficient Solvers for Incompressible Flow Problems.* An Algorithmic and Computational Approach. 1999. XVII, 352 pp, with CD-ROM. Hardcover. ISBN 3-540-65433-X

Vol. 7   R. von Schwerin, *Multi Body System SIMulation.* Numerical Methods, Algorithms, and Software. 1999. XX, 338 pp. Softcover. ISBN 3-540-65662-6

Vol. 8   H.-J. Bungartz, F. Durst, C. Zenger (eds.), *High Performance Scientific and Engineering Computing.* Proceedings of the International FORTWIHR Conference on HPSEC, Munich, March 16-18, 1998. 1999. X, 471 pp. Softcover. 3-540-65730-4

Vol. 9   T. J. Barth, H. Deconinck (eds.), *High-Order Methods for Computational Physics.* 1999. VII, 582 pp. Hardcover. 3-540-65893-9

Vol. 10   H. P. Langtangen, A. M. Bruaset, E. Quak (eds.), *Advances in Software Tools for Scientific Computing.* 2000. X, 357 pp. Softcover. 3-540-66557-9

Vol. 11   B. Cockburn, G. E. Karniadakis, C.-W. Shu (eds.), *Discontinuous Galerkin Methods.* Theory, Computation and Applications. 2000. XI, 470 pp. Hardcover. 3-540-66787-3

Vol. 12   U. van Rienen, *Numerical Methods in Computational Electrodynamics.* Linear Systems in Practical Applications. 2000. XIII, 375 pp. Softcover. 3-540-67629-5

Vol. 13   B. Engquist, L. Johnsson, M. Hammill, F. Short (eds.), *Simulation and Visualization on the Grid.* Parallelldatorcentrum Seventh Annual Conference, Stockholm, December 1999, Proceedings. 2000. XIII, 301 pp. Softcover. 3-540-67264-8

Vol. 14   E. Dick, K. Riemslagh, J. Vierendeels (eds.), *Multigrid Methods VI.* Proceedings of the Sixth European Multigrid Conference Held in Gent, Belgium, September 27-30, 1999. 2000. IX, 293 pp. Softcover. 3-540-67157-9

Vol. 15   A. Frommer, T. Lippert, B. Medeke, K. Schilling (eds.), *Numerical Challenges in Lattice Quantum Chromodynamics.* Joint Interdisciplinary Workshop of John von Neumann Institute for Computing, Jülich and Institute of Applied Computer Science, Wuppertal University, August 1999. 2000. VIII, 184 pp. Softcover. 3-540-67732-1

Vol. 16   J. Lang, *Adaptive Multilevel Solution of Nonlinear Parabolic PDE Systems.* Theory, Algorithm, and Applications. 2001. XII, 157 pp. Softcover. 3-540-67900-6

Vol. 17   B. I. Wohlmuth, *Discretization Methods and Iterative Solvers Based on Domain Decomposition.* 2001. X, 197 pp. Softcover. 3-540-41083-X

Vol. 18 U. van Rienen, M. Günther, D. Hecht (eds.), *Scientific Computing in Electrical Engineering.* Proceedings of the 3rd International Workshop, August 20-23, 2000, Warnemünde, Germany. 2001. XII, 428 pp. Softcover. 3-540-42173-4

Vol. 19 I. Babuška, P. G. Ciarlet, T. Miyoshi (eds.), *Mathematical Modeling and Numerical Simulation in Continuum Mechanics.* Proceedings of the International Symposium on Mathematical Modeling and Numerical Simulation in Continuum Mechanics, September 29 - October 3, 2000, Yamaguchi, Japan. 2002. VIII, 301 pp. Softcover. 3-540-42399-0

Vol. 20 T. J. Barth, T. Chan, R. Haimes (eds.), *Multiscale and Multiresolution Methods.* Theory and Applications. 2002. X, 389 pp. Softcover. 3-540-42420-2

Vol. 21 M. Breuer, F. Durst, C. Zenger (eds.), *High Performance Scientific and Engineering Computing.* Proceedings of the 3rd International FORTWIHR Conference on HPSEC, Erlangen, March 12-14, 2001. 2002. XIII, 408 pp. Softcover. 3-540-42946-8

Vol. 22 K. Urban, *Wavelets in Numerical Simulation.* Problem Adapted Construction and Applications. 2002. XV, 181 pp. Softcover. 3-540-43055-5

Vol. 23 L. F. Pavarino, A. Toselli (eds.), *Recent Developments in Domain Decomposition Methods.* 2002. XII, 243 pp. Softcover. 3-540-43413-5

Vol. 24 T. Schlick, H. H. Gan (eds.), *Computational Methods for Macromolecules: Challenges and Applications.* Proceedings of the 3rd International Workshop on Algorithms for Macromolecular Modeling, New York, October 12-14, 2000. 2002. IX, 504 pp. Softcover. 3-540-43756-8

Vol. 25 T. J. Barth, H. Deconinck (eds.), *Error Estimation and Adaptive Discretization Methods in Computational Fluid Dynamics.* 2003. VII, 344 pp. Hardcover. 3-540-43758-4

Vol. 26 M. Griebel, M. A. Schweitzer (eds.), *Meshfree Methods for Partial Differential Equations.* 2003. IX, 466 pp. Softcover. 3-540-43891-2

Vol. 27 S. Müller, *Adaptive Multiscale Schemes for Conservation Laws.* 2003. XIV, 181 pp. Softcover. 3-540-44325-8

Vol. 28 C. Carstensen, S. Funken, W. Hackbusch, R. H. W. Hoppe, P. Monk (eds.), *Computational Electromagnetics.* Proceedings of the GAMM Workshop on "Computational Electromagnetics", Kiel, Germany, January 26-28, 2001. 2003. X, 209 pp. Softcover. 3-540-44392-4

Vol. 29 M. A. Schweitzer, *A Parallel Multilevel Partition of Unity Method for Elliptic Partial Differential Equations.* 2003. V, 194 pp. Softcover. 3-540-00351-7

Vol. 30 T. Biegler, O. Ghattas, M. Heinkenschloss, B. van Bloemen Waanders (eds.), *Large-Scale PDE-Constrained Optimization.* 2003. VI, 349 pp. Softcover. 3-540-05045-0

Vol. 31 M. Ainsworth, P. Davies, D. Duncan, P. Martin, B. Rynne (eds.), *Topics in Computational Wave Propagation.* Direct and Inverse Problems. 2003. VIII, 399 pp. Softcover. 3-540-00744-X

Vol. 32 H. Emmerich, B. Nestler, M. Schreckenberg (eds.), *Interface and Transport Dynamics.* Computational Modelling. 2003. XV, 432 pp. Hardcover. 3-540-40367-1

Vol. 33 H. P. Langtangen, A. Tveito (eds.), *Advanced Topics in Computational Partial Differential Equations.* Numerical Methods and Diffpack Programming. 2003. XIX, 658 pp. Softcover. 3-540-01438-1

Vol. 34 V. John, *Large Eddy Simulation of Turbulent Incompressible Flows.* Analytical and Numerical Results for a Class of LES Models. 2004. XII, 261 pp. Softcover. 3-540-40643-3

Vol. 35 E. Bänsch (ed.), *Challenges in Scientific Computing - CISC 2002.* Proceedings of the Conference *Challenges in Scientific Computing,* Berlin, October 2-5, 2002. 2003. VIII, 287 pp. Hardcover. 3-540-40887-8

Vol. 36 B. N. Khoromskij, G. Wittum, *Numerical Solution of Elliptic Differential Equations by Reduction to the Interface.* 2004. XI, 293 pp. Softcover. 3-540-20406-7

Vol. 37 A. Iske, *Multiresolution Methods in Scattered Data Modelling.* 2004. XII, 182 pp. Softcover. 3-540-20479-2

Vol. 38 S.-I. Niculescu, K. Gu (eds.), *Advances in Time-Delay Systems.* 2004. XIV, 446 pp. Softcover. 3-540-20890-9

Vol. 39 S. Attinger, P. Koumoutsakos (eds.), *Multiscale Modelling and Simulation*. 2004. VIII, 277 pp. Softcover. 3-540-21180-2

Vol. 40 R. Kornhuber, R. Hoppe, J. Périaux, O. Pironneau, O. Wildlund, J. Xu (eds.), *Domain Decomposition Methods in Science and Engineering*. 2005. XVIII, 690 pp. Softcover. 3-540-22523-4

Vol. 41 T. Plewa, T. Linde, V.G. Weirs (eds.), *Adaptive Mesh Refinement – Theory and Applications*. 2005. XIV, 552 pp. Softcover. 3-540-21147-0

Vol. 42 A. Schmidt, K.G. Siebert, *Design of Adaptive Finite Element Software*. The Finite Element Toolbox ALBERTA. 2005. XII, 322 pp. Hardcover. 3-540-22842-X

Vol. 43 M. Griebel, M.A. Schweitzer (eds.), *Meshfree Methods for Partial Differential Equations II*. 2005. XIII, 303 pp. Softcover. 3-540-23026-2

Vol. 44 B. Engquist, P. Lötstedt, O. Runborg (eds.), *Multiscale Methods in Science and Engineering*. 2005. XII, 291 pp. Softcover. 3-540-25335-1

Vol. 45 P. Benner, V. Mehrmann, D.C. Sorensen (eds.), *Dimension Reduction of Large-Scale Systems*. 2005. XII, 402 pp. Softcover. 3-540-24545-6

Vol. 46 D. Kressner (ed.), *Numerical Methods for General and Structured Eigenvalue Problems*. 2005. XIV, 258 pp. Softcover. 3-540-24546-4

Vol. 47 A. Boriçi, A. Frommer, B. Joó, A. Kennedy, B. Pendleton (eds.), *QCD and Numerical Analysis III*. 2005. XIII, 201 pp. Softcover. 3-540-21257-4

Vol. 48 F. Graziani (ed.), *Computational Methods in Transport*. 2006. VIII, 524 pp. Softcover. 3-540-28122-3

Vol. 49 B. Leimkuhler, C. Chipot, R. Elber, A. Laaksonen, A. Mark, T. Schlick, C. Schütte, R. Skeel (eds.), *New Algorithms for Macromolecular Simulation*. 2006. XVI, 376 pp. Softcover. 3-540-25542-7

For further information on these books please have a look at our mathematics catalogue at the following URL: www.springer.com/series/3527

Texts in Computational Science and Engineering

Vol. 1 H. P. Langtangen, *Computational Partial Differential Equations*. Numerical Methods and Diffpack Programming. 2nd Edition 2003. XXVI, 855 pp. Hardcover. ISBN 3-540-43416-X

Vol. 2 A. Quarteroni, F. Saleri, *Scientific Computing with MATLAB*. 2003. IX, 257 pp. Hardcover. ISBN 3-540-44363-0

Vol. 3 H. P. Langtangen, *Python Scripting for Computational Science*. 2nd Edition 2006. XXIV, 736 pp. Hardcover. ISBN 3-540-29415-5

For further information on these books please have a look at our mathematics catalogue at the following URL: www.springer.com/series/5151